MW01075360

A Tour through Mathematical Logic

Library of Congress Catalog Card Number 2004113540

Complete Set ISBN 0-88385-000-1
Vol. 30 ISBN 0-88385-036-2

Printed in the United States of America

Current Printing (last digit):
10 9 8 7 6 5 4 3 2 1

The following Monographs have been published:

1. *Calculus of Variations,* by G. A. Bliss (out of print)
2. *Analytic Functions of a Complex Variable,* by D. R. Curtiss (out of print)
3. *Mathematical Statistics,* by H. L. Rietz (out of print)
4. *Projective Geometry,* by J. W. Young (out of print)
5. *A History of Mathematics in America before 1900,* by D. E. Smith and Jekuthiel Ginsburg (out of print)
6. *Fourier Series and Orthogonal Polynomials,* by Dunham Jackson (out of print)
7. *Vectors and Matrices,* by C. C. MacDuffee (out of print)
8. *Rings and Ideals,* by N. H. McCoy (out of print)
9. *The Theory of Algebraic Numbers,* second edition, by Harry Pollard and Harold G. Diamond
10. *The Arithmetic Theory of Quadratic Forms,* by B. W. Jones (out of print)
11. *Irrational Numbers,* by Ivan Niven
12. *Statistical Independence in Probability, Analysis and Number Theory,* by Mark Kac
13. *A Primer of Real Functions,* third edition, by Ralph P. Boas, Jr.
14. *Combinatorial Mathematics,* by Herbert J. Ryser
15. *Noncommutative Rings,* by I. N. Herstein
16. *Dedekind Sums,* by Hans Rademacher and Emil Grosswald
17. *The Schwarz Function and Its Applications,* by Philip J. Davis
18. *Celestial Mechanics,* by Harry Pollard
19. *Field Theory and Its Classical Problems,* by Charles Robert Hadlock
20. *The Generalized Riemann Integral,* by Robert M. McLeod
21. *From Error-Correcting Codes through Sphere Packings to Simple Groups,* by Thomas M. Thompson
22. *Random Walks and Electric Networks,* by Peter G. Doyle and J. Laurie Snell
23. *Complex Analysis: The Geometric Viewpoint,* by Steven G. Krantz
24. *Knot Theory,* by Charles Livingston
25. *Algebra and Tiling: Homomorphisms in the Service of Geometry,* by Sherman Stein and Sándor Szabó
26. *The Sensual (Quadratic) Form,* by John H. Conway assisted by Francis Y. C. Fung
27. *A Panorama of Harmonic Analysis,* by Steven G. Krantz
28. *Inequalities from Complex Analysis,* by John P. D'Angelo
29. *Ergodic Theory of Numbers,* by Karma Dajani and Cor Kraaikamp
30. *A Tour through Mathematical Logic,* by Robert S. Wolf

MAA Service Center
P. O. Box 91112
Washington, DC 20090-1112
800-331-1MAA FAX: 301-206-9789

Preface

This book is supposed to be about "mathematical logic," so it seems appropriate to begin with a definition of that phrase. On the surface, this is an easy task: we can say that mathematical logic is the branch of mathematics that studies the logic and the methods of deduction used in mathematics. Chapter 1 is an introduction to mathematical logic in this narrow sense.

A bit of reflection should convince you that this is not a typical branch of mathematics. Most branches of mathematics study mathematical "objects" or "structures": numbers, triangles, functions of a real variable, groups, topological spaces, etc. But mathematical logic studies something very different—essentially, mathematical reasoning.

A couple of points come to mind on the basis of this characterization. One is that mathematical logic could be among the most abstract branches of mathematics, since it is particularly difficult to give a concrete description of its subject material. (On the other hand, mathematical logic also studies the symbolic language used in mathematics, which is a concrete object.) Another is that there is something circular, perhaps even paradoxical, in the nature of this field. For its subject matter is mathematical reasoning, but since it is a branch of mathematics its method of inquiry must also consist of mathematical reasoning. The term "metamathematics" was introduced by David Hilbert to describe this unusual situation: a discipline that is mathematical in nature but at the same time beyond or above ordinary mathematics because it treats mathematics as an object of study. (Another branch of mathematics with a somewhat similar status is category theory.)

However, there are compelling reasons to broaden what we mean by mathematical logic. It turns out to be of limited value to study mathematical reasoning without also addressing the nature of mathematical objects. After all, reasoning is usually *about* something, and if it is, understanding what the reasoning is about may be nearly essential to analyzing the reasoning itself. One of the great achievements of modern mathematics has been the unification of the many types of objects studied in mathematics. This unification began in the seventeenth century, when René Descartes and Pierre de Fermat showed how geometric objects (points, lines, circles, etc.) could be represented numerically or algebraically, and culminated in the twentieth century with the sophisticated use of set theory to represent all the objects normally considered in mathematics. In common parlance (including the title of this book), the term "mathematical logic" is used as a synonym for "foundations of mathematics," which includes both mathematical logic in the stricter sense defined above, and the study of the fundamental types of mathematical objects and the relationships among them.

Accordingly, this book includes two chapters on set theory, one on the basics of the subject and the other on the explosion of methods and results that have been discovered since the 1960s. It also includes a chapter on model theory, which uses set theory to show that the theorem-proving power of the usual methods of deduction in mathematics corresponds perfectly to what must be true in actual mathematical structures. The remaining chapters describe several other important branches of the foundations of mathematics. Chapter 3 is about the important concepts of "effective procedures" and "computable functions," and Chapter 4 describes Kurt Gödel's (and others') famous incompleteness theorems. Chapter 7 is on the powerful tool known as nonstandard analysis, which uses model theory to justify the use of infinitesimals in mathematical reasoning. The final chapter is about various "constructivist" movements that have attempted to convince mathematicians to give "concrete" reasoning more legitimacy than "abstract" reasoning.

The word "tour" in the title of this book also deserves some explanation. For one thing, I chose this word to emphasize that it is not a textbook in the strict sense. To be sure, it has many of the features of a textbook, including exercises. But it is less structured, more free-flowing, than a standard text. It also lacks many of the details and proofs that one normally expects in a mathematics text. However, in almost all such cases there are references to more detailed treatments and the omitted proofs. Therefore, this book is actually quite suitable for use as a text at the university level (undergraduate or graduate), provided that the instructor is willing to provide supplementary material from time to time.

The most obvious advantage of this omission of detail is that this monograph is able to cover a lot more material than if it were a standard textbook of the same size. This de-emphasis on detail is also intended to help the reader concentrate on the big picture, the essential ideas of the subject, without getting bogged down in minutiae. Could this book have been titled "A Survey of Mathematical Logic"? Perhaps it could have, but my choice of the word "tour" was deliberate. A survey sounds like a rather dry activity, carried out by technicians with instruments. Tours, on the other hand, are what people take on their vacations. They are intended to be fun. On a tour of Europe, you might learn a lot about the cathedrals of Germany or Impressionist art, but you wouldn't expect to learn about them as thoroughly as if you took a course in the subject. The goal of this book is similar: to provide an introduction to the foundations of mathematics that is substantial and stimulating, and at the same time a pleasure to read.

Here are some suggestions for how to proceed on the tour that this book provides. It is self-contained enough that a reader with no previous knowledge of logic and set theory (but at least a year of mathematics beyond calculus) should be able to read it on his or her own, without referring to outside sources and without getting seriously stuck. This is especially true of the first two chapters, which (except for Section 1.7) consist of basic material that is prerequisite to the later chapters. Before proceeding to the later chapters, it is highly advisable to read and understand at least Sections 1.1 through 1.5 and 2.1 through 2.3.

Several appendices covering some important background material are provided for the reader with limited experience in higher mathematics. The main topics included in these appendices, other than those from foundations, are ordering relations and basic concepts from abstract algebra (groups, rings, and fields). Some topics that are not included are basic concepts from linear algebra (e.g., vector spaces), analysis, and topology. These topics are occasionally mentioned in the book, but the reader who is unfamiliar with them should not be significantly hindered.

The remaining chapters are distinctly more advanced than the first two. Attempting to digest them without first understanding the basics of logic and set theory could prove frustrating, but the savvy reader may be able to read one or more of these later chapters, needing to refer only occasionally to particular material in Chapters 1 and 2. Most of these later chapters depend only slightly on each other. So, for instance, someone who decides to read Chapters 1, 2, and 5 will not be significantly hindered by having skipped Chapters 3 and 4. The most significant exceptions to this are that Chapter 4 is based essentially on Chapter 3, and Chapters 6 and 7 use many concepts from Chapter 5.

Of course, even in the introductory chapters, if you want to learn the material thoroughly, then you must be prepared to reread certain parts, do the exercises, and perhaps fill in some details from outside sources. A book can be made pleasant to read, but it cannot turn abstract mathematics into a walk in the park. As Euclid supposedly said to King Ptolemy, "There is no royal road to geometry."

This book includes about a dozen biographies, in boxes, of people who have made important contributions to mathematical logic and foundations. In order to fend off any bewilderment about whom I chose for these and why, let me clarify two points. First, I decided not to include biographies of any living mathematicians. Second, my choices were motivated only partly by the importance, quality, quantity, brilliance, etc., of the person's contributions. My other main criterion was to what extent the person's life was noteworthy or unusual, perhaps even heroic or tragic. So I have omitted the biographies of some true "giants" in the field whose lives were relatively straightforward.

I sincerely hope that you will find this "tour" interesting, informative, and enjoyable. I welcome any type of feedback, including questions, corrections, and criticisms, sent to rswolf@calpoly.edu.

Acknowledgments

One of the few pleasurable activities that normally occurs during the frantic process of finalizing a book is thanking the people who helped the author(s) along the way. In the present situation, an unusually large number of people have aided me in one way or another, and I am genuinely grateful to all of them.

First, I want to thank the staff of the Mathematical Association of America, specifically Don Albers, Elaine Pedreira, and Bev Ruedi. They have been consistently helpful and encouraging during the long process of writing this book, and their technical expertise has been extremely valuable.

I also want to sincerely thank the editorial board of the Carus Monograph series for their time and their many constructive comments: Joseph Auslander, Harold P. Boas, Robert E. Greene, Dennis Hejhal, M. Susan Montgomery, Bruce A. Reznick, Ken Ross, Clarence Eugene Wayne, and David Wright.

Space does not allow me to thank all the mathematicians and mathematics students who have read part or all of this manuscript and provided valuable feedback. So, at the risk of omitting some deserving ones (and with apologies in advance, if that occurs), I want to thank the following people who have kindly donated a substantial amount of time to provide comments and guidance: Michael Beeson, Jim Henle, Robert Solovay, Charlie Steinhorn, Todor Todorov, and Dan Velleman.

I am especially indebted to Ken Ross, chair of the Carus Monograph editorial board during the writing of this book. Ken gave an extraordinary amount of his time to reading every chapter of this manuscript, most of them more than once, and providing thoughtful and detailed comments. He also provided much encouragement and helped me deal with more than one thorny issue. Ken, you're swell!

Finally, I want to thank my friends and family for their patience and encouragement during these several years of distraction and A.R.-ness on my part. Special kudos go to my wife, Laurie, for suppressing her impulse to laugh out loud when the number of times I said, "I've just finished my book!" hit double digits.

Contents

CHAPTER 1
Predicate Logic

1.1 Introduction

Why should students of mathematics want to know something about predicate logic? Here is one answer: predicate logic helps one understand the fine points of mathematical language, including the ambiguities that arise from the use of *natural* language (English, French, etc.) in mathematics. For research mathematicians, these ambiguities are usually easy to spot, and their intuition will normally prevent any errors. Yet, occasionally, such ambiguities can be very subtle and present real pitfalls. For instructors of mathematics, awareness of this type of ambiguity is indispensable for analyzing students' mistakes and misconceptions.

Let's illustrate this point with a rather fanciful example: imagine that you are an eminent mathematician with an international reputation. Your life is wonderful in every way except that, after many years, your marriage has become stale. Reluctantly, you decide to have a fling, hoping that a brief one will be enough and your marriage will survive. You approach your friend Lee, whose marriage is also stale, and the two of you agree to have a tryst. (For the record, I would never do such a thing. Remember, it's *you* I'm talking about. If you wouldn't do such a thing either, then I must be thinking of a different reader.)

To be on the safe side, you and Lee decide to go to another state for your tryst. You drive there and go to a nondescript motel. To your surprise, the desk clerk asks, "Are you folks married? We have to ask that, and you're required by state law to answer truthfully." Without batting an eyelash, you reply, "Of course we are." You get the room and have your little fling, but the next morning you are arrested—not for adultery (since nothing can be proved about that), but for perjury!

From your jail cell, you plan your clever defense: you will simply say that you didn't lie at all. You and Lee are married, but not to each other! Your answer was literally quite correct, so no perjury has occurred.

Will your defense succeed? I doubt it, but that's not what interests us anyway. Among the many types of ambiguities in the English language, this one is of a particularly interesting sort. Note that the statement "I am married" is not ambiguous. The statement (or **predicate**, since it involves a variable) "x is married" means that x is married to *someone*. But the statement "x and y are married" somehow becomes ambiguous. It would normally mean that x and y are married to each other, but it could mean that x and y are, individually, married persons. To analyze the source of this ambiguity, one should realize that the **binary** (two-variable) predicate "x and y are married to each other" is really the basic or **atomic** predicate here. If we use $M(x, y)$ to denote this sentence, then the **unary** (one-variable) predicate "x is married" actually means "There is a person y such that $M(x, y)$." It is not atomic because it includes a **quantifier** ("there is"), which is tacit in its usual English rendition. Thus the less common meaning of "x and y are married" is in a sense simpler than the usual meaning (because it's a statement about the individuals x and y separately), but in another sense more complicated (because it involves two quantifiers).

We are done with this example, but one could concoct other interesting questions. How would you normally interpret "Fred and Jane and Bill and Mary are married"? How about "Fred, Jane, Bill and Mary are married"? How about "Fred, Jane, and Bill are married"? Of course, we usually rely on context to interpret such statements, but sometimes there is little or no context to rely on.

Another example: Suppose you were told that "two sisters entered the room." Normally, this would mean that two women who are each other's sisters entered the room (binary predicate of sisterhood). It would be quite artificial to make this statement to express that the women both had siblings. But suppose you were told that "two sisters entered the convent." Suddenly a different meaning presents itself, in which sisterhood is a unary predicate, completely unrelated to the usual binary predicate.

As a final nonmathematical example, I recently found myself at lunch in a restaurant with two married couples. When the waitress asked if we wanted a single check or separate ones, I spontaneously responded, "Those two are together, and so are those two, but I'm not very together."

Does this sort of ambiguity occur in mathematics? Yes it does, frequently and in many guises. An example that I like to use as a pedagogical tool is "Let C be a collection of nonempty, disjoint sets." The professional mathematician has no problem understanding this sentence. But the beginning student can easily forget that nonemptiness is a property of individual sets, a unary predicate, while disjointness is a property of pairs of sets, a binary predicate. Sometimes, the word "pairwise" is inserted before "disjoint" to clarify this difference. Also, note that the statement that two sets are disjoint is symmetric, so it is essentially a predicate on unordered pairs of (distinct) sets rather than ordered pairs of sets. If this were not the case, the notion of a collection of disjoint sets would make no sense (but the notion of a *sequence* of disjoint sets could be meaningful).

A particularly complex example of this sort is the statement that a set B of vectors in an inner product space is an orthonormal basis. One part of this statement is that each member of B is a unit vector (a unary predicate). Another part is that the vectors are orthogonal (a binary predicate, about unordered pairs of distinct vectors). Finally, the property of being a basis is essentially a statement about the whole set B, as opposed to a statement about all individuals in B, or all pairs in B, etc. But a bit more can be said here: the linear independence of B is equivalent to the linear independence of every finite subset of B. So, if one desires, one may view linear independence as a predicate that is initially defined for finite sets of vectors. But the statement that B spans the space is not equivalent to saying that some (or every) finite subset of B spans the space. It has the more complex form, "For every vector v, there is a finite subset of B such that v is a linear combination of the vectors in that finite subset."

Of course, mathematical logic is about a lot more than analyzing ambiguities. The study of logic and the foundations of mathematics has led to many powerful and versatile methods, and many deep and beautiful results. Some of these methods and results are relevant within foundations only, but many of them are useful and important in other branches of mathematics and in fields outside of mathematics. The goal of this book is to show you some of the power, depth, and beauty of this subject.

A very brief history of mathematical logic

The basic principles of logic and its use in mathematics were well understood by the philosophers and mathematicians of the classical Greek era, and Aristotle provided the first systematic written treatment of logic (propositional logic, primarily). Aristotle's description of logic was carried out in the context of natural language, and so it is considered *informal* rather than *symbolic*. It was not until two thousand years later (almost exactly!) that the seeds of modern mathematical logic were sown, with Aristotle's work as a basis. In the late 1600s Gottfried Leibniz, one of the co-inventors of calculus, expressed and worked toward the goal of creating a symbolic language of reasoning that could be used to solve all well-defined problems, not just in mathematics but in science and other disciplines as well. We have just seen several examples of the ambiguities that are inherent in English and other natural languages. Leibniz understood clearly that such ambiguities could be problematic in a discipline as exacting as mathematics.

Almost two more centuries passed before Augustus De Morgan and George Boole extended Leibniz's work and began to create modern symbolic logic. De Morgan is best remembered for the two laws of propositional logic that bear his name, while **Boolean algebra** found many uses in the twentieth century, notably the essential design principles of computer circuitry. Finally, predicate logic as we know it today was created in the latter part of the 1800s by Gottlob Frege and, a few years later but independently, by Giuseppe Peano. Frege gave the first rigorous treatment of the logic of quantifiers, while Peano invented many of the symbols currently used in propositional logic and set theory. Both of them applied their work to create formal theories of arithmetic. Of these, it is Peano's theory that has since found wide acceptance (see Example 16 in Section 1.5), but Frege's work had more influence on the development of modern mathematical logic.

The contradictions that were discovered in set theory at the end of the nineteenth century (to be discussed in Chapter 2) heightened awareness of the potential dangers of informal reasoning in mathematics. As a result, one of the goals of those wanting to free mathematics from such contradictions was to reduce the importance of natural language in mathematics. One monumental effort toward this end was the **logicist** program of Bertrand Russell and Alfred North Whitehead, which extended the work of Peano and Frege. Their 2000-page *Principia Mathematica* [RW], which took a decade to write, attempted to formulate all of mathematics in terms of logic. They hoped to make logic not just a tool for doing mathematics, but also the underlying subject matter on which all mathematical concepts would be based. They did not quite succeed in this, but they provided much of the groundwork for the formalization of mathematics within set theory. (So mathematics can't be based solely on logic, but it can be based on logic and set theory. In a sense, this is what Russell and Whitehead were trying to do, since their idea of logic included some set-theoretic concepts.)

In the wake of these achievements of Peano, Frege, Russell, and Whitehead, another ambitious plan was proposed by Hilbert. The goals of his **formalist** program were to put all of mathematics into a completely symbolic, axiomatic framework, and then to use **metamathematics** (mathematics applied to mathematics), also called **proof theory**, to show that mathematics is free from contradictions. Earlier, in the 1890s, Hilbert had published his *Foundations of Geometry*, con-

Bertrand Russell (1872–1970) was not primarily a mathematician but continued in the ancient tradition of philosophers who have made important contributions to the foundations of mathematics. He was born into a wealthy liberal family, orphaned at age four, and then raised by his grandmother, who had him tutored privately.

Russell made two great contributions to modern set theory and logic. One was his co-discovery of the inconsistency of naive set theory, discussed in Section 2.2. The other was the monumental *Principia Mathematica* [RW], written with his former professor, Alfred North Whitehead. As a philosopher, Russell was one of the main founders of the modern analytic school. He was a prolific writer and wrote many books intended for the general public, notably the best-selling *A History of Western Philosophy* (1950), for which he won the Nobel Prize in Literature.

Outside of academia, Russell is best known for his political and social activism. During World War I, his pacifism led to his dismissal from Trinity College and a six-month prison sentence. Half a century later, he vehemently opposed nuclear weapons, racial segregation, and the U. S. involvement in Vietnam. He advocated sexual freedom and trial marriage as early as the 1930s, causing a court of law to nullify a faculty position that had been offered to him by the City College of New York in 1940.

By the way, the web site

http://www-history.mcs.st-andrews.ac.uk/history/index.html

run by the University of St. Andrews is an excellent source for biographical information about mathematicians.

sidered to be the first version of Euclidean geometry that was truly axiomatic, in the sense that there were no hidden appeals to spatial intuition. Referring to this work, Hilbert made the point that his proofs should stay completely correct even if the words "point," "line," and "plane" were replaced throughout by "table," "chair," and "beer mug."

Because of the famous negative results obtained by Kurt Gödel, there is a tendency to think of the formalist program as a failure. But in fact two major parts of Hilbert's program—the translation of mathematical *language* and mathematical *proofs* into a purely formal, symbolic format—succeeded extremely well.

Chapter 4 will provide a more detailed discussion of Hilbert's formalist program and how Gödel's work affected it. An excellent source for the history of mathematics and its foundations is [Bur]. See [Hei] or [BP] for important writings in foundations since 1879, in their original form.

In this chapter we will outline the key concepts of mathematical logic, the logic that underlies the practice of mathematics. Readers who want a more thorough introduction to this material at a fairly elementary level may want to consult [Wolf], [Men], [Ross], or [End]. A more advanced treatment is given in [Sho].

1.2 Propositional logic

This section and the next will cover the most elementary ideas of mathematical logic. Many people who have studied mathematics at the post-calculus level should be familiar with this material, either through a specific course (in logic, or perhaps an "introduction to higher mathematics" course) or through general experience. Readers in this category are encouraged to skip or skim these sections.

Informal treatments of propositional logic (also known as **sentential logic** or the **propositional calculus**) usually begin by defining a **proposition** to be simply "a declarative sentence that is either true or false." Frankly, this definition is rather imprecise. Certainly, questions and commands are not propositions. Statements such as "Snow is white" and "$2 + 2 = 5$" are often given as examples of propositions, while "$x + 3 = 7$" is not a proposition because it is not true or false as it stands. Its truth or falsity depends on the value of x, and so it is called a **propositional function** or a **predicate**.

But there are all sorts of declarative sentences whose inclusion as propositions is debatable. These include statements about the future ("The Cubs will win the World Series before 2099"), certain statements about the past ("The continent of Atlantis sank into the ocean"), value judgments ("Life is good"), abstract mathematical statements ("The continuum hypothesis is true"), and statements in which a noun or a pronoun is not clearly specified and thus is similar to a mathematical variable ("Bill is not the president"). Indeed, one could even question the classic "Snow is white," since in reality snow is not always white.

However, these philosophical problems are not germane to our mathematical treatment of logic. Let us use the word **statement** to mean any declarative sentence (including mathematical ones such as equations) that is true or false or could become true or false in the presence of additional information.

We will use the **propositional variables** P, Q, and R (possibly with subscripts) to stand for statements.

Propositional logic studies the meaning of combinations of statements created by using certain words called **connectives**. The most commonly used connectives, and the symbols used to abbreviate them, are: "and" ($\wedge$), "or" ($\vee$), "not" ($\sim$), "implies" or "if ... then" ($\rightarrow$), and "if and only if" ($\leftrightarrow$), commonly abbreviated "iff." The grammatical rules for the use of the connectives are simple: if P and Q are statements, then so are $P \wedge Q$, $P \vee Q$, $\sim P$, $P \rightarrow Q$, and $P \leftrightarrow Q$.

There is nothing sacred about this set of connectives, or the number five. One could use many more connectives (e.g., "neither ... nor," "not both," "unless"); or one could be very economical and express all these meanings with just two connectives (such as "and" and "not") or even with just one connective. In other words, there is plenty of redundancy among the five standard connectives.

There is some useful terminology associated with the connectives. A statement of the form $P \wedge Q$ is called the **conjunction** of the two **con-**

juncts P and Q. Similarly, $P \vee Q$ is the **disjunction** of the **disjuncts** P and Q. $\sim P$ is called the **negation** of P, while $P \to Q$ is called a **conditional** statement or an **implication**. In $P \to Q$, P is the **hypothesis** or the **antecedent** of the implication, while Q is its **conclusion** or **consequent**. Finally, $P \leftrightarrow Q$ is a **biconditional** statement or an **equivalence**.

If a statement contains more than one connective, parentheses may be used to make its meaning clear. In ordinary mathematics the need for parentheses is often eliminated by an understood priority of operations. For example, in algebra multiplication is given higher priority than addition, so $a + b \cdot c$ means $a + (b \cdot c)$ as opposed to $(a + b) \cdot c$. Similarly, we give the connectives the following priorities, from highest to lowest:

$$\sim, \wedge, \vee, \to, \leftrightarrow .$$

So $P \to Q \wedge R$ would mean $P \to (Q \wedge R)$ rather than $(P \to Q) \wedge R$. Parentheses can be used to convey the other meaning. Similarly, $\sim P \to Q$, with no parentheses, would mean $(\sim P) \to Q$ rather than $\sim (P \to Q)$.

A statement is called **atomic** if it has no connectives or quantifiers, and **compound** otherwise. (Quantifiers are introduced in the next section.)

In **classical** or **Aristotelian** logic, which is almost universally accepted as the appropriate type of logic for mathematics, the meaning of the connectives is defined by **truth functions**, also called **truth tables**. These are literally functions whose inputs and outputs are the words "true" and "false" (or T and F) instead of numbers. Specifically, $P \wedge Q$ is true if and only if both P and Q are true. $P \vee Q$ is true if and only if at least one of P and Q is true. $\sim P$ is true if and only if P is false. $P \to Q$ is true *unless* P is true and Q is false. $P \leftrightarrow Q$ is true if and only if P and Q are both true or both false.

Two of these truth tables deserve some discussion. The truth table for $P \vee Q$ defines the so-called **inclusive or**, in which $P \vee Q$ is considered true when both P and Q are true. Thus, for instance, "$2 + 2 = 4$ or $5 > \pi$" is a true statement. One could also define the **exclusive or**, where "P exor Q" is true if and only if exactly one of P or Q is truc.

Every mathematics student needs to develop an understanding of the logical meaning of implication. Note that $P \to Q$ is true in three of the four possible cases. In particular, it is *automatically* true if P is false. For example, "If $2 + 2 = 3$, then Mars has seven moons" is true. In ordinary speech, a statement of the form "P implies Q" normally asserts that there is some causal relationship between P and Q, that the truth of P somehow causes the truth of Q. This is not required in mathematics or logic (or in sarcastic implications such as "If you can run a six minute mile, then I'm the Queen of England").

It is also worth noting that, among the four basic connectives that involve more than one substatement, implication is the only one that is not symmetrical in P and Q. It is very important not to confuse an implication $P \to Q$ with its **converse** $Q \to P$.

A statement that is built up from simpler ones using connectives only is called a **propositional combination** or a **Boolean combination** of those statements. The truth functions of the connectives allow us to determine the truth function of any propositional combination of statements. For readability, truth functions are usually written for statements that are built up from propositional variables. It is easy to show that if a symbolic statement has n propositional variables, then the domain of its truth function consists of 2^n combinations of Ts and Fs.

Example 1. Here is a truth table for the symbolic statement

$$(P \to Q) \leftrightarrow (R \land P).$$

P	Q	R	$P \to Q$	$R \land P$	$(P \to Q) \leftrightarrow (R \land P)$
T	T	T	T	T	T
T	T	F	T	F	F
T	F	T	F	T	F
T	F	F	F	F	T
F	T	T	T	F	F
F	T	F	T	F	F
F	F	T	T	F	F
F	F	F	T	F	F

Note that since this statement has three propositional variables, its truth table has eight rows. The first three columns of the truth table show all possible combinations of the inputs T and F, and for uniformity they are arranged in a way that is equivalent to counting from 0 to 7 in base 2 (with T and F representing 0 and 1, respectively). The next two columns provide the truth values of the substatements $\mathrm{P} \rightarrow \mathrm{Q}$ and $\mathrm{R} \wedge \mathrm{P}$, and the last column gives the final output for the whole statement.

Exercise 1. Write out the truth tables for these statements:

(a) $(\mathrm{P} \wedge \mathrm{Q}) \leftrightarrow (\mathrm{Q} \rightarrow \sim \mathrm{P})$.

(b) $\mathrm{P} \rightarrow (\mathrm{Q} \rightarrow (\mathrm{P} \wedge \mathrm{Q}))$.

(c) $(\mathrm{P} \vee \mathrm{Q}) \leftrightarrow (\mathrm{R} \wedge \mathrm{P})$.

Here is the most important concept of propositional logic:

Definition. A statement P is called a **tautology** or **law of propositional logic** if there is a set of substatements of P such that:

(a) P is a propositional combination of those substatements, and

(b) P is true for every combination of truth values that is assigned to these substatements.

Example 2. If P and Q are any statements, then $\mathrm{P} \vee \sim \mathrm{P}$ and $\mathrm{P} \rightarrow (\mathrm{P} \vee \mathrm{Q})$ are tautologies. In particular, "Either it's raining or it isn't" and "$x < 2$ implies $x < 2$ or $y > 7$" are tautologies.

Example 3. The equation $2 + 2 = 4$ is not a tautology. Its only substatement is itself, and we must assign both T and F as truth values for that substatement when we construct a truth table for the statement. In other words, it's not a tautology because its form is simply "P," with no shorter substatements. This may be a law of arithmetic, but is it not a law of propositional logic.

Similarly, "For every number x, $x = x$" is not a tautology. Simply put, it cannot be a tautology because it includes no connectives. We will see that this is a law of *predicate* logic, but that is quite different from a law of propositional logic.

Definitions. A statement whose negation is a tautology is called a **contradiction**. Two statements P and Q are called **propositionally equivalent** if $\mathrm{P} \leftrightarrow \mathrm{Q}$ is a tautology. And a statement Q is said to be a **propositional consequence** of statements $\mathrm{P}_1, \mathrm{P}_2, \ldots, \mathrm{P}_n$ if

$$(\mathrm{P}_1 \wedge \mathrm{P}_2 \wedge \cdots \wedge \mathrm{P}_n) \rightarrow \mathrm{Q}$$

is a tautology.

Example 4. In general, an implication $\mathrm{P} \rightarrow \mathrm{Q}$ is not propositionally equivalent to its converse $\mathrm{Q} \rightarrow \mathrm{P}$. On the other hand, an implication $\mathrm{P} \rightarrow \mathrm{Q}$ is always propositionally equivalent to its **contrapositive** $\sim \mathrm{Q} \rightarrow \sim \mathrm{P}$. From this it follows that the converse of $\mathrm{P} \rightarrow \mathrm{Q}$ is propositionally equivalent to $\sim \mathrm{P} \rightarrow \sim \mathrm{Q}$, which is called the **inverse** of $\mathrm{P} \rightarrow \mathrm{Q}$.

Example 5. De Morgan's laws are the following simple but useful propositional equivalences (that is, biconditional tautologies):

$$\sim (\mathrm{P} \vee \mathrm{Q}) \leftrightarrow (\sim \mathrm{P} \wedge \sim \mathrm{Q}),$$

$$\sim (\mathrm{P} \wedge \mathrm{Q}) \leftrightarrow (\sim \mathrm{P} \vee \sim \mathrm{Q}).$$

The definition of propositional consequence may be restated as follows: Q is a propositional consequence of a finite set of statements S iff Q must be true (in a truth table) whenever all the statements in S are true. When restated in this way, the definition can be generalized by dropping the requirement that S be finite. It can then be proved that if Q is a propositional consequence of an infinite set of statements S, then Q is in fact a propositional consequence of some finite subset of S. This is an important result called the **compactness theorem** for propositional logic, because it is related to the compactness of a certain topological space. These ideas will be discussed more thoroughly in Chapter 5, in the context of predicate logic.

Here are a couple of results that describe standard forms for statements of propositional logic:

Proposition 1.1. *Suppose that* P *is a Boolean combination of statements* $\mathrm{Q}_1, \mathrm{Q}_2, \ldots \mathrm{Q}_n$. *Then there is a statement that is proposition-*

ally equivalent to P *and is in* ***disjunctive normal form*** *with respect to* $Q_1, Q_2, \ldots Q_n$*, meaning a disjunction of conjunctions of the* Q_i*'s and their negations.*

Proof. Here is a very informal proof: essentially, the disjunctive normal form of a statement comes directly from its truth table. Each row of the truth table with output T indicates one of the conjunctions whose disjunction must be taken. For instance, the truth table of Example 1 has just two T's in its output column. Therefore, the disjunctive normal form for that statement (with respect to P, Q, and R) is simply $(P \wedge Q \wedge R) \vee (P \wedge \sim Q \wedge \sim R)$. ∎

Proposition 1.2. *Suppose that* P *is a Boolean combination of statements* $Q_1, Q_2, \ldots Q_n$*. Then there is a statement that is propositionally equivalent to* P *and is in* ***conjunctive normal form*** *with respect to* $Q_1, Q_2, \ldots Q_n$*, meaning a conjunction of disjunctions of the* Q_i*'s and their negations.*

Exercise 2. Prove this proposition, using the previous proposition and de Morgan's laws.

Example 6. The statement $[(P \rightarrow \sim Q) \leftrightarrow (R \vee \sim P)]$ has the conjunctive normal form equivalent

$$[(P \vee Q \vee R) \wedge (\sim P \vee Q \vee R) \wedge (\sim P \vee \sim Q \vee \sim R)].$$

In the course of a mathematical proof, it seems intuitively clear that we should be allowed to assert any statement that is a propositional consequence of previous steps in the proof. Several important "proof methods" in mathematics are based on this principle, notably:

- Modus ponens: From P and $P \rightarrow Q$, we may conclude Q.
- Modus tollens: From $\sim Q$ and $P \rightarrow Q$, we may conclude $\sim P$.
- Indirect proof, proof by contradiction, or "reductio ad absurdum": From $(\sim P) \rightarrow (Q \wedge \sim Q)$, we may conclude P.
- Proof by cases: From $P \vee Q$, $P \rightarrow R$, and $Q \rightarrow R$, we may conclude R.

- Proof of a biconditional using separate "forward" and "reverse" directions: From $P \to Q$ and $Q \to P$, we may conclude $P \leftrightarrow Q$.

In fact, the only important proof method that is based on propositional logic but does not follow immediately from this principle is the method of **conditional proof**, often called "direct proof" of an implication: if we can produce a proof of Q from the assumption P, then we may conclude $P \to Q$. See Theorem 1.3 at the end of Section 1.4.

In some mathematical proofs, the statement being proved is a propositional consequence of given statements (axioms, previous theorems, and hypotheses of that proof). If there are only a finite number of givens, propositional consequence can be verified or refuted by a truth table, which requires only a finite amount of computation. Therefore, proofs of this type are generally straightforward. However, most proofs in mathematics require some reasoning involving quantifiers and so cannot be carried out entirely within propositional logic.

1.3 Quantifiers

Propositional logic captures only a part of the language and reasoning used in mathematics. To complete the picture, quantifiers are needed. Quantifiers are used in conjunction with **mathematical variables** (as opposed to propositional variables): variables that stand for real numbers, sets, vectors, functions, or other mathematical objects. If we want to say that every real number satisfies a certain inequality, or that there is a function with a particular property, quantifiers are needed. (From now on, the unmodified word "variable" will always mean "mathematical variable.")

For the most part, only two quantifiers are viewed as basic to mathematical language. The **universal quantifier** corresponds to the words "for all," "for every," "for any," or "for each" and is represented by the symbol $\forall$. The **existential quantifier** corresponds to the words "there exists" or "for some" or "there is a" (meaning that there is *at least one*), and is represented by the symbol $\exists$.

The grammatical rules for the use of quantifiers are simple: if P is any statement, and x is any mathematical variable, then $\forall x\text{P}$ and $\exists x\text{P}$ are also statements. It is not required that P include the variable x, but there is no reason to attach the quantifier unless it does.

Quantifiers are given higher priority than any of the connectives. So, for example, $\forall x\text{P} \vee \text{Q}$ means $(\forall x\text{P}) \vee \text{Q}$ rather than $\forall x(\text{P} \vee \text{Q})$. As usual, parentheses can be used to clarify the meaning of a symbolic statement with quantifiers.

When the same type of quantifier is repeated, it is common to replace subsequent occurrences of it by commas. So instead of $\forall x\, \forall y\text{P}$, we would usually write $\forall x, y\text{P}$. There are also abbreviations for quantified variables that are restricted to a certain domain: $(\forall x \in A)\text{P}$ stands for $\forall x(x \in A \rightarrow \text{P})$, and $(\exists x \in A)\text{P}$ stands for $\exists\, x(x \in A \wedge \text{P})$. Similar abbreviations are used for restrictions involving symbols such as $\subset$, $\subseteq$, $<$, $>$, $\leq$, and $\geq$ in place of $\in$.

A more thorough description of the formal grammar of mathematics will be provided in the next section.

Example 7. Suppose $U \subseteq \mathbb{R}$ and $f : U \rightarrow \mathbb{R}$ (in other words, f is a real-valued function with domain U). Then, for any number x in U, the rigorous definition of what it means for f to be continuous at x is

$$\forall \epsilon > 0\, \exists \delta > 0\, \forall v \in U(|v - x| < \delta \rightarrow |f(v) - f(x)| < \epsilon).$$

So this definition requires three quantifiers, all of which are restricted quantifiers in the sense mentioned above. For f to be a continuous function means that for every x in U, f is continuous at x, so this statement requires four quantifiers. The stronger condition that f is *uniformly* continuous is obtained simply by moving the quantifier on x to the right of the quantifier on δ:

$$\forall \epsilon > 0\, \exists \delta > 0\, \forall x, v \in U(\ldots),$$

where the inner statement $(\ldots)$ is as above.

These definitions, and the related definitions for limits of functions and sequences, are due to Karl Weierstrass and were among the most important pieces of the **arithmetization** of calculus and analysis that took place in the nineteenth century—that is, the rigorous development of these subjects without reliance on geometric intuition.

Example 8. The intermediate value theorem states that every continuous real-valued function that is defined on a closed interval, and takes values of opposite sign at the endpoints, has a zero. If we analyze the connective and quantifier structure of this statement, we see that it has the form

$$\forall a, b, f[(a \in \mathbb{R} \wedge b \in \mathbb{R} \wedge a < b \wedge f \text{ is a continuous function from } [a, b] \text{ to } \mathbb{R} \wedge f(a) \cdot f(b) < 0) \rightarrow \exists c(a < c < b \wedge f(c) = 0)].$$

So the statement includes a total of eight quantifiers, including the ones required to say that f is continuous. Note that seven of the quantifiers involve real number variables, but one of them, $\forall f$, refers to all functions.

Definitions. Every occurrence of the variable x in a statement (or substatement) of the form $\forall x\mathrm{P}$ or $\exists x\mathrm{P}$ is called **bound**. An occurrence of a variable is said to be **free** if it is not bound. A statement with no free variables (that is, no free occurrences of any variable) is called **closed**, or a **sentence**.

Example 9. In the statement $\exists y(y = x^2) \vee y > 3$, the last occurrence of y and the occurrence of x are free. So this statement has the free variables x and y. The first and second occurrences of y are bound. Thus y is both free and bound in this statement. Generally, this situation can and should be avoided, as we will soon see.

Definitions. Given a statement P, a **generalization** of P is formed by putting any finite (possibly empty) sequence of universal quantifiers in front of P. A **universal closure** of P is a generalization of P that is closed.

Notation. Notation such as $\mathrm{P}(x)$ and $\mathrm{P}(x, y)$ is used to convey that the variables shown in parentheses might be free in the statement P. However, the use of this notation actually guarantees nothing. The variables shown are not required to be free in P, and usually P is also allowed to have free variables other than those shown. A similar situation exists with the usual function notation ($f(x)$, etc.) of mathematics.

It is essential to understand the difference between free and bound variables. A free variable represents a genuine unknown, whose value must be specified before the given statement's truth or falsity can be determined (unless that free variable's appearance in the statement is trivial, such as x in the equation $x + 3 = x + y$). A bound variable is really a "dummy variable," like the variable of integration in a definite integral or the indexing variable of a summation. It is never necessary to know the value of a bound variable; rather, one needs to know the *domain* of bound variables in order to know whether a statement is true or false.

Example 10. Suppose that, in the context of basic algebra, you are asked whether the equation $x + 5 = 3$ is true. Here x is free, so you would want to know the value of x. The statement is true if and only if $x = -2$.

Now suppose you are asked whether the statement $\exists x(x + 5 = 3)$ is true. This time x is bound, so it makes no sense to ask the value of x. Rather, it makes sense to ask the domain of x. For instance, the statement is true if the domain of x is the set $\mathbb{R}$ of real numbers or the set $\mathbb{Z}$ of integers, but false if its domain is the set $\mathbb{N}$ of natural numbers (nonnegative integers).

By the way, many books define $\mathbb{N}$ to be the set of *positive* integers, so that $0 \notin \mathbb{N}$. In logic and set theory, it is more convenient to let 0 be a natural number, and we will stick with this convention.

Since a bound variable is just a dummy variable, we can replace any bound variable of a statement with a variable that does not appear in the original statement, without changing the meaning of that statement. For example, the usual additive inverse axiom $\forall x\, \exists y(x + y = 0)$ could just as well be written $\forall x\, \exists z(x + z = 0)$ or $\forall u\, \exists v(u + v = 0)$. More precisely, when we make such a change, the new statement will always be **logically equivalent** to the old statement; we will define logical equivalence in the next section.

It is not grammatically incorrect to have a variable appear both free and bound in the same statement, as in Example 9, but the result is usually awkward and/or confusing. By the previous paragraph, it is always possible to avoid this situation, and we will consistently do so from now on. For example, the statement $x > 3 \wedge \exists x(x^3 - 3x + 1 = 0)$ becomes much more readable, without any change in meaning, if it is rewritten as $x > 3 \wedge \exists u(u^3 - 3u + 1 = 0)$.

Notation. Suppose that $\mathrm{P}(x)$ is some statement in which x might occur free. Then we write $\mathrm{P}(t)$ to denote the statement that results by replacing every free occurrence of x in P by the **term** t. A term could be a variable, but it could also be a more complex expression denoting a mathematical object, such as -3.276 or $5yz^2$ (in the context of the real numbers). This convention is analogous to the standard practice of replacing a variable by terms in function notation such as $f(x)$. When we use this notation, we require that t denotes the same sort of mathematical object that x does, and that no free variable of t becomes bound in $\mathrm{P}(t)$. The next example illustrates the latter restriction:

Example 11. Suppose that we want $\mathrm{P}(n)$ to mean that the integer n is even. In symbols, $\mathrm{P}(n)$ could be $\exists m(n = 2m)$. (Let us assume that all variables in this example have the integers as their domain.) Then suppose that we want to express formally that m is even. We can write $\mathrm{P}(m)$ for this statement, but we cannot use m for the quantified variable in it. For instance, we could let $\mathrm{P}(m)$ be $\exists k(m = 2k)$. If we want to formalize the statement that m and n are both even, then m and n are free in this statement and so neither of them should be quantified. So we might take $\mathrm{P}(m) \wedge \mathrm{P}(n)$ to be $\exists k(n = 2k) \wedge \exists j(m = 2j)$. It would be acceptable to use the same bound variable in both conjuncts, but most people would probably prefer different variables, since k and j obviously represent different numbers unless $n = m$.

Uniqueness

The quantifier $\exists$ is given the interpretation "there is at least one...." In mathematics, one also needs to be able to express the existence of a *unique* object with some property $P(x)$. There are several equivalent ways to express this, for example

$$\exists x[P(x) \wedge \forall y(P(y) \rightarrow y = x)].$$

We will use the abbreviation $\exists! x P(x)$ for this statement. Many other types of statements can be expressed using the basic quantifiers. For instance, "There are at least two objects such that P" is expressed by $\exists x, y[P(x) \wedge P(y) \wedge x \neq y]$. On the other hand, there is no way in *pure logic* to write a single statement expressing that there are an infinite number of objects with a certain property.

Exercise 3. Show that the statement $\exists x[P(x) \wedge \forall y(P(y) \rightarrow y = x)]$ really does say that there is exactly one object for which P is true. What's intended here is an intuitive verbal explanation, not a rigorous proof.

Exercise 4. Write symbolic statements expressing:

(a) "There are at least three objects such that P."

(b) "There are exactly two objects such that P."

Proof methods based on quantifiers

The previous section defined propositional consequence and listed several "proof methods" based on this idea. The definition of the analogous concept of **logical consequence**, which includes reasoning based on quantifiers as well as connectives, must wait until the next section. What we can do now is list some standard proof methods that are based on quantifiers and are known to be logically correct (in the sense that if the assumptions or "givens" are true, the conclusion must also be true). The correctness of some of these proof methods requires the assumption that the domain of a mathematical variable must be nonempty, which is a standard convention.

- $\sim \forall x P$ and $\exists x \sim P$ are logically equivalent (that is, each is a logical consequence of the other). So we can always go from either of these to the other in a proof.

- Similarly, $\sim \exists x\mathrm{P}$ and $\forall x \sim \mathrm{P}$ are logically equivalent. (By negating both sides of (1) and (2), we also get that $\forall x\mathrm{P}$ is logically equivalent to $\sim \exists x \sim \mathrm{P}$, and $\exists x\mathrm{P}$ is logically equivalent to $\sim \forall x \sim \mathrm{P}$. In other words, either quantifier can be defined in terms of the other.)
- Universal specification: From $\forall x\mathrm{P}(x)$ we may conclude $\mathrm{P}(t)$, where t is an appropriate term in the sense described earlier.
- Universal generalization: If we can prove $\mathrm{P}(x)$, with no assumptions about x other than what sort of object it represents, then we may conclude $\forall x\mathrm{P}(x)$, where the quantified variable has the same restriction as in the assumption.
- Existential generalization, or "proof by example": From $\mathrm{P}(t)$, where t is an appropriate term, we may conclude $\exists x\mathrm{P}(x)$.

Universal specification is the simple principle that if some property is true for all members of a certain domain, then it is true for any particular one that we specify. For instance, since $\sin^2 x + \cos^2 x = 1$ holds for every real number x (being a "trigonometric identity"), we may assert that $\sin^2 \pi + \cos^2 \pi = 1$.

Universal generalization refers to the type of proof that begins "Let x be an arbitrary ... " and then proves something about x, concluding that this "something" is true for all members of the intended domain.

Existential generalization says that if we have found a particular object satisfying some property, then we can assert that there exists an object satisfying that property. Mathematicians don't usually use the phrase "proof by example," but they frequently talk about "proof by counterexample." This refers to a method of *disproving* a statement of the form $\forall x\mathrm{P}(x)$: first, note that $\sim \forall x\mathrm{P}(x)$ is equivalent to $\exists x \sim \mathrm{P}(x)$, by the first equivalence listed above. So if we can produce a term t (in essence, a particular number or other object) for which we can prove $\sim \mathrm{P}(t)$, we may conclude $\exists x \sim \mathrm{P}(x)$ and hence $\sim \forall x\mathrm{P}(x)$.

Example 12. Suppose you were asked to prove or find a counterexample to the assertion that $n^2 - n + 41$ is prime for every natural number n. To prove this statement, you could try to prove it for an arbitrary n (that is, by universal generalization). Or, you could try to prove it by mathematical induction. But no such proof will work, because the statement is not true. Only one counterexample is needed to disprove the assertion, but every nonzero multiple of 41 (including 41 itself) is easily seen to provide a counterexample.

The fascinating thing about this polynomial, which was discovered by Leonhard Euler, is that it gives prime outputs for every n between 0 and 40, so it is tempting to guess that all of its outputs are prime. But it is not hard to show that no nonconstant polynomial gives prime outputs for every natural number input. In fact, the outputs must include an infinite set of composite numbers.

The terminology introduced above ("universal specification," etc.) is not commonly used by mathematicians, but the methods themselves are extremely common and useful.

Translating statements into symbolic form

Both natural language and symbolic language, separately or together, are used to express mathematical statements. It is essential for mathematicians to be able to "translate" back and forth between verbal and symbolic versions of statements. We conclude this section with several examples of this translation process. We'll begin with some nonmathematical statements, but even though the content of these statements may seem frivolous, the translation process is still a valuable mental exercise.

It is important to bear in mind that when a statement is translated from one form to another, *the original version and the translated version must have exactly the same free variables.*

Example 13. Translate the statement "Every person likes somebody who doesn't like him or her" into symbols, by defining a propositional variable for each atomic substatement and specifying the domain of every mathematical variable used.

Solution. The statement is based on the property of one person liking another. So let $L(x, y)$ stand for "x likes y," where x and y are "people variables"—that is, their domain is the set of all people. The statement can then be written symbolically as

$$\forall x\, \exists y[L(x, y) \wedge \sim L(y, x)].$$

Exercise 5. Translate the following statements into symbolic form, as in the previous example.

(a) All crows are black, but not all black things are crows.

(b) If everybody complains, no one will get help.

(c) Everybody loves somebody sometime.

Now let's consider some mathematical statements. Generally, these can be translated into symbolic form using standard mathematical symbols:

Example 14. Let j, k, l, m, and n be integer variables. Then the statement "n is even" translates to $\exists m(n = 2m)$ in symbols, while "n is odd" becomes $\exists m(n = 2m + 1)$. The (true, but not at all obvious) statement that every nonnegative integer can be written as the sum of four squares becomes

$$\forall n \geq 0\, \exists j, k, l, m(n = j^2 + k^2 + l^2 + m^2).$$

By the way, can every nonnegative integer be written as the sum of three squares?

Exercise 6. Translate these statements into completely symbolic form. Use m, n, and k as integer variables and x, y, and z as real variables:

(a) n is a prime number. (Hint: a prime number is an integer greater than 1 that cannot be written as the product of two such integers.)

(b) There is no largest real number.

(c) For any two distinct integers, there is a real number between them.

(d) For any two integers, there's an integer between them if and only if they differ by more than 1.

1.4 First-order languages and theories

We will now give a more rigorous and detailed treatment of the ideas introduced in the previous two sections. First, we must explain what is meant by a **first-order language**. Any particular first-order language is determined by its *symbols*. These consist of:

1. a denumerable list of (mathematical) **variables** $v_0, v_1, v_2, \ldots$. For readability, we will also use arbitrary letters, with or without subscripts, as variables.
2. for each natural number n, a set of n**-ary relation symbols** (also called **predicate symbols**).
3. for each natural number n, a set of n**-ary function symbols**.
4. a set of **constant symbols**.
5. the **equality symbol** $=$.
6. the **connectives** $\vee$, $\wedge$, $\sim$, $\rightarrow$, and $\leftrightarrow$.
7. the **quantifiers** $\forall$ and $\exists$.
8. parentheses and the comma.

Note that (1), (5), (6), (7), and (8) are the same for all first-order languages (although it is permissible, and occasionally fruitful, to consider a first-order language without the equality symbol). Therefore, all that distinguishes one first-order language from another are the sets described in (2), (3), and (4). There is no restriction on the cardinalities of these sets; in particular, they may be empty or uncountable.

There are many well-known ways to make the above list more economical: one may choose to have a shorter list of connectives. Either of the quantifiers can be eliminated. One may view equality as a particular binary relation symbol (albeit a special one). One may view constant symbols as 0-ary function symbols. And, for most purposes, one may use the usual definition of a function to replace each n-ary function symbol by an $(n+1)$-ary relation symbol.

It is harmless, and often convenient, to allow a first-order language to have more than one **sort** of variable, with a denumerable list of variables of each sort. Such a language is called a **many-sorted** first-order language. The sort of every constant symbol and of every argument of every relation symbol and function symbol must be specified. For example, for the first-order theory of a vector space, we would presumably want two sorts of variables: one for scalars and one for vectors. When doing real analysis, we might want separate variables for real numbers and integers; but the same effect can be achieved with predicate symbols, or with set notation if sets are allowed in the discussion.

The definition of a first-order language must specify the *syntax* of the language as well as the symbols. This part of the definition is the same for all first-order languages: first we define what is meant by a **term**. Every variable and every constant symbol is a term. Also, if $t_1, t_2, \ldots, t_n$ are terms and f is an n-ary function symbol, then the string of symbols $f(t_1, t_2, \ldots, t_n)$ is also a term. (Note that this is an inductive definition, or more precisely a recursive definition.)

Having defined terms, we can then define the **well-formed formulas**, or simply **formulas**, of a first-order language. First of all, if $t_1, t_2, \ldots, t_n$ are terms and R is an n-ary relation symbol, then the string of symbols $\mathrm{R}(t_1, t_2, \ldots, t_n)$ is a formula. Equations of the form $t_1 = t_2$ are also formulas. (These two categories comprise the atomic formulas of the language.) Inductively, if P and Q are any two formulas, then so are $(\mathrm{P} \wedge \mathrm{Q})$, $(\mathrm{P} \vee \mathrm{Q})$, $\sim \mathrm{P}$, $(\mathrm{P} \rightarrow \mathrm{Q})$, $(\mathrm{P} \leftrightarrow \mathrm{Q})$, $\forall v_n \mathrm{P}$, and $\exists v_n \mathrm{P}$ (where n can be any natural number).

In practice, many binary relation and function symbols are written between their arguments rather than in front of them. For instance, we write $x + y$ instead of $+(x, y)$, and similarly for $\cdot$, $<$, $\in$, etc. There are also more specialized notational conventions such as xy for multiplication and x^y for exponentiation.

When writing formulas, we may use the priority of connectives and quantifiers described in the previous two sections in order to eliminate some of the parentheses required by these syntactical rules. We may also drop parentheses when all the syntactically different interpretations of a certain string of symbols are propositionally equivalent. For instance, since $(\mathrm{P} \wedge \mathrm{Q}) \wedge \mathrm{R}$ and $\mathrm{P} \wedge (\mathrm{Q} \wedge \mathrm{R})$ are propositionally equivalent we would normally shorten both of these to $\mathrm{P} \wedge \mathrm{Q} \wedge \mathrm{R}$.

Every formula of a first-order language is a statement in the sense of Section 1.2, but not conversely.

Notation. We will use the letter $\mathcal{L}$ to denote a first-order language, or, more specifically, the set of formulas of a first-order language. An $\mathcal{L}$-formula simply means a formula of the language $\mathcal{L}$.

Example 15. The **first-order language of an ordered ring** (see Appendix D for basics) has one binary relation symbol $<$ (in addition to equality), a unary function symbol $-$, two binary function symbols $+$ and $\cdot$, and a constant symbol 0. The terms of the language are the variables, the constant 0, and all expressions of the forms $-t_1$, $t_1 + t_2$, and $t_1 \cdot t_2$, where t_1 and t_2 are terms. The atomic formulas are all expressions of the forms $t_1 = t_2$ and $t_1 < t_2$, where t_1 and t_2 are terms.

This language is rather "sparse," but we can make it richer by defining a variety of standard abbreviations such as $t_1 - t_2$, $t_1 > t_2$, $t_1 \leq t_2$, and $t_1 \geq t_2$.

If we omit the symbol $<$, we obtain the first-order language of a ring. If we want to consider rings with unity (ordered or not), we could have another constant symbol 1.

If we want to consider fields, then we might want a symbol for division or reciprocals. Here we run into a snag because division by 0 is not allowed, and under the usual interpretation of first-order logic, function symbols represent operations that are defined for all members of the intended domain. There are several solutions to this difficulty. One is to relax this requirement and allow function symbols to be interpreted as "partial" operations, as in [Bee, pp. 97–99]. Another solution is to define division rather than having a symbol for it in the language, so that $x = y/z$ by definition means $xz = y$ and $z \neq 0$. But then it requires some care to define exactly what is meant when fractions are used as terms in equations such as $(u/v) + (v/x) = y/z$. A third approach to this difficulty is mentioned at the end of Section 1.6.

First-order languages provide a workable framework for translating mathematical statements into a completely symbolic form. This was the first goal of Hilbert's program. To carry out the second part of the formalist program requires a rigorous and accurate definition of what is meant by a *proof* within a first-order language.

The notion of a mathematical proof or deduction, and the criteria for judging the correctness of such proofs, have changed remarkably little in the last two-and-a-third millenia. Although modern mathematicians have found flaws in some of Euclid's definitions and proofs, the vast majority of his deductive arguments are sound by modern standards. Conversely, Euclid, Archimedes, and even Pythagoras would be able to grasp the deductive processes used in today's mathematics and would be quite capable of understanding today's proofs, once they had caught up on the subject material. This is in stark contrast to all the natural sciences, in which the methodology of even three hundred years ago is quite primitive by modern standards and bears little resemblance to today's methodology.

There are several equivalent ways of defining what is meant by a formal proof in a first-order language. The most common ways are based on the concept of the **axiomatic method** in mathematics. This term was coined by Hilbert to describe part of his formalist program, but the essentials of the method were developed in ancient Greece. Certain formulas are taken to be **axioms**. There are **logical axioms** that are universal, independent of the particular first-order language in use. These usually include some or all tautologies, the usual equality axioms, and some simple laws involving quantifiers. One usually also has some **proper axioms**, which are specific to the first-order language and the subject under consideration. These correspond to the Euclidean notion of **postulates**, except that, classically, postulates were supposed to be evident truths, whereas truth does not enter into the formalist viewpoint.

Technically, any set of formulas can be taken to be proper axioms, but in practice proper axioms are almost always sentences. However, when these axioms are listed, it is common to omit universal quantifiers at the beginning of them. Then the usual convention is that some universal closure of the formula shown is the actual axiom. For instance, it is common to write one of the field axioms (the commutative law of addition) as $x + y = y + x$, with the intention that either $\forall x, y(x + y = y + x)$ or $\forall y, x(x + y = y + x)$ is "officially" an axiom. It doesn't matter which universal closure is chosen. We could instead

Euclid was one of the most famous mathematicians of antiquity, and yet almost nothing is known of his life. Not even the years of his birth and death or his birthplace are known. In his youth, he probably studied geometry at Plato's academy in Athens. It is known that he spent many years in Alexandria and reached his prime there around 300 B.C. He is best known for his *Elements*, a monumental compendium of thirteen books. Most of the books deal with plane geometry, but number theory and solid geometry are also covered.

The Elements is the oldest surviving work in which mathematical subjects were developed from scratch in a thorough, rigorous, and axiomatic way. However, the vast majority of the results in *The Elements* were first proved by someone other than Euclid. He is remembered less for his original contributions than for the impressive organization and rigor of his work. *The Elements* was viewed as *the* model of mathematical rigor for over two thousand years, has been printed in over a thousand editions, and is still used as a geometry textbook in some places. No one, before or since, has made nearly as much of an impact on the methodology of mathematics.

A couple of interesting quips are attributed to Euclid. One was mentioned in the preface. On another occasion, when a student asked what he would gain from studying geometry, Euclid supposedly told his servant to "give this man a coin, since he must make a profit from what he learns."

assume that all universal closures or even all generalizations of this formula are axioms.

The formalist program goes beyond the classical axiomatic approach by explicitly defining not just the language and axioms to be used, but also the **rules of inference**. A rule of inference is a precise rule that states how one or more steps in a proof may be used or combined to create a new step that can be asserted. Modus ponens and most of the other "proof methods" listed in the two previous sections are rules of inference. In the usual approach to first-order logic, two features of rules of inference are worth noting: first, they are based entirely on logic, and second, they are the only way of generating new steps (i.e., theorems) from the axioms. If one accepts this approach, then it follows that logic is the only *formal* mechanism by which theorems are proved in mathematics. However, it is important not to overstate this point of view. All mathematicians, including logicians, know that the creative process of mathematics is based on far more than logic. Nor is it essential that all rules of inference be based on logic. For example, mathematical induction is an extremely important proof method; shouldn't it be considered a rule of inference? It certainly could be, but since it states a property of a particular structure, $\mathbb{N}$, it is more common to view it as a proper axiom, stated as an implication.

One standard list of logical axioms and rules of inference for first-order logic is provided in Appendix A. One appealing feature of this version is that there is only one rule of inference, modus ponens.

Definition. Let T be a set of first-order formulas. A **proof** from (proper axioms) T is a finite sequence of formulas ("steps") such that every step is either a logical axiom, a member of T, or the result of applying a rule of inference to previous steps in the proof. A proof of a formula P from T is a proof from T whose last step is P.

More precisely, this is the definition of a **formal proof**. Of course, mathematicians never actually write formal proofs; their proofs are informal in many ways. For one thing, proofs normally are written at least partly in a natural language such as English, rather than a first-order language. Furthermore, they include all sorts of shortcuts, such as definitions, and citing previously proved statements rather than reproving them. Still, most logicians (though perhaps not most mathematicians) are convinced that all correct proofs in mathematics could, with enough effort, be translated into formal proofs of first-order logic.

A **first-order theory** technically consists of two parts: a first-order language, and a set of formulas (usually, sentences) of that language to be used as proper axioms. There is no need to specify the logical axioms or the rules of inference since they are standard, except for inconsequential variations. For the most part, we will use the word **theory** a bit more loosely to mean any set of formulas T in a first-order language.

Notation. We write $T \vdash \mathrm{P}$ ("T **proves** P") to mean that there is a proof of P from T. There are several other ways to read this notation: P is **provable** from (or in) T, P is **deducible** from T, P is **derivable** from T, P is a **theorem** of T, or P is a **logical consequence** of T.

There are also numerous variants of this notation:

$T_1 \vdash T_2$ means that $T_1 \vdash \mathrm{P}$ for every $\mathrm{P} \in T_2$.

$\mathrm{Q} \vdash P$ means $\{\mathrm{Q}\} \vdash \mathrm{P}$.

$\vdash \mathrm{P}$ means $\emptyset \vdash \mathrm{P}$, that is, P is derivable from logical axioms and rules of inference alone. A formula with this property is called a **law of logic**.

Finally, P and Q are called **logically equivalent** if $\vdash (\mathrm{P} \leftrightarrow \mathrm{Q})$.

Notation. The set of theorems of T will be denoted $Thm(T)$.

Definition. If $Thm(T_1) \subseteq Thm(T_2)$ (which is the same as saying $T_2 \vdash T_1$), we say that T_1 is a **subtheory** of T_2, and T_2 is an **extension** of T_1. If $Thm(T_1) = Thm(T_2)$, we say that T_1 and T_2 are **equivalent** theories.

The notions of "logical consequence," "law of logic," and "logical equivalence" are analogous to the corresponding propositional notions defined in Section 1.2. The important qualitative difference is that the propositional versions are all based on truth tables and therefore are **computable** or **decidable**—a finite amount of straightforward computation always suffices to determine whether or not they hold, provided that the sets of formulas involved are finite. This is not the case for the concepts that we just defined. There is no "effective procedure" (essentially, a computer program) that can even determine whether or not any given first-order sentence is a law of logic. We should perhaps be thankful for this, since if there were such a computer program, humans would hardly ever (if ever) be needed for proving theorems! (Computability and decidability will be thoroughly discussed in Chapter 3.)

Even though the *notions* of "logical consequence," "law of logic," and "logical equivalence" are analogous to notions defined in Section 1.2, the *definitions* themselves are not analogous. The definitions in Section 1.2 are **semantic**, meaning that they are based on some concept of truth—in this case, truth tables. A law of propositional logic is a statement that is true under all possible interpretations of certain substatements. It is very natural to use semantic definitions for propositional logic because truth tables are so simple to understand and use.

By contrast, the definitions we have just given for first-order logic are **syntactic**—they are based on provability in some formal system. We can obtain syntactic definitions of the corresponding propositional notions simply by removing those parts of Appendix A that mention quantifiers. It is also fruitful to give semantic definitions of these notions for first-order logic; again, these are based on the idea that a law of logic is a statement that is true under all possible interpretations. However, it requires some "machinery" to make this precise so we will defer these definitions until Chapter 5.

Are the semantic and syntactic definitions equivalent? Yes they are, which is reassuring. This tells us that a formula P is provable from a theory T if and only if P must be true whenever all the statements in T are true. A detailed discussion of this equivalence in the case of first-order logic will also be given in Chapter 5. This equivalence provides a compelling argument that the formalists succeeded in codifying, precisely and compactly, the 2300-year-old notion of a correct mathematical deduction.

We will occasionally refer to reasoning "informally." Usually, this will mean using the fact that provability corresponds to truth, in order to avoid a tedious formal proof. For instance, it can certainly be proved formally that $\forall x\, \forall y \mathrm{P}(x, y)$ is logically equivalent to $\forall y\, \forall x \mathrm{P}(x, y)$. But the obvious truth of this equivalence may be considered a nonrigorous proof, acceptable in most circumstances.

Definitions. A theory T is called **consistent** if no contradiction can be derived from it. A formula P is said to be **independent** of T if neither P nor $\sim$ P can be proved from T. T is called **complete** if it is consistent and no *sentence* of its language is independent of it. In other words, T is complete if $T \vdash$ P or $T \vdash \sim$ P, but not both, for every sentence P of the language of T. (The language of T is the smallest first-order language that contains T.)

The subject that deals with first-order languages and theories is called **first-order predicate logic**, or simply **first-order logic**. We conclude this section by stating two of the most important **metatheorems** of first-order logic, that is, theorems *about* first-order logic. Their proofs can be found in most logic texts such as [End] and [Sho]:

Theorem 1.3 (Deduction Theorem).

$$\textit{If } T \cup \{\mathrm{P}\} \vdash \mathrm{Q}, \textit{ then } T \vdash (\mathrm{P} \rightarrow \mathrm{Q}).$$

The converse of this theorem also holds; essentially, it is the rule of inference modus ponens. The deduction theorem is the formal justification of the method of conditional proof (direct proof of implications). Similarly, the next result is the formal justification of the method of universal generalization:

Theorem 1.4 (Generalization Theorem). *If $T \vdash \mathrm{P}(x)$ and the variable x does not occur free in any formula in T, then $T \vdash \forall x \mathrm{P}(x)$.*

Exercise 7. Let T be a theory and P a sentence. Prove:

(a) If T is inconsistent, then every formula is derivable from T.

(b) $T \vdash$ P if and only if $T \cup \{\sim \mathrm{P}\}$ is inconsistent.

(c) P is independent of T if and only if $T \cup \{\mathrm{P}\}$ and $T \cup \{\sim \mathrm{P}\}$ are both consistent.

1.5 Examples of first-order theories

In the previous section it was claimed that first-order languages form an adequate framework for the translation of mathematical statements into a purely symbolic form. We will now illustrate, by means of several examples, precisely what we mean by this claim. As we will see, the claim is not completely unproblematic.

Example 16 (Peano Arithmetic). One of the first successful formalizations of a part of mathematics was the axiomatization of arithmetic, first carried out by Richard Dedekind and then refined by Peano in the 1890s. The intended domain of this theory is the set $\mathbb{N}$, but one can use the theory to define and study $\mathbb{Z}$ and $\mathbb{Q}$ (the set of rational numbers) as well. The most common first-order language $\mathcal{L}$ used for this theory has two binary function symbols, $+$ and $\cdot$, a unary function symbol S ("successor"), and a constant symbol $\overline{0}$. (We write $\overline{0}$ rather than 0 to emphasize that this is a formal symbol, not the number 0. However, most logic books—including this one—are not consistently careful to make this type of distinction.) As usual, the operators $+$ and $\cdot$ are written between their arguments. The proper axioms of Peano arithmetic, PA for short, include the following straightforward ones:

1. $S(x) \neq \overline{0}$. ($\overline{0}$ is not any number's successor.)
2. $S(x) = S(y) \rightarrow x = y$. ($S$ is one-to-one.)
3. $x + \overline{0} = x$.
4. $x + S(y) = S(x + y)$.
5. $x \cdot \overline{0} = \overline{0}$.
6. $x \cdot S(y) = (x \cdot y) + x$.

In addition to the above, PA also needs some sort of principle of mathematical induction. The most straightforward statement of induction is

7. $[\overline{0} \in A \wedge \forall n(n \in A \rightarrow S(n) \in A)] \rightarrow \forall n(n \in A)$.

The intention here is that the variable n ranges over natural numbers, while the variable A ranges over *sets* of natural numbers. But $\mathcal{L}$ does not have variables for sets. One obvious solution to this difficulty is to expand $\mathcal{L}$ to a two-sorted language with natural number variables and set variables, and add the binary relation symbol $\in$ to $\mathcal{L}$.

But this type of two-sorted first-order language, with variables for elements and for subsets of an intended domain, does not provide the intended meaning unless the domain of the set variables consists of all subsets of the domain of the element variables. As we will see in Chapter 5, the rules for interpreting first-order theories do not require this. If we want this version of induction to mean what it ought to mean, we need to go beyond first-order logic and instead use a **second-order** version of formal arithmetic. Second-order logic, and the reasons for using it in this type of situation, will be discussed in the next section.

In order to complete the axiomatization of *first-order* PA, we must replace the concise form of induction given above with the so-called **predicate form**: for each $\mathcal{L}$-formula $\mathrm{P}(n)$ with the free variable n (and possibly other free variables), we include the axiom

7′. $[\mathrm{P}(\overline{0}) \wedge \forall n(\mathrm{P}(n) \rightarrow \mathrm{P}(S(n)))] \rightarrow \forall n \mathrm{P}(n)$.

This works well for many purposes, but it does have two drawbacks. One is that the single induction axiom has been replaced by an infinite list of axioms (a so-called **axiom schema**), a situation that cannot be avoided if we stay in the language $\mathcal{L}$. That is, there is no finite set of axioms that is equivalent (in terms of the theorems obtained) to axioms (1) through (6) above plus the schema (7′). We express this limitation by saying that first-order PA is not **finitely axiomatizable**. (Because P is a propositional variable, rather than a mathematical variable within $\mathcal{L}$, it cannot be quantified in first-order PA.)

The second, more serious, drawback is that this axiom schema might have less "power" than the version involving sets. To see this, note that our axiom schema of induction may be viewed as the set version restricted to sets that are **definable** by a formula of $\mathcal{L}$, that is, sets of the form $\{n : \mathrm{P}(n)\}$. The set of formulas of $\mathcal{L}$ is countable, by Proposition 1(e) of Appendix C. Therefore, since there are uncountably many subsets of $\mathbb{N}$, it is quite plausible that the predicate form of induction could be inadequate for proving some important theorems.

In summary, we have the following situation, which turns out to be quite common: there is a sensible first-order theory that seems to provide a correct formalization of arithmetic. However, this first-order theory does not have as much theorem-proving power as one would like, especially for more advanced purposes, and this limitation cannot be remedied in the original first-order language. At the same time, those who want to maintain that all of mathematics can be carried out within first-order logic need not admit defeat: they can specify ZFC set theory (see Example 18 below) as the first-order theory to be used for all of mathematics.

From now on, unless stated otherwise, the terms "Peano arithmetic" and "PA" refer to *first-order* Peano arithmetic. The discussion above emphasizes the limitations of PA, but in fact a surprising amount of mathematics can be carried out in subtheories of PA in which the induction axiom schema is severely restricted, instead of being allowed for all $\mathcal{L}$-formulas $\mathrm{P}(n)$. The investigation of these "weak" subsystems of PA has proven to be a very fruitful area of research. Section 4.4 will provide a more detailed treatment of what can and cannot be proved in PA.

Exercise 8. To get a feel for PA, you might want to prove a few basic arithmetical facts from its axioms (not too formally, or you'll drive yourself crazy!). Reasonable choices might be the commutative laws of addition and multiplication and the distributive laws. Also, you might want to show how to define the predicate $m < n$ within PA, and then prove irreflexivity, transitivity, etc. Almost all of these proofs require induction (on just one variable, even if there is more than one variable in the statement).

Example 17 (The First-Order Theory of Rings and Fields). Let $\mathcal{L}$ be the first-order language of a ring, as described in Example 15. In the context of ring theory, it is not necessary to have the symbols 0 and $-$ in the language, because they are definable. However, for most purposes it is more convenient to include these symbols in $\mathcal{L}$.

In $\mathcal{L}$, it is simple to write down the usual axioms of a ring, a commutative ring, a ring with unity, a field, etc. (See Appendix D for basics.) In this context, the axioms are just the defining properties of these algebraic structures, rather than basic, assumed truths. Many simple theorems of ring and field theory can be stated in this language and proved from the appropriate axioms: the uniqueness of identity elements and inverses, the fact that any number times 0 equals 0, the fact that a field has no zero-divisors, etc. On the other hand, it is easy to see that $\mathcal{L}$ is not adequate for a full treatment of rings and fields. Among other things, it provides no way of discussing arbitrary subrings or subsets of a ring, or mappings between rings.

Furthermore, the language $\mathcal{L}$ lacks the means to express some rather basic facts about a single ring or field. For example, consider the idea of characteristic of a field. If we want to state in $\mathcal{L}$ that the characteristic of a field is 3, it is easy to do so: $1 + 1 + 1 = 0$. (As usual, the associative law enables us to omit parentheses on the left side of this equation.)

Exercise 9. How would we state this property of a field in $\mathcal{L}$ if there were no symbols for the identity elements?

But now suppose we want to axiomatize the theory of a field of characteristic zero. If you look in a standard abstract algebra text, the definition given for this is something like "There is no positive integer n such that $n \cdot 1 = 0$, where $n \cdot 1$ is an abbreviation for $1 + 1 + \cdots + 1$ (n times)." It's tempting to think that this can be formalized in $\mathcal{L}$ as $\forall n > 0(n \cdot 1 \neq 0)$. But the problem is that the variable n in this formula denotes a natural number, not a member of the field in question, so this formula is not within the language $\mathcal{L}$.

So how can we formalize, in $\mathcal{L}$, that a field has characteristic zero? The standard way is to use an axiom schema, as in Example 16: start with the usual field axioms and add, for each n, the formula $1 + 1 + \cdots + 1 \neq 0$, where there are n 1's on the left side of the equation. We will prove in Chapter 5 that there is no finite set of axioms of $\mathcal{L}$ that is equivalent to this infinite list of axioms. In other words, the first-order theory of a field of characteristic zero, like first-order Peano arithmetic, is not finitely axiomatizable. Furthermore, there is absolutely no way in $\mathcal{L}$, even using an infinite set of axioms, to express that a field has *finite* (that is, nonzero) characteristic! We will also see that these limitations are not just an esoteric curiosity; they lead to some questions that are of genuine interest to algebraists.

If we want a theory in which we can work with concepts such as finite characteristic, subrings, and homomorphisms, we need to go beyond the first-order theory of rings and fields. As in Example 16, we could use a second-order theory, or we could use the full power of set theory.

Example 18 (Zermelo–Fraenkel Set Theory). One of the most important mathematical achievements of the early part of the twentieth century was the development of versions of set theory that apparently avoid the paradoxes of "naive" set theory (to be discussed in Chapter 2), and yet do not significantly diminish the freedom to define abstract and infinite sets, as envisioned by the founders of set theory. The most important of these theories is called Zermelo–Fraenkel (ZF) set theory; with the addition of the axiom of choice, it is simply called ZFC set theory.

ZFC is a remarkable first-order theory. All of the results of contemporary mathematics can be expressed and proved within ZFC, with at most a handful of esoteric exceptions. Thus it provides the main support for the formalist position regarding the formalizability of mathematics. In fact, logicians tend to think of ZFC and mathematics as practically synonymous.

On the other hand, ZFC is in many ways an extremely simple theory. This is especially true of its language. The language of set theory has just one binary relation symbol $\in$. It is not even necessary to include the equality symbol, since equality of sets can be defined (two sets being equal if and only if they have exactly the same elements). It is worth noting that ZFC is a "pure" set theory: all the objects under discussion are technically sets. There are not even variables or axioms for the natural numbers; every mathematical object must be a set. We will examine ZFC in much more detail in Chapters 2 and 6.

Example 19 (A First-Order Theory of Family Relationships). We conclude this section with a nonmathematical example. But this example provides a good setting to practice translating English statements into a formal symbolic language, with careful use of quantifiers.

Let $\mathcal{L}$ be a first-order language with one unary relation symbol $\mathrm{W}(x)$ and one binary relation symbol $\mathrm{P}(x, y)$, in addition to equality. The intended interpretation is that the variables of $\mathcal{L}$ denote people (living or dead), $\mathrm{W}(x)$ means "x is female," and $\mathrm{P}(x, y)$ means "x is a parent of y." (There is no requirement to mention any interpretation when defining a first-order language or theory, but doing so is often very helpful to the reader.) All of the usual blood relationships for which we have words can be expressed in this language. For example, the statements that one person is another's mother, grandparent, uncle, brother, half-sibling, or first cousin can all be formalized in $\mathcal{L}$.

On the other hand, not everything one might want to say about family relationships can be expressed in $\mathcal{L}$. If you give it some thought, you should be able to convince yourself that there's no way to express "x is a descendant of y" or "x and y are blood relatives" in $\mathcal{L}$. This limitation is rather similar to the situation we discussed in first-order field theory, where certain statements regarding the characteristic of a field could not be formalized. As an exercise, see if you can expand $\mathcal{L}$ in a way that makes it possible to say anything about blood relationships that can be expressed in English. Would it suffice to add integer variables to the language, and a relation symbol $\mathrm{D}(x, y, n)$ that means "y is an nth level descendant of x"? (So $\mathrm{D}(x, y, 1)$ would mean y is x's child, etc.) Is it possible to express "x and y are blood relatives" in this expanded language?

So we have a couple of possibilities for a first-order language for the formalization of family relationships. What about axioms? In the original language with only "people variables," the most obvious axioms to include are that each person has a unique mother and father. From this, we can prove many simple facts, for example, that each person has exactly four grandparents (if there's no incest!), and that siblinghood is an equivalence relation (if we agree that each person is his or her own sibling).

But this simple axiomatization is inadequate if we want to prevent circular relationships, such as someone being his or her own parent or grandparent. What kind of axioms would prevent these? Surprisingly, there is no finite set of axioms in $\mathcal{L}$ that will do so. If we go to an expanded language as above, or a language in which we can refer to a person's time of birth, and assert (as an axiom) that every person is born after his or her parents were, then we can guarantee noncircularity. Essentially, it would be desirable to be able to express the "descendant" (or "ancestor") binary relation in whatever formal language we choose, and then have axioms that state that these relations are partial orderings. You might find it instructive to carry this out in some detail.

1.6 Normal forms and complexity

In this section we present two more metatheorems of logic that give useful ways of rewriting arbitrary first-order formulas in simple, standard forms. Both of these are well known to logicians and understood intuitively by all mathematicians. These normal forms will be useful at various points in the following chapters.

Definition. A first-order formula is said to be in **prenex normal form**, or simply in **prenex form**, if it consists of a string (possibly empty) of quantifiers, followed by a quantifier-free subformula.

In a prenex formula, the entire quantifier-free subformula must be the scope of the initial quantifiers. So if P and Q are quantifier-free, then $\forall x(\mathrm{P} \rightarrow \mathrm{Q})$ is prenex, but $(\forall x \mathrm{P}) \rightarrow \mathrm{Q}$ is not, because $\mathrm{P} \rightarrow \mathrm{Q}$ is not a subformula of the latter formula.

Example 20. The usual additive inverse axiom of ring theory, $\forall x\, \exists y(x + y = 0)$ is in prenex form. But the multiplicative inverse axiom, $\forall x[x \neq 0 \rightarrow \exists y(x \cdot y = 1)]$ is not, because one of the quantifiers is to the right of a connective.

Theorem 1.5. *For any first-order formula* P, *there is a prenex formula* Q *that is logically equivalent to* P.

Proof. We will take this opportunity to illustrate an extremely common way of proving results in logic, namely, by induction on the total number of connectives and quantifiers appearing in P (which we will abbreviate from now on as "induction on the structure of P").

The claim is trivial if P is atomic. So now assume that the claim has been proved for every formula with n or fewer connectives and quantifiers, and let P be given with $n + 1$ connectives and quantifiers. So now we have seven cases to consider, based on the "outermost" connective or quantifier of P:

If P has the form $\forall x$ Q then, by the induction hypothesis, Q has a prenex equivalent Q′. But then $\forall x$ Q′ is a prenex formula that is equivalent to P. The same reasoning applies if P has the form $\exists x$ Q.

If P has the form $\sim$ Q then, by the induction hypothesis, Q has a prenex equivalent Q′. So P is equivalent to $\sim$ Q′. The usual logical equivalences used to move negations through quantifiers ($\sim \forall x \mathrm{P} \leftrightarrow \exists x \sim \mathrm{P}$, and $\sim \exists x \mathrm{P} \leftrightarrow \forall x \sim \mathrm{P}$) are derivable in first-order logic, and by applying them repeatedly to $\sim$ Q′ we get a prenex formula that is equivalent to P.

If P has the form $\mathrm{Q}_1 \wedge \mathrm{Q}_2$ then, by the induction hypothesis, Q_1 and Q_2 have prenex equivalents Q'_1 and Q'_2, respectively. By changing some bound variables if necessary (which is allowable, as explained in Section 1.3), we may assume that none of the bound variables of Q'_1 appears in Q'_2, and vice-versa. But if this is so, all of the quantifiers of both Q'_1 and Q'_2 can simply be moved to the front of the formula, creating a prenex equivalent of P.

Rather than prove this point formally, let's illustrate it with an example. Suppose P has the form $\mathrm{Q}_1 \wedge \mathrm{Q}_2$, where Q_1 has the prenex equivalent $\forall x \exists z \mathrm{R}_1$, and Q_2 has the prenex equivalent $\exists u \forall w \forall y \mathrm{R}_2$. Under the (harmless) assumption that x and z do not appear in R_2, and u, w, and y do not appear in R_1, P has many prenex equivalents such as $\forall x\,\exists z\,\exists u\,\forall w\,\forall y(R_1 \wedge R_2)$, $\exists u\,\forall w\,\forall y\,\forall x\,\exists z(R_1 \wedge R_2)$, $\exists u\,\forall x\,\forall w\,\exists z\,\forall y(R_1 \wedge R_2)$, etc. The quantifiers within each conjunct must stay in the same order, but the two strings of quantifiers can be "meshed" arbitrarily.

There are three more cases to consider: P having the forms $Q \vee R$, $Q \rightarrow R$, and $Q \leftrightarrow R$. But there is no need to consider these cases, because these connectives are redundant: $Q \vee R$ is equivalent to $\sim(\sim Q \wedge \sim R)$, $Q \rightarrow R$ is equivalent to $\sim(Q \wedge \sim R)$, and $Q \leftrightarrow R$ is equivalent to $(Q \rightarrow R) \wedge (R \rightarrow Q)$. In proofs of this sort, one almost always uses such equivalences in order to deal with only a couple of connectives instead of five. The most convenient combinations are $\{\wedge, \sim\}$, $\{\vee, \sim\}$, and $\{\rightarrow, \sim\}$. In the same way, it suffices to consider either one of the quantifiers rather than both of them. ■

The ability to put statements into prenex form is useful in mathematics generally. For the most part, we like our quantifiers at the front of statements, not in the middle, and Theorem 1.5 guarantees that we can always put them there. The process of putting the negation of a prenex statement back into prenex form is well known, but it's not much harder to do the same with implications, conjunctions, etc.

On the other hand, the prenex form of a statement is sometimes awkward. For example, consider this special case of the intermediate value theorem: Whenever f is a continuous function from $[0, 1]$ to $\mathbb{R}$ such that $f(0) < 0$ and $f(1) > 0$, then f has a root in $(0, 1)$. Formally, this has the form

$$\forall f[P \rightarrow \exists x(0 < x < 1 \wedge f(x) = 0)],$$

where P is a statement about f that can be written without the variable x. So this theorem has the equivalent prenex form

$$\forall f\,\exists x[P \rightarrow (0 < x < 1 \wedge f(x) = 0)].$$

Placing the quantifier on x in front of the antecedent P obscures the intended meaning of an existence theorem such as this. Still, it doesn't hurt to be aware that the theorem can be written this way without changing its logical meaning.

Prenex form also provides an important way of judging the syntactic **complexity** of a mathematical statement. Given a formula in prenex form, the number of *alternations* of quantifiers (that is, switches between $\forall$ and $\exists$, in either order) turns out to be the most useful measure of its complexity. Even two alternations of quantifiers, as in the definition of continuity (Example 7, Section 1.3), can make a formula complicated and difficult to grasp. Mathematical statements with more than three or four alternations of quantifiers are unwieldy and hard to comprehend but, fortunately, they are uncommon.

Here is the most important scheme for categorizing statements according to their complexity:

Definition. A quantifier-free formula is called a Σ_0 formula or a Π_0 formula. A prenex formula that begins with an existential (respectively, universal) quantifier and has exactly n alternations of quantifiers is called a Σ_{n+1} (respectively, Π_{n+1}) formula.

For example, a statement of the form $\forall u, v\, \exists x, y, z\mathrm{P}$, where P is quantifier free, is only Π_2, in spite of the five quantifiers. This is deliberate, primarily because with many important domains (e.g., integers, real numbers, or sets), a finite string of quantifiers of the same type can be rewritten as a single quantifier.

The categories of formulas that we have just defined are technically disjoint. For instance, a Π_1 formula is neither Π_2 nor Σ_2. But we can always use "dummy quantifiers" to find more complex equivalents of any given formula. For example, the commutative law of multiplication, a Π_1 sentence, is logically equivalent to the Π_2 sentence $\forall x, y\, \exists z(xy = yx)$ and the Σ_2 sentence $\exists z\, \forall x, y(xy = yx)$.

Exercise 10. Show that the symbolic statement that a real-valued function is continuous (or uniformly continuous) is logically equivalent to a Π_3 formula. (The only reason this exercise is not trivial is that the restricted quantifiers in these statements need to be "unabbreviated." Essentially, the point of this exercise is to show that restricted quantifiers should be counted as if they were ordinary quantifiers, when determining syntactic complexity.)

Second-order logic and Skolem form

Perhaps the main advantage of writing statements in prenex form is that their "content" becomes very clear. If a statement in prenex form is to be true, every existentially quantified variable represents a *function* that must exist, specifically a function of all the universally quantified variables to its left. All mathematicians know this, regardless of whether they are familiar with the logicians' terminology. For instance, suppose a formula has the form $\exists u\, \forall v\, \exists w\, \forall x, y\, \exists z \mathrm{P}$. For this statement to be true, there must exist a specific value of u (that is, a constant, a function of no variables), a function of v that defines a value of w, and a function of v, x, and y that defines a value of z, such that the formula P is true for all values of v, x, and y. If P is quantifier-free, then the existence of these three functions (that define u, w, and z) can be viewed as what the given statement is really "asserting." Functions that interpret existential quantifiers in this way are called **Skolem functions**.

Example 21. Consider again the formal statement that a function f is continuous at a number x: $\forall \epsilon > 0\ \exists \delta > 0\ \forall v\ (\dots)$, where the formula in parentheses is quantifier-free. Essentially, this statement is in prenex form in spite of its restricted quantifiers. For beginning students of analysis, it is helpful to realize that this statement asserts the existence of a function (called a **modulus of continuity**) that specifies δ in terms of ϵ, such that the inner part of the statement holds for every positive ϵ and every v in the domain of f. If f is continuous on its entire domain, then δ becomes a function of x and ϵ. If f is uniformly continuous, then δ must again be a function of ϵ alone. For both of these last two statements, the inner quantifier-free formula must hold for every positive ϵ and every v and x in the domain of f.

How can this sort of "function variant" of a statement be formalized? Given a prenex formula in a first-order language, there is no way in general to express this alternative version in the same first-order language. For instance, if we are working in the first-order language of ring theory, then we have no variables available that represent functions.

With this in mind, let us define the notion of a **second-order language**. In addition to all the components of a first-order language, such a language has an infinite list of **n-ary relation variables** for each natural number n. It is convenient (but not strictly necessary) to assume that there are also **n-ary function variables** for each n. These two new types of variables are called second-order variables, as opposed to the usual first-order variables. Relation variables and function variables are used just like the relation symbols and function symbols of first-order logic to form terms and formulas. The only difference is that relation variables and function variables may be quantified. Also, if $t_1, \dots, t_n$ are first-order terms and A is an n-ary relation variable, we may write $(t_1, \dots, t_n) \in A$ instead of $A(t_1, \dots, t_n)$.

Example 22. The "straightforward" version of induction that was mentioned in Example 16 would be an axiom in second-order Peano arithmetic. Here, A is a unary relation variable; in other words, we think of it as representing an arbitrary set of natural numbers. So, technically, this axiom is prefaced by "$\forall A$."

Example 23. Suppose that we have a language with a binary relation symbol $*$. The usual definition of what it means for $*$ to be an equivalence relation is first-order. Similarly, the statement that $*$ is a partial order or a total order is first-order. The property of being a minimal, maximal, least, or greatest element is also first-order. But the statement that $*$ is well-founded or a well-ordering is unavoidably second-order: it refers to all subsets of the domain, and it can be shown that there is no way to achieve the intended meaning without such a quantifier. (The definitions of all of these types of relations and orderings are given in Appendix B.)

Example 24. The standard way of describing the structure of $\mathbb{R}$ is that it is a complete ordered field. The defining properties of an ordered field are all first-order, but the **completeness property**—that every set of reals with an upper bound has a least upper bound—is unavoidably second-order, as in the previous example. We will expand on this point in Section 5.1.

From a purely syntactic standpoint, anything that can be expressed in a second-order language could also be expressed in a many-sorted first-order language. What makes second-order logic special is the interpretation given to the variables: once the domain of the first-order variables is chosen, the n-ary relation variables (respectively, n-ary function variables) must range over all n-ary relations (respectively, n-ary functions) on the first-order domain. We will see the significance of this requirement in Section 5.7.

If one wants to axiomatize second-order logic, then in addition to the axioms of first-order logic, there must be rules for the existence of relations and functions on the first-order domain. In particular, if A is a unary relation variable and $\mathrm{P}(x)$ is any formula that does not contain A free, then

$$\exists A\, \forall x[x \in A \leftrightarrow \mathrm{P}(x)]$$

is a standard second-order axiom. There are similar axioms for formulas $\mathrm{P}(x, y)$, $\mathrm{P}(x, y, z)$, etc. Also, whenever $\mathrm{P}(x, y)$ is a formula that does not contain the function variable F free,

$$\forall x\, \exists ! y \mathrm{P}(x, y) \rightarrow \exists F\, \forall x \mathrm{P}(x, F(x))$$

is generally included as a second-order axiom.

It is also plausible to accept the previous axiom with the symbol "!" removed. Since this makes the left side of the implication weaker, the axiom becomes stronger. If you are familiar with the axiom of choice, which will be discussed in chapters 2 and 6, you can see that this strengthened axiom is clearly a form of the axiom of choice. We will use the phrase "second-order logic with choice" to indicate that this axiom is included. It's fine to make this axiom a biconditional if desired, since the converse is trivial.

Definition. A second-order formula is said to be in **Skolem normal form**, or simply in **Skolem form**, if it consists of a string (possibly empty) of existentially quantified variables of any types, followed by a string (possibly empty) of universally quantified *first-order* variables, followed by a quantifier-free subformula.

Theorem 1.6. *For any first-order formula* P, *there is a second-order formula in Skolem form that is equivalent to* P, *in second-order logic with choice.*

Proof. Given P, let P′ be an equivalent first-order formula in prenex form. If any variable is quantified more than once in P′, delete all the quantifiers except the rightmost one for that variable, obtaining a formula P″. (These extra quantifiers are meaningless.) Then, replace each existentially quantified variable that is to the right of one or more universally quantified variables with an existentially quantified function variable whose arguments are precisely those universally quantified variables. The order of these existentially quantified function variables is immaterial, as long as they are placed at the front of the final formula. Naturally, each occurrence of an existentially quantified variable in the quantifier-free part of P′ must also be replaced by the appropriate function term. Call the resulting formula Q; we claim that Q is a Skolem equivalent of P.

It is easy to see that Q is in Skolem form, since P′ was already in prenex form, every existential quantifier has been moved to the front of Q, and no second-order universal quantifiers have been introduced.

It remains to prove that Q is equivalent to P in second-order logic with choice. First of all, we know from Theorem 1.5 that P′ is equivalent to P. The equivalence of P″ and P′ is a simple exercise in first-order logic. Finally, we must show that Q is equivalent to P″. We will not do this rigorously, but we have already mentioned the main idea. In a typical case, if x and y are first-order variables, then

$$\forall x\, \exists y \mathrm{R}(x, y) \leftrightarrow \exists F\, \forall x \mathrm{R}(x, F(x))$$

is an axiom of second-order logic with choice (or only the forward direction is an axiom, and the reverse is trivial). ∎

Example 25. Returning to Example 21: the true prenex form for the property of continuity at a point is $\forall \epsilon\, \exists \delta\, \forall v[\epsilon > 0 \rightarrow (\delta > 0 \wedge (\dots))]$.

So the Skolem form is $\exists G\,\forall \epsilon, x[\epsilon > 0 \rightarrow (G(\epsilon) > 0 \wedge (\ldots))]$, where every occurrence of δ in (...) is also replaced by $G(\epsilon)$. If ϵ and δ are real variables, then the modulus of continuity G is technically a function from $\mathbb{R}$ to $\mathbb{R}$ whose values on nonpositive values of ϵ are irrelevant. But it's harmless and more natural to think of G as a function with domain $\mathbb{R}^+$.

Exercise 11. Put each of the following first-order formulas into prenex form and then into Skolem form, assuming that P and Q are quantifier-free:

(a) $\exists u\,\forall v\,\exists w\,\forall x, y\,\exists z \mathrm{P}(u, v, w, x, y, z)$.

(b) $\sim\ \exists u\,\forall v\,\exists w\,\forall x, y\,\exists z \mathrm{P}(u, v, w, x, y, z)$.

(c) $\forall x\,\exists y \mathrm{P}(x, y) \rightarrow \forall x, y\,\exists z \mathrm{Q}(x, y, z)$.

An alternative approach to Skolem form is to introduce additional function symbols into the given first-order language, instead of second-order (function) variables. However, there are good reasons not to overdo this practice. Certainly, one does not want to introduce a new function symbol into the language every time an existential quantifier appears in the statement of a theorem.

But a purpose for which it often *is* convenient and useful to introduce new function symbols is the elimination of existential quantifiers from *axioms*. For example, consider the inverse axioms of field theory. The additive inverse axiom $\forall x\,\exists y(x + y = 0)$ has the Skolem form $\forall x[x + (-x) = 0]$, where the function $f(x) = -x$ is viewed as a Skolem function. Similarly, the multiplicative inverse axiom has the Skolem form $\forall x(x \neq 0 \rightarrow x \cdot x^{-1} = 1)$. Once the existential quantifiers are removed from axioms in this way, it becomes convenient to omit the universal quantifiers from them as well. Thus the use of this first-order Skolem form allows all mathematical axioms to be written without quantifiers. This is often an appealing simplification, especially when the existentially quantified variables in the axioms can be proved to be unique, as is the case with the inverse axioms.

One awkward consequence of viewing reciprocation as a Skolem function is that it requires x^{-1} to be defined even when $x = 0$. This seems unnatural, but there is no real problem with it; 0^{-1} can be given some arbitrary value, with no harmful effects; of course, we cannot expect $0^{-1} \cdot 0 = 1$ to hold! (Recall the related remarks at the end of Example 15 in Section 1.4.) By the way, one can also view 0 and 1 as "Skolem constants," that is, Skolem functions of no variables.

1.7 Other logics

This chapter has emphasized that first-order predicate logic is a framework that successfully embodies the reasoning used in mathematics. We have also discussed second-order logic, one of the most important extensions of first-order logic. Every theorem or deduction of first-order logic is also valid in second-order logic, and the underlying logic of both is the usual Aristotelian or classical logic.

In this section we will list a few alternatives to classical logic. We will not discuss them in detail, in part because they are of limited applicability to mathematics. However, it can't hurt for mathematicians to be aware of some of these alternatives and the types of reasoning they are intended to formalize. Most of them are important in philosophy, linguistics, computer science, or the foundations of mathematics.

The most mathematically relevant alternative to classical logic is **intuitionistic logic**. Developed by L. E. J. Brouwer in the early 1900s, intuitionism was the first serious challenge to the use of Aristotelian logic in mathematics. But since this topic will be discussed thoroughly in Chapter 8, we will not say more about it now.

Two excellent sources for all the types of logic discussed in this section (and many more) are volumes 2 and 3 of [GG], and the online encyclopedia [SEP].

Many-valued logic

In 1920, not long after the introduction of intuitionism, Jan Łukasiewicz proposed a three-valued system of logic. Łukasiewicz was probably influenced not only by Brouwer but also by various earlier thinkers, including Aristotle himself, who was quite aware of the arguments against the claim that every proposition is true or false. (Recall the discussion of propositions at the beginning of Section 1.2.) In Łukasiewicz's logic, a statement may have a truth value of 1, 0, or 1/2, corresponding to "true," "false," and "possible" (or "undefined"). Here are the most sensible truth tables for three of the basic connectives in this type of logic:

P	Q	$\sim$P	P $\wedge$ Q	P $\vee$ Q
T	T	F	T	T
T	F	F	F	T
T	U	F	U	T
F	T	T	F	T
F	F	T	F	F
F	U	T	F	U
U	T	U	U	T
U	F	U	F	U
U	U	U	U	U

Exercise 12. Write out reasonable truth tables for the remaining basic connectives, $\rightarrow$ and $\leftrightarrow$, in three-valued logic.

In many-valued systems, it often makes sense to have a greater variety of connectives than in ordinary logic. For instance, some many-valued systems have two or more conjunction connectives, all of which act the same on the values F and T but act differently on "intermediate" truth values.

Łukasiewicz and others went on to devise a great variety of many-valued logics, including ones with an infinite set of possible truth values. One such proposal, due to Max Black, was a type of logic in which the truth value of each statement is a real number between 0 and 1 (inclusive), viewed as the probability that the statement is true. Black's system was the immediate forerunner of fuzzy logic, which we will discuss in the next subsection. Another outgrowth of the development of many-valued logic was the theory of **Boolean-valued models**, which became an extremely important tool in set theory in the 1960s. Many-valued logic has also been applied to linguistics, philosophy, and hardware design, among other fields.

Fuzzy logic

Fuzzy logic, a type of many-valued logic, was invented by Lotfi Zadeh in the 1960s (see [Zad]). In the most common form of fuzzy logic, the truth value of a statement is a real number in the interval [0, 1], which is to be thought of as the *degree* (not the probability!) of truth of the statement. Usually, the statements to which one assigns these truth values are statements about membership in a set, and this leads to the notion of a **fuzzy set** and the subject of **fuzzy set theory**.

For example, suppose we want to model the notion of what it means for a person to be "young." In classical logic, a person is either young or not young. This would suggest that each person (except for those who die young) is young for a while and then suddenly ceases to be young the next day or even the next second. Clearly, this doesn't seem quite right. Youth ought to be a matter of degree, so that a person becomes less young over a period of years. So we might decide, say, that a person who is less than 18 years old is totally young, a person who is more than 50 years old is not young at all, and from 18 to 50 a person's youth decreases linearly from 1 to 0.

To translate this into fuzzy set theory, we would start with the ordinary or **crisp** set U of all people. A crisp subset A of U can be defined by its **characteristic function**, the function from U to $\{0, 1\}$ whose output is 1 on members of A only. Analogously, a fuzzy subset of U is defined to be a function from U to the interval [0, 1]. Thus, the fuzzy set Y of all young people, according to the above criterion, is defined as the function

$$\mu_Y(x) = \begin{cases} 1, & \text{if } x\text{'s age is 18 or less,} \\ 0, & \text{if } x\text{'s age is 50 or more,} \\ (50 - A)/32, & \text{otherwise, where } A \text{ is } x\text{'s age.} \end{cases}$$

There are more sophisticated notions of fuzziness as well. In the example we have been considering, a person is exactly 50% young on his or her 34th birthday. That doesn't sound fuzzy at all! Maybe, for some purposes, we would rather say that a 34-year-old is between 40% and 60% young, depending on other factors. This leads to consideration of **interval-valued fuzzy sets**, in which the truth value of each membership statement is a subinterval of [0, 1] rather than a specific number.

The main practical purpose of fuzzy logic and set theory is to model shades of gray into computer systems. In the real world, few things are absolute, and our everyday rules for living reflect this reality. If the house is *a little cold*, we turn the thermostat up *a little bit*. If the house is *very cold* we would probably turn the thermostat up *a lot*. We make these decisions without turning the phrases in italics into black-and-white criteria or exact quantities. But how does the heating system itself operate? A simple thermostat is a two-state device. If the temperature it records is less than a specific cut-off, the heat is on full blast. Otherwise, there is no heat. For most home situations, this may be good enough. But for many computerized control systems, this would not be satisfactory. It may be necessary for the system to decide to what degree a certain condition exists, and to what degree it should respond. Situations of this sort lend themselves to fuzzy logic and set theory.

The jury is still out on the importance and usefulness of fuzzy methods. They have been most favorably received and fruitfully applied in Japan, but in the 1990s their use in the rest of the world increased substantially. To learn more about fuzzy logic and set theory, see [Zad] or [YOTN].

Modal logic

Modal logic is not a type of many-valued logic, but it is motivated by similar philosophical considerations. In modal logic, the usual syntax of propositional or first-order logic is supplemented by an additional connective $\Box$, read "necessarily." Thus $\Box$ P may be read "Necessarily P" or "P is necessary" or "P is necessarily true." Usually there is also a symbol $\Diamond$, read "possibly," but this connective is redundant because $\Diamond P$ is equivalent (by definition or axiom) to $\sim \Box \sim P$. Of course, there is also redundancy among the usual connectives and quantifiers.

The idea of modal logic, which may be traced back to Leibniz, is that our "real world" is just one of many "possible worlds." The statement $\Box P$ is thought of as meaning that P is true in all possible worlds. So, in a way, the modal operators $\Box$ and $\Diamond$ are analogous to the quantifiers $\forall$ and $\exists$, except that they are evaluated on a set of possible worlds rather than domains of numbers, sets, etc. There are many statements P that happen to be true but can easily be imagined to be false under different circumstances. For such statements, $\Box P$ would be false. Presumably, every law of logic is necessarily true. But it is plausible that the converse of this could fail; in particular, some basic laws of mathematics and even of physics may be deemed to be necessarily true.

A great variety of formal modal systems have been considered. Some thorny problems arise when modal operators are combined with quantifiers, so most elementary treatments of modal logic concentrate on propositional logic. One common system of modal propositional logic is called S4 and has the following axioms and rule of inference, in addition to the usual ones of propositional logic:

1. $\Box P \to P$.
2. $\Box(P \to Q) \to (\Box P \to \Box Q)$.
3. $\Box P \to \Box\Box P$.
4. Rule of Inference: If $\vdash P$, then $\vdash \Box P$.

It is important to understand the relationship between (1) and (4). Axiom 1 says that if P is necessarily true, then P is true. This makes perfect sense, while the converse would not. But (4) says that if P has been *proved*, then we may assert that P is necessarily true, so that $\Box P$ is also proved. Note that this rule cannot be applied as it stands to proofs from assumptions or proper axioms; that is, we can't put "T" before the $\vdash$ on both sides of this rule. In particular, we always have $P \vdash P$, but we don't want $P \vdash \Box P$. Technically, this makes (4) a so-called **meta-rule** rather than a true rule of inference.

Axiom 3 of S4 is also interesting. It may be informally justified as follows: if P is necessary, then the fact that P is necessary does not depend on any particular possible world and so is true in every possible world. Therefore, $\Box$P is necessary.

Another common system of modal propositional logic, called S5, is identical to S4 except that axiom 3 is replaced with:

3′. $\Diamond P \rightarrow \Box\Diamond P$.

Axiom 3′ is perhaps less clear than (3), but it can be justified with a similar argument. To demonstrate how some simple formal proofs look in S5, we will show that S5 is an extension of S4. In other words, axiom 3 is provable in S5.

Lemma 1.7. *The theorems of* S5 *are closed under propositional consequence. That is, if* $S5 \vdash P_i$ *for* $1 = 1, 2, \ldots, n$, *and* $(P_1 \wedge P_2 \wedge \cdots \wedge P_n) \rightarrow Q$ *is a tautology, then* $S5 \vdash Q$.

Proof. The statements $(P_1 \wedge P_2 \wedge \cdots \wedge P_n) \rightarrow Q$ and $[P_1 \rightarrow (P_2 \rightarrow (P_3 \rightarrow \cdots (P_n \rightarrow Q)\ldots)]$ are propositionally equivalent, as is easily verified with a truth table. So if the former is a tautology, then so is the latter, and both of them are axioms of S5. So if each P_i is a theorem of S5, we can prove Q in S5 by repeated applications of modus ponens on the second tautology. ■

Lemma 1.8. *For any statements* P *and* Q*:*

(a) *If* $S5 \vdash P \rightarrow Q$, *then* $S5 \vdash \Box P \rightarrow \Box Q$.

(b) *If* $S5 \vdash P \leftrightarrow Q$, *then* $S5 \vdash \Box P \leftrightarrow \Box Q$.

Exercise 13. Prove this lemma.

Proposition 1.9. *The following are provable in* S5, *for any statement* P*:*

(a) $P \rightarrow \Diamond P$.

(b) $P \rightarrow \Box\Diamond P$.

(c) $\Box P \leftrightarrow \sim\Diamond\sim P$ *(the definition of* $\Box$ *from* $\Diamond$*).*

(d) $\Diamond\Box P \leftrightarrow \sim\Box\Diamond\sim P$.

(e) $\Diamond\Box P \rightarrow \Box P$.

(f) $\Box P \rightarrow \Box\Box P$. *Therefore,* S5 *is an extension of* S4.

Proof.

(a) Working in S5, we have $\Box\sim P \rightarrow\sim P$, by axiom 1. So we have its contrapositive $P \rightarrow\sim\Box\sim P$, by Lemma 1.7. By definition, this is what we want.

(b) We have $P \rightarrow \Diamond P$ by part (a), and $\Diamond P \rightarrow \Box\Diamond P$ by axiom $3'$. So $P \rightarrow \Box\Diamond P$ follows by Lemma 1.7. ■

Exercise 14. Prove the rest of this proposition. If you skip parts, it is reasonable to assume the previous parts. (Hint: For part (f), you may need the previous lemma. Also, remember that the propositional variable P may be replaced by any statement you want, in any axiom or theorem.)

Modal logic is important in philosophy and linguistics and, as we will see in the next two subsections, it is a useful tool for developing other types of logic that have applications to computer science. It also has significant applications within the foundations of mathematics, as we will see in Section 4.3.

Nonmonotonic logic

Standard mathematics uses **deductive** logic, which has an absolute quality to it. We start with some axioms and then we derive theorems that follow with complete certainty from those axioms. Except in the case of outright mistakes in proofs, theorems never need to be retracted. People may decide to stop using a particular axiom system because it is no longer interesting or relevant, but that doesn't nullify the correctness of theorems proved from those axioms.

A bit of thought should convince you that "real world" reasoning is rarely this "clean." Outside of mathematics, in ordinary life and even in the natural sciences, one is usually not completely certain of facts or even of general principles. The reasoning in these realms is **defeasible**, meaning that the conclusions derived must potentially be subject to retraction in the light of new evidence.

This distinction can be made more precise by pointing out that deductive logic, including first-order logic and other types of logic used in mathematics, is **monotonic**. This means that we never retract a theorem on the basis of new givens. In other words, if $T \vdash \varphi$ and $T \subseteq S$, then $S \vdash \varphi$. This metatheorem is trivial to establish: if $T \subseteq S$, then any proof of φ from T is automatically a proof of φ from S.

By contrast, in order for us to function in the real world, our commonsense reasoning must be **nonmonotonic**. We use "general" principles that are not absolute, such as "Birds can fly" (the classic example). Note that this is not as strong as the statement "All birds can fly," which could easily be formalized in first-order logic. The word "Birds" in the first rule is an example of what linguists call a "generic" plural and is much harder to formalize than the phrase "all birds." Nonetheless, if we are told that Opus is a bird, we would naturally apply our general principle to infer that Opus can fly. But this conclusion is only tentative: if we subsequently learn that Opus is a penguin, we retract the previous conclusion and instead infer that Opus cannot fly.

In artificial intelligence, the term **knowledge representation** refers to efforts to formalize subtle types of knowledge and reasoning such as the above. Effective knowledge representation is crucial for expert systems and sophisticated databases. Nonmonotonic logic arose from the need to solve a variety of knowledge representation problems. The field was pioneered by John McCarthy, Raymond Reiter, and others in the 1970s and 1980s.

Nonmonotonic logic is a rich field within theoretical computer science, with many branches. The situation involving Opus illustrates a typical problem in **default logic**, one branch of nonmonotonic logic. The principle that birds can fly may be viewed as a **default rule**: if we know that something is a bird, the default conclusion—the reasonable conclusion in the absence of evidence to the contrary—is that it can fly. Formalizing such rules in a way that allows them to be used effectively by computer systems is a deep problem. For example, to reach the correct conclusion about Opus, a system must understand that the rule "Penguins cannot fly" is absolute and therefore takes precedence over the rule "Birds can fly." Of course, we could deal with this specific situation by replacing the rule "Birds can fly" with "Birds that are not penguins can fly," but systems become unwieldy if every exception to a general rule must be explicitly written into the rule in this way.

Modal logic is one of the important tools used to formalize default reasoning. There are several useful types of modal nonmonotonic logic. The most well-known of these, known as **autoepistemic logic**, uses a modal operator whose intended interpretation is something like "P is believed" rather than "P is necessary" or "P is possible." In a certain formal sense, autoepistemic logic is equivalent to the most common system of default logic.

Here is a situation that shows the need for a different type of nonmonotonic logic. Imagine a database that contains many facts, such as all regularly scheduled flights of a particular airline. The implicit working assumption about such a database is that it is complete. For instance, if the database does not list a flight from Boston to Chicago between 10:00 a.m. and 11:00 a.m., it is reasonable to conclude that there is no such flight. But this conclusion cannot be absolute: if a 10:14 flight from Boston to Chicago is added to the database, the previous negative conclusion must be retracted. Thus we are in the realm of nonmonotonic logic.

The standard formalization used to deal with situations like this is called the **closed world assumption**. In the case of a database consisting only of specific facts, this assumption means that whenever a (positive) fact fails to appear in the database, its negation may be inferred, tentatively (or temporarily). In the case of more sophisticated databases, the closed-world assumption can best be formalized with a type of nonmonotonic logic known as **model-preference logic**.

The concept behind model-preference logic is quite different from that behind default logic and modal nonmonotonic logic. But there have been attempts to unify them to some extent. One notable achievement was McCarthy's invention of **abnormality theories** to formalize default reasoning within model-preference logic. In an abnormality theory, the general rule that "Birds can fly, normally" would be formalized as "Every bird that is not abnormal can fly." For this to work, the abnormality predicate for birds (which would presumably address other issues in addition to flying) must be different from the abnormality predicates for cats, artichokes, etc. In order to represent default rules properly, the system must be geared to minimize abnormalities; that is, things are tentatively assumed to be normal until there is some evidence to suspect otherwise.

To learn more about nonmonotonic logic, see [BDK].

Temporal logic

Mathematicians normally think of their subject as a timeless activity. When we prove a theorem or become convinced in some other way that a mathematical statement is true, we believe that the statement has always been true and will always be true. For instance, even though "Fermat's last theorem" was not proved until 1994 by Andrew Wiles, almost nobody would question that it was already true when Fermat claimed it in the 1600s. We don't prove theorems that sound like, "If C is a closed and bounded set of real numbers one day, then C will be compact by the next day." Essentially, time is not a variable that is relevant to pure mathematics. (The intuitionist school provided one of the few exceptions to this way of thinking. See Chapter 8.)

But time and reasoning that involves time are clearly important in the "real world" and in many academic subjects, including all sciences. There have been many attempts to develop formalisms that embody the special role that time plays in reasoning. One of the first important ones was **tense logic**, created by Arthur Prior in the 1950s. The original version of tense logic includes four temporal operators, whose interpretations are "It has always been true that ... ," "It has at some point been true that ... ," and an analogous pair of operators that pertain to the future. There is redundancy within each pair of these operators, just as the quantifiers $\forall$ and $\exists$ can be defined from each other, as can the modal operators $\Box$ and $\Diamond$. For example, "It has at some point been true that ... " is equivalent to "It has not always been false that ... "

Prior and others have proposed a wide variety of axiom systems for tense logic. In addition to the usual axioms and inference rules of propositional or predicate logic, it is typical to include such principles as, "If P will always be true, then P will at some point be true," and "If P is true, then it will always be true that P has been true at some point." These systems have a similar flavor to modal logic, and in fact tense logic is usually considered to be a type of modal logic.

Since the introduction of tense logic, several interesting extensions of it have been proposed. For example, one could have connectives interpreted as "since" and "until," to be able to express "Q has been true since a time when P was true" and "Q will be true until a time when P will be true." Prior himself later considered **metric tense logic**, which provides the ability to refer to particular points in time, past or future. For instance, it is impossible to express "P has been true for the past five days" in basic tense logic, but this can be expressed in metric tense logic.

Another approach (perhaps *the* main other approach) to temporal logic is to dispense with special modal-type operators and instead try to carry out temporal reasoning entirely within first-order logic. In this approach, called the method of **temporal arguments**, each predicate whose truth might depend on time is augmented to allow an extra variable denoting a time. Thus, a typical predicate $P(x)$ is augmented to $P(x, t)$, interpreted as "$P(x)$ is true at time t." Time becomes a new sort of variable, with its own axioms. These axioms would presumably include that time is totally ordered, and perhaps that time is a dense linear ordering without endpoints. One could also include some algebraic axioms for time, and maybe even a completeness axiom, in order to make the time "axis" behave somewhat like the real numbers. For some purposes, it is useful to include a special constant symbol that denotes a particular time called "now." If this is done, it becomes straightforward to model most versions of tense logic using temporal arguments in first-order logic. For example, "P has always been true" may be defined as "$\forall t < \text{now}\ P(t)$."

Like most of the types of logic discussed in this section, temporal logic has applications in philosophy, linguistics, and computer science.

Within computer science, one use of temporal logic is to regulate parallel processing, where the essentially simultaneous actions of two or more processors must be carefully coordinated. Temporal logic also has applications to artificial intelligence, primarily to the so-called **frame problem**. This refers to an aspect of knowledge representation pertaining to inferences about various conditions *not* changing over time. In everyday life, people make all sorts of inferences of this type easily and subconsciously, but formalizing this type of reasoning in expert systems and other artificial intelligence programs has proven to be very challenging.

CHAPTER 2
Axiomatic Set Theory

2.1 Introduction

Why should students of mathematics want to know something about axiomatic set theory? Here is one answer: set theory provides a natural and efficient framework in which all of contemporary mathematics can be unified. We live in a time when the specialization within mathematics (and many other disciplines, of course) is mind-boggling, and even though that specialization is a tribute to the great success of mathematics and mathematicians, it can be alienating. Not too many centuries ago, the few professional mathematicians generally understood just about all of the mathematics of their day, and many of them also did important work in one or more branches of science and engineering. Now it is almost impossible to be this "broad." Mathematics has major branches such as real analysis, algebraic geometry, number theory, and topology, but it is quite difficult to master even one of these branches. Most mathematicians work in much more narrowly defined areas and, if they are motivated, may be able to understand most of the research in one of these major branches. Again, this observation is not intended as any sort of criticism! I believe the current situation is the almost inevitable consequence of the enormous progress that has been made in almost all areas of mathematics in the last hundred years or so.

Personally, I find it somewhat comforting to know that, formally at least, all the different branches and fragments of mathematics can be nicely packaged together. It may also be valuable to explain this to nonmathematicians who see our discipline as a complex hodgepodge of barely related subjects. Of course, one must not overstate this position. It would be absurd to claim that all of mathematics *is* set theory. The objects of study and the creative activity of most mathematicians are not about sets. Indeed, the great variety of concepts and methods in mathematics makes it all the more remarkable that they can all be embedded in a single, apparently simple theory.

We briefly discussed the elegant simplicity of ZFC set theory in Chapter 1, but the point bears repeating. In the intended interpretation of this first-order theory, the only objects are sets—the same basic "clumps" of things that one learns about in middle school or even elementary school. The only atomic formulas allowed, besides equations, are statements saying that one set is an element of another. With a couple of exceptions, the axioms of ZFC make completely plausible assertions about sets and are not difficult to understand. It seems almost magical that the deepest results of modern mathematics, in all of these varied branches of the subject, can be formally stated and proved in this "sparse" theory.

It is also possible to develop axiomatic set theory with additional variables for objects that are not sets. Such objects are called **urelements**, **individuals**, or **atoms**. For example, we might want variables for natural numbers and perhaps even for real numbers, with appropriate axioms, included as a basic part of our set theory. The obvious way to do this is with a many-sorted first-order theory. There is no harm in this approach, but it doesn't really gain anything either (once the strangeness of developing number systems within pure set theory wears off), so we will not discuss it further in our development of set theory.

Another reason for mathematicians to know something about set theory is that it induces us to think about the meaning of what we do. Specialists in most branches of mathematics do not need to think very often about foundational questions. Number theorists, analysts, and even algebraists have little incentive to spend much time wondering whether the objects they study are “real,” or what their theorems “really mean.” But when one starts thinking about set theory, and especially the independence results that tell us, for example, how unlikely it is that we will ever know whether the continuum hypothesis is “true,” it becomes natural to ask such questions about the more abstract objects of mathematics. And mental exercises of this sort, while they may be unsettling, are also a valuable philosophical endeavor.

A third reason to be familiar with set theory is that its history is so interesting and so intertwined with developments in other parts of mathematics. In order to highlight this, much of this chapter and Chapter 6 are arranged historically, outlining three major phases in the development of the subject while also presenting the main concepts and results of set theory.

For a more thorough introduction to set theory at an elementary level, see [Gol], [Vau], [Roi], or [Sup]. A more advanced treatment can be found in [Jech78], [JW], or [TZ].

2.2 “Naive” set theory

The first phase in the development of set theory, which extended from the 1870s until about 1900, was marked by the attempts of Dedekind, Georg Cantor, and others to gain acceptance of the use of infinite sets in mathematics. Through his early work on trigonometric series, Cantor came to realize that the efforts of the time to establish a rigorous theory of the real numbers, primarily by Dedekind and Weierstrass, were essentially based on infinite sets. From today’s perspective, it seems surprising just how much resistance this new subject sparked. But Carl Friedrich Gauss, certainly the most influential mathematician of the first half of the nineteenth century, shared the ancient Greek “horror of the infinite.” Thus he believed that infinite collections should be considered only as incomplete, potential objects, never as completed ones that can be “grasped as a whole.” Many mathematicians of the latter

Georg Ferdinand Cantor (1845–1918) is generally considered to be the main founder of set theory. Cantor's father wanted him to study engineering, but Georg was more interested in philosophy, theology and mathematics. Eventually, Cantor concentrated on mathematics and received his doctorate from the University of Berlin in 1867. In 1874, Cantor published one of the first papers that seriously considered infinite sets as actual objects, and he devoted the rest of his career to this subject.

Cantor's work encountered a degree of resistance that, in retrospect, seems quite unfair and regrettable. Kronecker in particular was often vicious in his criticisms of other mathematicians. His attacks on the free use of the infinite angered Weierstrass and Dedekind, but had a more profound effect on Cantor. Kronecker used his influence to block Cantor's applications for positions at Germany's most prestigious universities; thus Cantor spent his entire 44-year career at the relatively minor Halle University. Cantor became exhausted and discouraged by the resistance to his work, and began having bouts of severe depression and mental illness in 1884. Cantor did very little new research during the last thirty years of his life, and even though his work finally received proper recognition after the turn of the century, he died in a mental institution in Halle.

part of the century, notably Leopold Kronecker, shared Gauss's **finitist** philosophy and refused to accept Cantor's radical ideas.

Set theory during this period was based on two very simple axioms. One, called the **comprehension** axiom, says that any collection of objects that can be clearly specified can be considered to be a set. The other axiom, **extensionality**, asserts that two sets are equal if and only if they have the same elements. This theory was meant to be combined with the rest of mathematics, not to replace it.

Cantor not only viewed infinite sets as actual objects; he defined operations on them and developed an elaborate theory of their sizes, called **cardinal arithmetic**. This program was especially offensive to Kronecker and to many mathematicians of the next generation such as Henri Poincaré. When Russell's amazingly short "paradox" (1902) showed that set theory based on the full comprehension axiom is inconsistent, Poincaré was particularly pleased, stating, "Later mathematicians will regard set theory as a disease from which we have recovered." Hilbert, who supported the new subject, countered by saying that "no one will evict us from the paradise that Cantor has built for us."

Russell's paradox, which Ernst Zermelo actually discovered independently a bit before Russell, begins by letting A be the set of all sets that are not members of themselves. In symbols, $A = \{B \mid B \notin B\}$. By definition of this set, $A \in A$ if and only if $A \notin A$, and so we have a contradiction. (The word paradox, which usually means an *apparent* contradiction, understates the situation here.) While the Burali–Forti paradox (described in the next section) was discovered about five years earlier, it was based on more sophisticated concepts and was not viewed as a major threat to the subject. But Russell's paradox is so simple that it put an end to set theory as it was practiced at the time, which is now called *naive* set theory. Cantor had an inkling of this development, but Frege, who was also a pioneer of the subject, was crushed by Russell's discovery and did no serious mathematics thereafter. (To be fair, there were other important factors in Frege's depression and retirement, such as the death of his wife.)

Russell's paradox was later popularized as the **barber paradox**: in a certain town the barber, who is a man, shaves exactly those men who don't shave themselves. Who shaves the barber? This question has no consistent answer.

Two of the most important problems in modern set theory arose from Cantor's study of cardinality, so we will devote the rest of this section to this topic. Here are the fundamental definitions, which are among Cantor's major contributions. While parts of naive set theory had to be discarded, the definitions and theorems in the rest of this section are for the most part not in this category and are an essential part of contemporary set theory. More of the basics of cardinal arithmetic are outlined in Appendix C.

Definitions. For any sets A and B:

(a) $A \preceq B$ means there is a one-to-one function from A to B.

(b) $A \sim B$ means there is a **bijection** or **one-to-one correspondence** between A and B, that is, a one-to-one function from A *onto* B.

(c) $A \prec B$ means $A \preceq B$ but not $A \sim B$.

There are many ways of reading $A \sim B$: A and B are **equinumerous**, or **equipollent**, or A and B have the same **cardinality**, or the same **size**. It is clear that this defines an equivalence relation on all sets. Frege defined a **cardinal** as an equivalence class of $\sim$. For example, the cardinal 3 would be the class of all sets with three members. This definition is intuitively appealing, but it is not permissible in ZFC. (We'll say more about this in the next section.) Two other definitions of this term that are permissible in ZFC are given later in this chapter, and in Appendix C.

Cantor was very interested in the ordering on sets based on cardinality. The relation $\preceq$ is a preordering on sets, but it is more natural to think of $\preceq$ and $\prec$ as orderings on cardinals. One of the first nontrivial accomplishments of set theory was to show that this makes sense:

Theorem 2.1 (Cantor–Schröder–Bernstein (CSB) Theorem). *If $A \preceq B$ and $B \preceq A$, then $A \sim B$.*

Proof. Assume that $A \preceq B$ and $B \preceq A$. So there are one-to-one functions $f : A \to B$ and $g : B \to A$. Let $C = Rng(f)$ and $D = Rng(g)$. So $g^{-1} : D \to B$ is a bijection. We will define a bijection $h : A \to B$ such that, for every x in A, $h(x)$ is either $f(x)$ or $g^{-1}(x)$. We always let $h(x) = f(x)$ unless "forced" to do otherwise.

Suppose y is any element of $B - C$, and let $x_1 = g(y)$. If we let $h(x_1) = f(x_1)$, then y will not be in the range of h, because y is not in the range of f, and x_1 is the only element of A such that

$g^{-1}(x_1) = y$. Therefore, we must let $h(x_1) = g^{-1}(x_1)$. But then, by the same reasoning, h must be defined to be g^{-1} on the following elements of A: $x_2 = g(f(x_1))$, $x_3 = g(f(x_2))$, $x_4 = g(f(x_3))$, etc. So, for each element of $B - C$, we get an entire infinite sequence of elements of A on which h must be defined to be g^{-1}. (See Figure 2.1.) On all other elements of A, we let h be f. It is routine to show that this h is indeed a bijection between A and B. ■

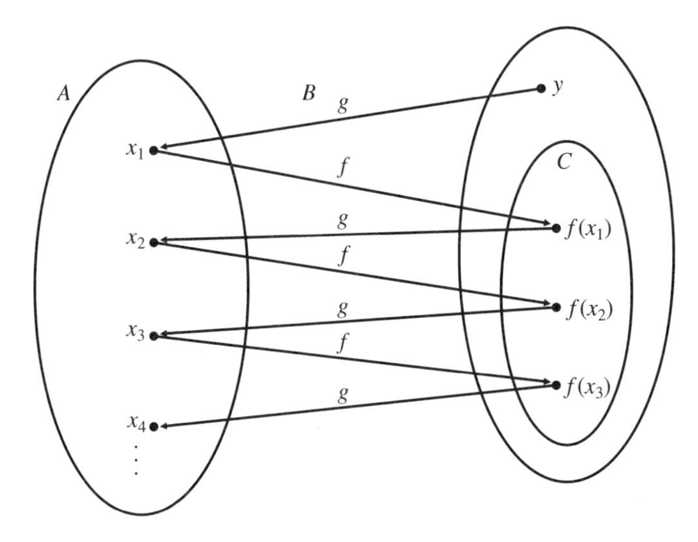

Figure 2.1. Construction of the sequence (x_n) in the proof of the CSB theorem

Example 1. Let's use the proof of the CSB theorem to define a bijection between the intervals $(-1, 1)$ and $[-1, 1]$. We have simple one-to-one functions $f : (-1, 1) \to [-1, 1]$ and $g : [-1, 1] \to (-1, 1)$ defined by $f(x) = x$ and $g(x) = x/2$. In the notation of the above proof, we have $B - C = \{-1, 1\}$. So we must set $h(1/2) = 1$, $h(1/4) = 1/2$, $h(1/8) = 1/4$, etc. Similarly, $h(-1/2) = -1$, $h(-1/4) = -1/2$, $h(-1/8) = -1/4$, etc. For all other elements of $(-1, 1)$, we let $h(x) = x$. It follows that $(-1, 1) \sim [-1, 1]$.

Note that this function h is not continuous. It is not hard to show that there cannot be a continuous bijection between an open interval and a closed interval.

Exercise 1. Complete the details of the proof of the CSB theorem.

The bulk of the proof of the CSB theorem was provided by Cantor's student Felix Bernstein at the age of nineteen. Without this theorem, there would be another natural equivalence relation based on size of a set, defined by ($A \preceq B$ and $B \preceq A$). That the two equivalence relations actually coincide is much more appealing. Another way of stating the CSB theorem is that the relation $\prec$ is strongly antisymmetric: $A \prec B$ and $B \prec A$ cannot hold simultaneously.

Other questions regarding cardinality were more elusive. Cantor naturally hoped to prove that the ordering on sets is total: $\forall A, B(A \preceq B \vee B \preceq A)$ or, equivalently, $\forall A, B(A \prec B \vee B \prec A \vee A \sim B)$. He was able to do this, but only by assuming the **well-ordering principle**, that every set can be well ordered. For some time Cantor claimed to have proved this principle from more elementary assumptions, but later he realized that he could not do so.

Here is another important achievement of Cantor's study of cardinality:

Definition. For any set A, its **power set**, denoted $\mathcal{P}(A)$, is the set of all subsets of A.

Theorem 2.2 (Cantor's Theorem). *For any set A, $A \prec \mathcal{P}(A)$.*

Proof. The function $f : A \to \mathcal{P}(A)$ defined by $f(u) = \{u\}$ is clearly one-to-one, so $A \preceq \mathcal{P}(A)$. Now we must show that $A \not\sim \mathcal{P}(A)$. Assume that g is any one-to-one function from A to $\mathcal{P}(A)$. Now let $B = \{u \in A \mid u \notin g(u)\}$. The set B is in $\mathcal{P}(A)$ but cannot be in the range of g, because if we assume that $g(u) = B$, we find that $u \in B$ if and only if $u \notin B$, a contradiction. Thus g is not a bijection, so we conclude that $A \not\sim \mathcal{P}(A)$. ■

This proof was the first example of a **diagonalization argument**, which has since become a powerful tool. Note the similarity to Russell's paradox, except that here we don't reach a contradiction. We just show that a certain function can't exist. We will encounter several more diagonalization arguments in this book, mostly in Chapters 3 and 4.

Cantor's theorem implies that there is no largest cardinal, and more specifically that there are **uncountable** sets, sets greater in size than $\mathbb{N}$. Cantor also proved that $\mathbb{R} \sim \mathcal{P}(\mathbb{N})$, so in particular the reals are uncountable. Now, if A is a finite set, say with n elements, then $\mathcal{P}(A)$ has 2^n elements. Unless $n = 0$ or 1, this means that there are sets that are strictly between A and $\mathcal{P}(A)$ in cardinality. But if A is infinite, no such "intermediate" sets present themselves. Cantor conjectured, but could not prove, that there are no sets that are between $\mathbb{N}$ and $\mathcal{P}(\mathbb{N})$ in cardinality. This conjecture is called the **continuum hypothesis** (CH). The more general conjecture obtained by replacing $\mathbb{N}$ with an arbitrary infinite set is called the **generalized continuum hypothesis** (GCH).

Cantor's theorem also provides an alternative proof of the inconsistency of naive set theory, almost as short as Russell's paradox: in naive set theory, we can define the set of all sets A. But then $\mathcal{P}(A)$ must be larger than A in size, which is absurd because $\mathcal{P}(A)$ is clearly a set of sets, and is therefore a subset of A.

Here is a direct proof of the uncountability of $\mathbb{R}$, by a modification of the proof of Cantor's theorem that makes it more clear where the term "diagonalization argument" comes from. In the following proof, we assume for notational simplicity that $0 \notin \mathbb{N}$. Only a slight modification is required if $0 \in \mathbb{N}$.

Proposition 2.3. $\mathbb{N} \prec \mathbb{R}$.

Proof. First of all, $\mathbb{N} \preceq \mathbb{R}$ because $\mathbb{N} \subseteq \mathbb{R}$. To complete the proof, we must show that there is no bijection between $\mathbb{N}$ and $\mathbb{R}$. We will prove a bit more, namely that if $f : \mathbb{N} \to \mathbb{R}$, then the range of f cannot contain the entire interval $[0, 1]$.

So let $f : \mathbb{N} \to \mathbb{R}$. We will construct a real number c between 0 and 1 that is not in the range of f. For each $n \in \mathbb{N}$, let the digit in the nth decimal place of c be obtained by increasing or decreasing the digit in the nth decimal place of $f(n)$ by 5. (This is one of many satisfactory procedures for constructing c.) For instance, if $f(1) = 17.374\ldots$ and

$$
\begin{aligned}
f(1) &= \quad 17.37742609\ldots \\
f(2) &= \quad -5.3974420\ldots \\
f(3) &= \quad 0.6402714\ldots \\
f(4) &= \quad -1.3372407\ldots \\
f(5) &= -804.7241036\ldots \\
\\
c &= \quad 0.84575\ldots
\end{aligned}
$$

Figure 2.2. Diagonalization argument used to define the number c

$f(2) = -5.397\ldots$, then c begins $0.84\ldots$. (See Figure 2.2.) Since c differs in at least one decimal place from each $f(n)$, c is not in the range of f. ■

This proof assumes the ability to represent real numbers as decimals and glosses over the fact that some reals have two different decimal forms. For instance, $0.999\ldots = 1$ and $7.47999\ldots = 7.48$. This is the only type of ambiguity in decimal representation: real numbers with terminating decimal expansions are the only ones with more than one decimal form.

Exercise 2. Show that the interval [0, 1] is uncountable.

Further discussion of cardinality will be given in Sections 2.5 and 3.2, as well as Appendix C. In Chapter 6 we will return to the two conjectures that eluded Cantor, the well-ordering principle and the continuum hypothesis, and see the prominent role they played in the development of set theory.

2.3 Zermelo–Fraenkel set theory

The second phase in the history of set theory began with efforts to free set theory, and hopefully all of mathematics, from contradictions such as Russell's paradox. Obviously, the thought that a branch of mathematics, especially such a simple-looking one, could turn out to be inconsistent was quite disturbing.

In the early years of the twentieth century, three movements emerged whose goals included ridding mathematics of contradictions: logicism, formalism, and intuitionism. We briefly mentioned logicism and formalism in Chapter 1. Intuitionism, founded by Brouwer, continued and expanded the tradition of Gauss and Kronecker by insisting that mathematical activity should be confined to "constructive" operations. We will discuss intuitionism further in Chapter 8. While all of these movements made important contributions to mathematics, none of them accomplished the main goal.

Attempts to fix set theory were much more successful. Zermelo was the first (in 1908) to create a set of axioms for set theory that replaced the unrestricted comprehension axiom with a more cautious list of principles for the existence of sets. His ideas were refined by Abraham Fraenkel, Thoralf Skolem, John von Neumann, and others in the 1920s, creating the theory ZF that has withstood eighty years of extensive use and scrutiny. Even though Gödel's incompleteness theorem creates a substantial obstacle to proving that ZF is consistent, almost all mathematicians are confident that it is.

We now list the axioms of ZF set theory. For the most part, the axioms are written completely formally, except that we use the standard abbreviation $x \subseteq y$ for $\forall u(u \in x \rightarrow u \in y)$, as well as the restricted quantifier notation introduced in Section 1.3. Also, starting with axiom 5, we will use "terms" to shorten the axioms.

Proper axioms of ZF set theory

1. **Extensionality:** $\forall x, y[x = y \leftrightarrow \forall u(u \in x \leftrightarrow u \in y)]$. (Two sets are equal if and only if they have the same elements.)
2. **Pairing:** $\forall x, y\, \exists z\, \forall u(u \in z \leftrightarrow u = x \vee u = y)$. (For any x and y, the set $\{x, y\}$ exists.)
3. **Union:** $\forall x\, \exists y\, \forall u(u \in y \leftrightarrow \exists w \in x(u \in w))$. (For any x, the union of all the sets in x exists. This union is denoted $\bigcup x$.)

4. **Empty Set:** $\exists x \, \forall y \sim (y \in x)$. (The empty set $\emptyset$ exists.)
5. **Infinity:** $\exists x[\emptyset \in x \wedge \forall y \in x((y \cup \{y\}) \in x)]$. (There exists an infinite set.) We will explain this axiom more fully in the next section. Note that the way we have written this axiom is not within the first-order language of set theory because it includes terms (in the sense of Section 1.4) like $\emptyset$ and $y \cup \{y\}$. This situation is discussed in the two examples following this list of axioms.
6. **Power Set:** $\forall x \, \exists y \, \forall u(u \in y \leftrightarrow u \subseteq x)$. (For any set x, its power set $\mathcal{P}(x)$ exists.)
7. **Replacement:**

$$[\forall x \in a \, \exists! y \mathrm{P}(x, y)] \rightarrow [\exists b \, \forall y(y \in b \leftrightarrow \exists x \in a \mathrm{P}(x, y))].$$

   (If the formula $\mathrm{P}(x, y)$, which cannot contain b as a free variable, defines a function on the domain a, then there is a set b that is the range of this function.) Replacement is an axiom schema since there are infinitely many choices for the formula P.
8. **Regularity or Foundation:** $\forall x[x \neq \emptyset \rightarrow \exists y \in x(x \cap y = \emptyset)]$. (A nonempty set must contain an element that is disjoint from it.) We will thoroughly discuss the significance of this axiom at the end of this section.

ZFC set theory is obtained from ZF by adding one more axiom:

9. **Axiom of Choice (AC):**

$$[\forall u \in x(u \neq \emptyset) \wedge \forall u, v \in x(u \neq v \rightarrow u \cap v = \emptyset)]$$
$$\rightarrow \exists y \, \forall u \in x \, \exists! w \in u(w \in y).$$

   (If x is a set of nonempty, pairwise disjoint sets, then there is a set y that consists of exactly one member of every set in x. Such a y is called a **choice set** for x.)

Example 2. The empty set axiom asserts the existence of a set with no members. By extensionality, this set is unique. Therefore, it is permissible and reasonable to introduce the term $\emptyset$ to denote this set. So the first-order Skolem form of the empty set axiom would be $\forall y(y \notin \emptyset)$.

When we write something like $\emptyset \in x$, as in the axiom of infinity, this is an abbreviation for what would be a much longer formula in the sparse language of ZF, namely: $\exists z[z \in x \wedge \forall y(y \notin z)]$. It would be extremely cumbersome to carry out the development of set theory without introducing terms for sets. The next example will continue the discussion of terms.

There are many other natural statements that are equivalent (in ZF) to the axiom of choice, such as the well-ordering principle and the totality of the ordering on cardinals, mentioned in the previous section. Another concise version of AC is that the Cartesian product of any family of nonempty sets is nonempty. The axiom of choice will be discussed further in Chapter 6.

We will not give a systematic development of basic set theory from the axioms of ZF or ZFC. To see how this is done in detail, refer to [Sup] or [Jech78]. We will just mention a few of the most useful basic results and then move on to more specific topics.

Set-builder notation $\{x \mid P(x)\}$ (read "the set of all x such that $P(x)$") is very convenient and commonly used throughout mathematics. Of course, the intended meaning is that if $y = \{x \mid P(x)\}$, then, for every x, $x \in y$ if and only if $P(x)$. Russell's paradox makes it clear that we cannot expect this notation to define a set in all cases. The more cautious viewpoint of modern set theory is that we should at least be able to assert the existence of any set of the form $\{x \in a \mid P(x)\}$, "the set of all x in a, such that $P(x)$." The idea is that since what we are asserting to exist in this way is a subset of some set a that already exists, we can't end up with a set that is "too big," such as the set of all sets. The principle that such subsets always exist is called the axiom (schema) of **separation**.

So, in ZF or ZFC, one cannot define the set of all sets, the set of all rings, etc. Informally, it's convenient and harmless to refer to such collections as **classes**. If a class is known to be too large to be a set, as these are, then it is called a **proper class**. Specifically, we can talk about the class of all sets x such that $P(x)$, for any formula $P(x)$. But there are no variables for such classes, and a proper class can never be a member of a set.

John von Neumann (1903–1957) was born to a well-to-do and intellectual family in Budapest. He was a true child prodigy who could divide eight-digit numbers in his head at age six, and he learned calculus at age eight. He would also show off his photographic memory by reading a page of a telephone book and then repeating all the names, addresses and phone numbers by heart. Von Neumann obtained a university degree in chemical engineering in 1925, and then got his PhD in mathematics just one year later. Like Albert Einstein and Gödel, he left Europe in the 1930s to become a permanent member of the Institute for Advanced Study in Princeton.

Von Neumann made important contributions to many fields within mathematics and science, from the very abstract to the very practical. In the 1920s, he worked with Hilbert and Paul Bernays on the formalist program and the foundations of set theory. When Gödel publicly announced his incompleteness theorem, von Neumann was the first member of the audience to grasp the significance of Gödel's accomplishment. During this period, von Neumann also studied the mathematical foundations of quantum mechanics, and in 1932 published a very successful textbook in that field. His next major achievement was in game theory; he collaborated with the economist Oskar Morgenstern to publish, in 1944, *The Theory of Games and Economic Behavior*, which is the primary reference for modern game theory. During World War II he made substantial contributions to the American atom bomb project, and much of his work in the forties and fifties was on military projects.

From a practical standpoint, von Neumann's most important achievement was his pioneering work on the development of the

(*continued*)

John von Neumann *continued*

computer. Toward the end of World War II, the army built a machine called ENIAC, considered to be the first digital computer. Not only was it huge—over 100 feet long—but it was also extremely awkward and complicated to instruct it what to do. Von Neumann realized that it would be more efficient to give the computer instructions using a "stored program" that one could create outside the computer and then insert into the machine. In other words, he essentially invented the notion of a computer program. Von Neumann built a computer at the Institute (a project that was very controversial at one of the world's "purest" ivory towers), and his work helped IBM develop the machines that launched the computer age. He also pioneered related fields such as cellular automata theory and the theory of self-reproducing machines.

Given his huge output of important work, von Neumann spent a surprising amount of time having fun. His personality was outgoing and friendly, and he loved money, fine wine, women, noise, dirty limericks, and fast cars—several of which he wrecked. He gave large parties, often more than one per week, which were legendary in the Princeton area. In short, von Neumann was a rare combination of genius and "party animal."

In another important version of axiomatic set theory, created by von Neumann, Bernays, and Gödel and therefore called VBG, the variables represent classes, which may be sets or proper classes. But only a set can be a member of a class. So, for instance, Frege's definition of a cardinal becomes acceptable in VBG: the cardinal of any set x is a legitimate object. Unless x is empty, it's a proper class. VBG proves exactly the same theorems about *sets* as ZFC does and, unlike ZF and ZFC, it is finitely axiomatizable. In spite of these desirable features of VBG, most contemporary treatments of set theory use ZFC exclusively, and we will also.

Proposition 2.4. *The full separation schema is derivable in ZF set theory. In other words, for any formula* Q(x) *in which* b *is not a free variable, the following formula is provable in ZF:*

$$\forall a\, \exists b\, \forall x[x \in b \leftrightarrow (x \in a \wedge \mathrm{Q}(x))].$$

Proof. We give a rather informal proof that can easily be formalized in ZF; technically, the formal proof consists of an infinite set of proofs, one for each Q.

Let a be given. We consider two cases. If there are no members of a for which Q(x) holds, then let $b = \emptyset$, and we are done.

If there are members of a for which Q(x) holds, let c be one of these. Now define the formula P(x, y) to be

$$(\mathrm{Q}(x) \wedge y = x) \vee ({\sim}\, \mathrm{Q}(x) \wedge y = c).$$

Then apply the replacement axiom to this P and a. The set b that must exist by replacement is easily shown to be $\{x \in a \mid \mathrm{Q}(x)\}$. ■

Zermelo's original version of set theory did not include the regularity axiom and had separation instead of replacement. By the previous proposition, Zermelo's theory is a subtheory of ZF, and it turns out to be a proper subtheory, but it is powerful enough to prove the great majority of important mathematical results outside of foundations.

With the exception of extensionality, all of the axioms of ZFC assert the existence of sets with certain properties. The existence of other familiar sets can easily be derived in ZF.

Example 3. The set that is asserted to exist by the pairing axiom is denoted $\{x, y\}$. From this, we can write $\{x\}$ for $\{x, x\}$. The ordinary union of two sets x and y, denoted $x \cup y$, is $\bigcup(\{x, y\})$, whose existence follows from pairing and union. Then $\{x, y, z\} = \{x, y\} \cup \{z\}$, and from this we can define $x \cup y \cup z$, etc. No special axiom is needed for intersections, since their existence follows from separation: $x \cap y = \{z \in x \mid z \in y\}$. Similarly, the set $x - y$, defined as $\{z \in x \mid z \notin y\}$, exists by separation. By the extensionality axiom, all of these terms denote sets that are unique, for any given values of the variables appearing in them. Therefore, it is natural to think of these terms as defining Skolem functions.

The notation $x - y$ may be read "x minus y," but this set is more correctly called the **relative complement of** y **in** x. There is also the related concept of the **symmetric difference** of any two sets x and y, denoted $x \Delta y$, and defined to be $(x \cup y) - (x \cap y)$ (or, equivalently, $(x-y) \cup (y-x)$. It is worth noting that, in ZF or ZFC, all complements are relative. That is, if x is a set, then $\{z : z \notin x\}$ cannot be a set; it is always a proper class.

Occasionally, one must resort to an artificial definition in order to "embed" some mathematical notion smoothly into ZFC. One such definition is Kazimierz Kuratowski's definition of the **ordered pair** of any two objects: $(x, y) = \{\{x\}, \{x, y\}\}$. The set on the right side of this equation has no conceptual connection with ordered pairs. It is used simply because it allows us to prove, in ZF, the two essential properties of ordered pairs: that the ordered pair of any two sets exists, and that $(x, y) = (u, v)$ if and only if $x = u$ and $y = v$.

Exercise 3. Prove these two properties of ordered pairs, in ZF. Don't try to make your proof too formal, but make sure your steps follow from the axioms of ZF.

Once ordered pairs are available, one can prove (in ZF) the existence of various other important sets, such as the Cartesian product $A \times B$ and the set B^A of all functions from A to B, for any sets A and B.

Exercise 4. Outline a proof (in ZF) that the Cartesian product of any two sets exists. You will need to use the replacement and union axioms twice each.

The regularity axiom

To conclude this section, we will examine the regularity axiom in some detail. What does it say? What is it about? Superficially, it asserts the existence of a certain type of set, just as all the other axioms of ZFC except extensionality do. But it really has a different flavor from the other axioms, in that the set asserted to exist is an element of a given set x. So, in an important sense, it doesn't assert the existence of any *new* sets.

The regularity axiom may be viewed as the result of the following line of thinking: naive set theory suffered from paradoxes, and paradoxes in logic and mathematics are almost always traceable to some sort of circular reasoning or definition. In set theory, one is constantly defining sets by specifying their members, and a prudent rule of thumb to avoid circular definitions is to require that all the members of a set must already be defined or "constructed" before we can define that set. This would imply, among other things, that a set cannot be a member of itself. For instance, note that the set of all sets, which we have shown cannot exist because it leads to paradoxes, would violate this principle.

With this in mind, let's consider what regularity says and some of its consequences. One immediate consequence is that no set is a member of itself. For if $y \in y$, then letting $x = \{y\}$ violates this axiom. Another consequence is that we cannot have $y \in z$ and $z \in y$ simultaneously, for then $x = \{y, z\}$ would violate regularity. Generalizing this, the regularity axiom guarantees that there cannot be any finite *cycles* in the relation $\in$, and this is clearly one desirable result if we are trying to eliminate circularity in the construction of sets.

Here is an even more significant consequence of regularity: imagine an infinite sequence of sets $x_0, x_1, x_2, \ldots$ such that $x_{n+1} \in x_n$ for every n. Then the set $\{x_0, x_1, x_2, \ldots .\}$ violates regularity. In other words, no such sequence can exist; we say that regularity prevents **infinite descending sequences** under $\in$. So if we start with any set x_0 and try to generate such a sequence, we inevitably find that $x_n = \emptyset$ for some n. (Remember, every object is a set.) This result corresponds to the notion that if sets may not be defined in a circular way, then they must be defined "from scratch," in "stages." "From scratch" can only mean from the empty set. And, in order for the idea of stages to make sense, it should not be possible to have an infinite sequence of earlier and earlier stages. More mathematically, what regularity says is precisely that $\in$ is a **well-founded** relation—hence the alternative name foundation. (It is important not to overstate this message. We are not saying that every set must be definable from $\emptyset$ in a finite number of stages. As we will see, there can be infinite *increasing* sequences of sets under $\in$.)

The well-foundedness of $\in$ has important ramifications in set theory. The fact that there are no infinite descending sequences in $\mathbb{N}$ is essentially equivalent to the principle of mathematical induction. In fact, Fermat's **method of infinite descent**, considered the first clear statement of induction, was based on the postulate that every decreasing sequence of natural numbers must terminate. We will soon see that the well-foundedness of $\in$ is useful for the development of the theory of ordinals as well as for embedding the theory of $\mathbb{N}$ in set theory.

2.4 Ordinals

In this section we outline an essential and fascinating part of set theory that is not well known to most mathematicians outside of foundations. In less theoretical treatments, an ordinal is usually defined to be an equivalence class of well-orderings. Here is the more rigorous definition that can be formalized in ZF:

Definitions. A set is **transitive** if every member of it is a subset of it (that is, every member of a member of it is a member of it). An **ordinal** is a transitive set, all of whose members are also transitive.

Example 4. The sets $\emptyset$, $\{\emptyset\}$, and $\{\emptyset, \{\emptyset\}\}$ are ordinals. But the set $\{\{\emptyset\}\}$ is not even transitive, because its only member is not a subset. Why such strange examples? Remember that in pure set theory, all sets must be built up from $\emptyset$.

We now present some basic facts about transitive sets and ordinals. We will usually omit the words "The following is provable in ZF" from the beginning of such results.

Proposition 2.5.

(a) *The power set of a transitive set is transitive.*

(b) *The union and intersection of a collection of transitive sets are transitive.*

Proof.

(a) Assume y is transitive. To show that $\mathcal{P}(y)$ is also transitive, consider $x \in \mathcal{P}(y)$. That means $x \subseteq y$. So if $u \in x$, then $u \in y$. Since y is transitive, this implies $u \subseteq y$, so $u \in \mathcal{P}(y)$. So we have $x \subseteq \mathcal{P}(y)$, as desired. ■

Exercise 5. Prove part (b) of this proposition.

Notation. Lower case Greek letters are used to denote ordinals. So a statement of the form $\forall\alpha \mathrm{P}(\alpha)$ means that P holds for all ordinals.

Proposition 2.6.

(a) *Every member of an ordinal is an ordinal.*

(b) *$\emptyset$ is an ordinal.*

(c) *For any ordinal α, $\alpha \cup \{\alpha\}$ is also an ordinal.*

Proof.

(a) Assume α is an ordinal, and $x \in \alpha$. Then x is transitive by definition of ordinals. To show x is an ordinal, we must also show that every member of x is transitive. But if $u \in x$, then u is a member of a member of α, and thus a member of α since α is transitive. So u must be transitive because α is an ordinal. ■

Exercise 6. Prove parts (b) and (c) of this proposition.

The ordinal $\alpha \cup \{\alpha\}$ referred to in (c) of this proposition is called the **successor** of α, denoted $S(\alpha)$. If β is of the form $S(\alpha)$, then we say β is a **successor ordinal**, written $\mathrm{Suc}(\beta)$. If λ is neither empty nor a successor, then we say λ is a **limit ordinal**, written $\mathrm{Lim}(\lambda)$. We also write 0 for the ordinal $\emptyset$.

Exercise 7. Prove that the successor operation is one-to-one, not just on ordinals but on arbitrary sets, that is: $S(x) = S(y) \to x = y$.

Lemma 2.7 (Trichotomy). *Any two ordinals are comparable under* $\in$, *that is,*

$$\forall \alpha, \beta (\alpha \in \beta \vee \alpha = \beta \vee \beta \in \alpha).$$

Proof. Let's abbreviate what we want to prove as $\forall \alpha, \beta\, \mathrm{C}(\alpha, \beta)$. Assuming it's false, choose α_0 such that $\exists \beta \sim \mathrm{C}(\alpha_0, \beta)$. Then let $A = \{\alpha \in S(\alpha_0) \mid \exists \beta \sim \mathrm{C}(\alpha, \beta)\}$. A is a set by separation, and $A \neq 0$ because $\alpha_0 \in A$. So by regularity, there is an α_1 that is an $\in$-minimal member of A. So every member of α_1 is comparable with every ordinal.

Since α_1 is incomparable to some β, we can choose β_0 that is incomparable to α_1. Just as in the previous paragraph, we can then get an $\in$-minimal β_1 such that $\sim \mathrm{C}(\alpha_1, \beta_1)$. So every member of β_1 is comparable with α_1.

We claim that $\beta_1 \subset \alpha_1$. Assume $\gamma \in \beta_1$. Then γ is an ordinal by Proposition 2.6(a). By definition of β_1, $\mathrm{C}(\gamma, \alpha_1)$. But either $\gamma = \alpha_1$ or $\alpha_1 \in \gamma$ contradicts the fact that $\sim \mathrm{C}(\beta_1, \alpha_1)$. Thus $\gamma \in \alpha_1$. So we have shown that $\beta_1 \subseteq \alpha_1$. Since $\sim \mathrm{C}(\alpha_1, \beta_1)$, we know that $\beta_1 \neq \alpha_1$, so $\beta_1 \subset \alpha_1$.

Now let $\gamma \in (\alpha_1 - \beta_1)$. By definition of α_1, every member of it is comparable to everything. In particular, $\mathrm{C}(\gamma, \beta_1)$. Since $\gamma \notin \beta_1$, we must have $\gamma = \beta_1$ or $\beta_1 \in \gamma$. But each of these possibilities contradicts $\sim \mathrm{C}(\alpha_1, \beta_1)$, so our original assumption must be false. ∎

We have included this rather technical proof because it illustrates the power of regularity in a very typical way. Note that regularity is used twice: to define α_1, and then to define β_1 from α_1.

Notation. When α and β are ordinals, we write $\alpha < \beta$ to mean $\alpha \in \beta$.

Exercise 8. Prove:

(a) $\alpha < \beta \leftrightarrow \alpha \subset \beta$.

(b) $\alpha \leq \beta \leftrightarrow \alpha \subseteq \beta$.

(c) $S(\alpha)$ really is the successor of α. That is, $\alpha < S(\alpha)$, but $\sim \exists\beta(\alpha < \beta < S(\alpha))$.

Theorem 2.8.

(a) *The class of all ordinals is well ordered by* $<$.

(b) *If* $\exists\alpha \mathrm{P}(\alpha)$, *then there is a least* α *such that* $\mathrm{P}(\alpha)$.

Proof.

(a) What we mean by this rather informal statement is that the defining properties of an (irreflexive) well-ordering, with the variables ranging over ordinals, can be proved about $<$, in ZF. So we need to show that $<$ is a partial ordering (irreflexive and transitive), well-founded (every nonempty set of ordinals has a minimal element under $<$), and total for ordinals.

By regularity, we know that $\alpha < \alpha$ is always false, so $<$ is irreflexive. The well-foundedness of $<$ on ordinals also follows immediately from the regularity axiom. If $\alpha < \beta$ and $\beta < \gamma$, then $\alpha < \gamma$ because γ is a transitive set. This shows $<$ is transitive on ordinals. Finally, to establish that $<$ is a total ordering on ordinals rather than just a partial ordering, we need trichotomy, which was proved in Lemma 2.7.

(b) Assume $\exists\alpha \mathrm{P}(\alpha)$. We can't form the set $\{\alpha \mid \mathrm{P}(\alpha)\}$, but we can proceed as in the proof of Lemma 2.7: choose a particular β_0 such that $\mathrm{P}(\beta_0)$, and form the set $\{\alpha \leq \beta_0 \mid \mathrm{P}(\alpha)\}$. This set has a least element, by (a). ■

Corollary 2.9. *Each ordinal is well ordered by* $<$.

Proof. Every initial segment of a well-ordering is a well-ordering. ■

Exercise 9. Prove the following near-converse of this corollary: if x is transitive and is totally ordered by $\in$, then x is an ordinal.

Proposition 2.10. *If* x *is any set of ordinals, then* $\bigcup x$ *is an ordinal, which is also the least upper bound of* x.

Exercise 10. Prove this proposition.

This proposition is quite useful. It tells us that every set of ordinals is bounded above in the class of all ordinals, and in fact has a least upper bound, which is simply its own union.

Corollary 2.11. *There is no set that contains all ordinals.*

Proof. Given any set x, let y be the set of all ordinals in x. Then let $\alpha = S(\bigcup y)$. By the previous proposition, α is an ordinal that contains every ordinal in x. Since $\alpha \notin \alpha$ by regularity, we have $\alpha \notin x$. So x does not contain all ordinals. ∎

The paradox that results from assuming the existence of the set of all ordinals and then arguing as above is called the **Burali–Forti Paradox**. Note that this corollary provides yet another proof, the third we have seen, that there is no set of all sets.

So the ordinals form a very large collection, a proper class (denoted *Ord*), but they are naturally well ordered by the simplest possible binary relation, $\in$. We will see that the ordinals are perfectly suited to represent the "stages" in the construction of sets mentioned earlier. Also, it is not hard to show that every well-ordering of a set is isomorphic to a unique ordinal. So there is a natural bijection between ordinals and equivalence classes of well-orderings (which are the ordinals in the intuitive sense).

We have not shown that there are any limit ordinals; it's time to fix that.

Theorem 2.12. *There exists a limit ordinal.*

Proof. Let x be a set satisfying the axiom of infinity, and define y to be the set of all ordinals in x. We have $0 \in y$ and $\forall\alpha(\alpha \in y \rightarrow S(\alpha) \in y)$. Now let $\beta = \bigcup y = \text{LUB}(y)$, as in Proposition 2.10. Since $0 \in \beta$, we know that $\beta \neq 0$. And whenever $\alpha < \beta$, we also have $S(\alpha) < \beta$. Therefore, $\beta \neq S(\alpha)$. In other words, β cannot be a successor. Thus β is a limit ordinal. ∎

It follows from this result that there is a least limit ordinal, which is called ω ("omega"). The members of ω are called **finite** ordinals or

natural numbers. In other words, to a set theorist, $\omega = \mathbb{N}$. Of course, ω and all larger ordinals are called **infinite** ordinals.

We are already writing 0 to mean $\emptyset$. Similarly, 1 means $S(0)$ or $\{0\}$, 2 means $S(1)$ or $\{0, 1\}$, etc. (So the three ordinals mentioned in Example 4 are in fact 0, 1, and 2.) Since each ordinal is the set of all smaller ordinals, each natural number is the set of all smaller natural numbers.

Since the ordinals are well ordered, a principle of proof by induction should hold for them. Here are the two main versions of it:

Theorem 2.13 (Transfinite Induction). *For any formula* $\mathrm{P}(\alpha)$*:*

(a) $\forall\alpha[(\forall\beta < \alpha)\mathrm{P}(\beta) \rightarrow \mathrm{P}(\alpha)] \rightarrow \forall\alpha\mathrm{P}(\alpha)$.

(b) $[\mathrm{P}(0) \wedge \forall\alpha(\mathrm{P}(\alpha) \rightarrow \mathrm{P}(S(\alpha)) \wedge \forall\lambda((\mathrm{Lim}(\lambda) \wedge (\forall\alpha < \lambda)\mathrm{P}(\alpha)) \rightarrow \mathrm{P}(\lambda))] \rightarrow \forall\alpha\mathrm{P}(\alpha)$.

Proof. (a) is just the contrapositive of Theorem 2.8, and is the same principle that we often use for $\mathbb{N}$: if there's no least counterexample to a certain statement, then there's no counterexample at all. Part (b) just takes (a) and breaks it down according to the three types of ordinals (0, successors, and limits). ■

Just as with $\mathbb{N}$, a principle of proof by induction always gives rise to a procedure for defining functions by induction. With ordinals, we may use Theorem 2.13(a) and give a single condition defining $f(\alpha)$ in terms of f's values on the entire domain α. Or we may use Theorem 2.13(b) and give a three-part definition. We may inductively define a function whose domain is some ordinal, or we may (informally, or in VBG) define a proper class function whose domain is all ordinals.

Here are two important binary operations defined on all ordered pairs of ordinals, with a three-part inductive definition on the second variable:

Definition. Ordinal **addition** is defined by induction as follows:

(a) $\alpha + 0 = \alpha$.

(b) $\alpha + S(\beta) = S(\alpha + \beta)$.

(c) For $\mathrm{Lim}(\lambda)$, $\alpha + \lambda = \bigcup_{\beta<\lambda}(\alpha + \beta)$.

Setting $\beta = 0$ in clause (b) implies that $\alpha + 1 = S(\alpha)$ for every α, so we will no longer use the special notation $S(\alpha)$.

Definition. Ordinal **multiplication** is defined by induction as follows:

(a) $\alpha \cdot 0 = 0$.

(b) $\alpha \cdot (\beta + 1) = (\alpha \cdot \beta) + \alpha$.

(c) For $\text{Lim}(\lambda)$, $\alpha \cdot \lambda = \bigcup_{\beta<\lambda}(\alpha \cdot \beta)$.

It is natural to think of α as fixed in these definitions. Parts (a) and (b) of both definitions are the standard inductive definitions for $+$ and $\cdot$ on $\mathbb{N}$, starting from the successor operation. Part (c) just extends this by taking the least upper bound of all previous values when a limit ordinal is reached.

The existence of a set like $\bigcup_{\beta<\lambda}(\alpha + \beta)$ follows from the replacement and union axioms.

What does it mean to define, in ZF, functions like these whose domain is the proper class of all ordinals? Certainly, the function we define is not a set. Technically, when we write such a definition, we are asserting (in ZF) that for any ordinals α and γ, there is a unique function with domain γ that satisfies all the clauses of the definition, for all β in γ (and that one fixed α).

Ordinal arithmetic is an interesting topic in its own right; for one thing, neither $+$ nor $\cdot$ is commutative. We will not go into it further, except to mention that $\alpha + \beta$ is the ordinal whose order type looks like "a copy of α followed by a copy of β," and $\alpha \cdot \beta$ is the ordinal whose order type looks like "β copies of α."

Exercise 11.

(a) Prove that $1 + \omega = \omega$, while $\omega + 1 > \omega$. Thus ordinal addition is not commutative.

(b) Prove that $2 \cdot \omega = \omega$, while $\omega \cdot 2 = \omega + \omega > \omega$. Thus ordinal multiplication is not commutative.

Exercise 12.

(a) Define **ordinal exponentiation**. The definition of α^β should be by transfinite induction on β. The clauses for $\beta = 0$ and $\text{Suc}(\beta)$

should be correct for natural numbers, and the clause for $\text{Lim}(\beta)$ should be the same as for addition and multiplication.

(b) Prove that $\alpha^{\beta} \cdot \alpha^{\gamma} = \alpha^{\beta+\gamma}$.

(c) Prove that $(\alpha^{\beta})^{\gamma} = \alpha^{\beta\cdot\gamma}$.

It can be shown in ZF that ω, together with the operations $+$, $\cdot$, and S (restricted to ω, of course) and the ordinal 0, satisfies all the axioms of Peano arithmetic. In other words, all the usual mathematics of $\mathbb{N}$ can be carried out in ZF. From there, we can give the usual constructions of $\mathbb{Z}$, $\mathbb{Q}$, and $\mathbb{R}$, and all the usual mathematics of these number systems can also be carried out in ZF or ZFC. So the ability to carry out all of standard mathematics within axiomatic set theory depends crucially on the theory of ordinals.

2.5 Cardinals and the cumulative hierarchy

In axiomatic set theory, ordinals are also useful for defining some important notions relating to cardinality:

Definitions. A set x is called **finite** if $x \prec \omega$; **denumerable** if $x \sim \omega$; **countable** if $x \preceq \omega$; **infinite** if $\omega \preceq x$; and **uncountable** if $\omega \prec x$.

Note that we have not defined "infinite" and "uncountable" simply as the negations of "finite" and "countable." The next few results will clarify these terms. We will indicate those results whose proofs require the axiom of choice. Make sure not to confuse the relations $<$ and $\leq$ between ordinals with the relations $\prec$ and $\preceq$ between arbitrary sets.

Lemma 2.14. *For any $n \in \omega$, $n \prec n + 1$.*

Proof. Since $n \subseteq n + 1$, we have $n \preceq n + 1$. We also need $n \not\sim n + 1$, which we will prove by ordinary induction on n.

Since $0 = \emptyset$ and $1 = \{\emptyset\}$, the only function from 0 to 1 is the function $\emptyset$. This function is not onto 1, so $0 \not\sim 1$.

Now assume $n \not\sim n + 1$. We want to show that $n + 1 \not\sim n + 2$. Assume on the contrary that f is a bijection between $n + 1$ and $n + 2$.

If $f(n) = n + 1$, then $f - \{(n, n + 1)\}$ is clearly a bijection between n and $n + 1$, contradicting the induction hypothesis. If $f(n) \neq n + 1$, let $k = f(n)$, and $m = f^{-1}(n + 1)$. Now define g to be the relation $[f \cup \{(m, k)\}] - \{(n, k), (m, n+1)\}$. It is easy to see that g is a bijection between n and $n + 1$, again contradicting the induction hypothesis. ■

Corollary 2.15. *If $m < n < \omega$, then $m \prec n \prec \omega$.*

Lemma 2.16. *Let $x \subseteq \omega$. Then:*

(a) *x is finite if and only if it is bounded above.*

(b) *x is denumerable if and only if it is unbounded above.*

Exercise 13. Prove the previous corollary and lemma.

Proposition 2.17. *For any set x,*

(a) *x is finite if and only if $x \sim n$ for some $n \in \omega$.*

(b) *x is infinite if and only if it has a proper subset of the same cardinality as x itself. (Dedekind used this as the definition of infinite sets.)*

(c) *If x is infinite, then it is not finite.*

(d) *(AC) x is infinite if and only if it is not finite.*

(e) *If x is uncountable, then it is not countable.*

(f) *(AC) x is uncountable if and only if it is not countable.*

Proof.

(a) This follows easily from the previous corollary and lemma, and is left as an exercise (see below).

(b) Assume x is infinite. So there is a one-to-one function f from ω to x. Clearly, ω itself is "Dedekind infinite." For example, the function $g(n) = 2n$ is a bijection between ω and the set of even numbers in ω. Now define $h : x \to x$ by

$$h(u) = \begin{cases} u, & \text{if } u \notin Rng(f), \\ f(g(f^{-1}(u))), & \text{otherwise.} \end{cases}$$

It is easy to show that h is a bijection between x and a proper subset of x. (See the next exercise.)

For the converse, assume $f : x \to y$, where f is one-to-one and $y \subset x$. Let c be any element of $x - y$. Define $g : \omega \to x$ inductively by $g(0) = c$, and $g(n + 1) = f(g(n))$. It is easy to show that g is one-to-one. Hence $\omega \preceq x$, making x infinite. (See the next exercise.)

Parts (c) and (e) are trivial. Parts (d) and (f) follow immediately, assuming the fact (which we will not prove) that AC implies that all sets are comparable by cardinality. ■

Exercise 14. Prove part (a) of this proposition, and complete the proofs of parts (b) and (c).

The phenomenon described in part (b) of this proposition was first described by Galileo in the early 1600s, and for over two centuries thereafter it was viewed as paradoxical and an argument against the use of infinite sets. Nowadays, this strange property is viewed as a fact of mathematical life.

Example 5. Since $\mathbb{N}$ (that is, ω) is infinite, it must have proper subsets of the same cardinality. In fact, Lemma 2.16(b) tells us that the set of even natural numbers, the set of primes, the set of perfect squares, etc., are just as big as all of $\mathbb{N}$.

Exercise 15. Appendix C shows that $\mathbb{N} \times \mathbb{N} \sim \mathbb{N}$. (See the discussion following Proposition C.1.) Use this fact and the CSB theorem to prove that $\mathbb{Q} \sim \mathbb{N}$. Many people find this especially surprising because, on a number line, there are an infinite number of rationals between each pair of whole numbers.

Exercise 16. This amusing scenario was concocted by Hilbert to illustrate the surprises that are inherent in the study of infinite sets. You are the desk clerk at Hilbert Hotel, which has a denumerable number of single rooms (numbered 1, 2, 3, etc.), and is currently full.

(a) Suddenly a man comes in, desperately wanting a room. At first you tell him that he can't have one because the hotel is full, but then you realize you can give him a room, provided that you are willing to move people around (but not force people to share rooms). How do you do that?

(b) Later, an even bigger problem occurs. There is another Hilbert Hotel across the street, and it burns down. Suddenly a denumerable set of customers arrives, all wanting rooms in your hotel. How can that be done?

(c) Now comes the true disaster. Across town, there is an infinite sequence of Hilbert Hotels, all full, and they all burn down. All the customers from all those hotels appear at your desk, wanting rooms. How can you accommodate them?

Example 6. The function e^x from $\mathbb{R}$ to $\mathbb{R}$ is one-to-one but not onto; it is a bijection between $\mathbb{R}$ and its proper subset $\mathbb{R}^+$, and so $\mathbb{R} \sim \mathbb{R}^+$. The function $x^3 - x$ from $\mathbb{R}$ to $\mathbb{R}$ is onto but not one-to-one. Such functions cannot exist from a finite set to itself.

Exercise 17. Give examples (or prove the nonexistence) of functions from $\mathbb{N}$ to $\mathbb{N}$ that are:

(a) one-to-one and onto
(b) neither one-to-one nor onto
(c) one-to-one but not onto
(d) onto but not one-to-one.

Exercise 18.

(a) Find a bijection between $\mathbb{R}$ and a bounded open interval. There is at least one such function that you have been familiar with since high school.

(b) Using part (a), the CSB theorem, and other familiar functions, show that all intervals of the forms (a, b), $[a, b]$, $(a, b]$, $[a, b)$, (a, ∞), $[a, \infty)$, $(-\infty, b)$, and $(-\infty, b]$ have the same cardinality as $\mathbb{R}$, provided that $a < b$.

Proposition C.1(c) of Appendix C also tells us that $\mathbb{R} \times \mathbb{R} \sim \mathbb{R}$, and therefore $\mathbb{R}^k \sim \mathbb{R}$ for every positive integer k. This special case does not require AC, and in fact the idea behind the required bijection is straightforward: since a real number is basically a decimal, two reals can be "coded" into one simply by alternating digits. That is, if we want to define $f : \mathbb{R} \times \mathbb{R} \to \mathbb{R}$, we might let $f(2.5, 1/3) = 20.530303\ldots$, since $2.5 = 2.5000\ldots$ and $1/3 = 0.3333\ldots$. However, there are some sticky points here, such as the treatment of negative numbers and the ambiguity of decimal representation mentioned in Section 2.2. If we allow ourselves to use the fact that $\mathbb{R} \sim 2^{\mathbb{N}}$, then these difficulties disappear, since we can code two infinite sequences of bits into one sequence in the same way.

If we combine the content of the previous paragraph and the previous exercise, we reach a conclusion that seems geometrically absurd: a line segment one millimeter long has the same "number of points" (cardinality) as all of three-dimensional space. Results like this one and the fact that $\mathbb{Q} \sim \mathbb{N}$ contributed to the resistance that Cantor faced in the early years of set theory. Perhaps even more bizarre is the existence of a **space-filling curve**, devised by Peano: a *continuous* function from the unit interval $[0, 1]$ *onto* the unit square $[0, 1] \times [0, 1]$. Such a function cannot be one-to-one, but even so, most people find it difficult to imagine how such a curve could exist.

Exercise 19. Prove that $\mathbb{N}^{\mathbb{N}} \sim 2^{\mathbb{N}}$, and therefore $k^{\mathbb{N}} \sim 2^{\mathbb{N}}$ for every natural number $k > 1$. The main thing you need to show is how to "code" an infinite sequence of natural numbers into an infinite sequence of 0's and 1's.

Here are a few other basic facts involving cardinality, which are restated as parts (f), (h), (i) and (j) of Proposition C.1 in Appendix C. These facts are somewhat abstract in that they deal with functions on functions, but otherwise they are straightforward.

Proposition 2.18.

(a) *For any set A, $\mathcal{P}(A) \sim 2^A$. Here, 2^A has its literal meaning: the set of all functions from A to the ordinal* $\{0, 1\}$.

(b) *If B and C are disjoint, then $A^B \times A^C \sim A^{B \cup C}$.*

(c) *For any sets A, B, and C,* $(A^B)^C \sim A^{B\times C}$.

(d) *For any sets A, B, and C,* $A^C \times B^C \sim (A \times B)^C$.

Proof.

(a) To define a bijection F from $\mathcal{P}(A)$ to 2^A, let $F(B)$ be the characteristic function of B (with domain A). It is then routine to show that F is one-to-one and onto 2^A. (See the exercise below).

(b) Here, we need to define a mapping F that takes as input an ordered pair (g, h), where $g \in A^B$ and $h \in A^C$, and outputs a function from $B \cup C$ to A. What is the obvious way to do that if B and C are disjoint?

(c) This is the most notationally confusing part of this proposition. We want to define a bijection F between these sets. Now, an element of $(A^B)^C$ is by definition a function g such that, for any $y \in C$, $g(y)$ is a function from B to A. So let $F(g)$ be the function with domain $B \times C$ such that $[F(g)](x, y) = [g(y)](x)$, for every $x \in C$ and $y \in C$. Again, it is routine to show that F produces the required bijection.

(d) See the exercise. ∎

Exercise 20. Complete the proof of this proposition. For parts (b) and (d), this requires defining the appropriate mapping. For all four parts, show that the mapping that's been defined really does yield the desired bijection.

Von Neumann cardinals

Here is another use of ordinals. Recall that Frege's definition of a cardinal as an equivalence class of sets under $\sim$ is unsatisfactory in ZFC. Von Neumann gave this more rigorous definition:

Definition. An ordinal α is called a **cardinal** if, for every $\beta < \alpha$, we have $\beta \prec \alpha$. Such an ordinal is also called an **initial** ordinal.

Under this definition, every finite ordinal is also a cardinal, and ω is the first infinite cardinal. Obviously, there are no other countable cardinals. In ZFC, we can show that every set "has" a unique von Neumann cardinal(ity):

Theorem 2.19 (AC). *For every set x there's a unique von Neumann cardinal α such that $x \sim \alpha$.*

Proof. Let x be given. By AC, x can be well ordered. But then, given a well-ordering on x, there's a unique ordinal β such that this well-ordering is isomorphic to the ordering of β under $\in$. (This fact was mentioned without proof in the previous section.) It follows that $x \sim \beta$. Therefore, by Theorem 2.8(b) there's a smallest ordinal α such that $x \sim \alpha$. So this α is the unique initial ordinal of the same size as x. ■

This theorem, in combination with transfinite induction, is a powerful tool for proving things in "ordinary mathematics." This method can be used to prove many results that are usually proved by Zorn's lemma or the Hausdorff maximal principle (defined in Appendix B). Here is a typical example, with two proofs; we will encounter more examples in Sections 6.4 and 6.5.

Theorem 2.20 (AC). *Every vector space has a basis.*

Proof. (By transfinite induction) Given a vector space V over some field, let α be the von Neumann cardinal of V, and then let f be a bijection between α and V. By transfinite induction, we define a subset B of V: for each $\beta < \alpha$, include $f(\beta)$ in B if and only if $f(\beta)$ is not a linear combination of the vectors that have already been included in B for $\gamma < \beta$. It is very easy to show that B is a basis for V. ■

Proof. (By Zorn's lemma) Given a vector space V over some field, let A be the collection of all linearly independent subsets of V. Partially order A by the subset relation, so $S_1 \leq S_2$ means $S_1 \subseteq S_2$. Since the union of any chain of linearly ordered sets of vectors is still linearly independent, that union is the least upper bound of the chain. Therefore, every chain in this partial ordering has an upper bound. By Zorn's lemma, there is a maximal element B, which is easily shown to span V and so must be a basis. ■

Exercise 21. Complete both of these proofs by showing that B is a basis for V: every vector in V is a linear combination of vectors in B, and no nontrivial linear combination of vectors in B is the zero vector.

Is there a rigorous definition of cardinality for all sets that "works" without the axiom of choice? Yes, there is, but we have to wait until the end of the section to be able to present it.

You might wonder whether the existence of any uncountable ordinals can be proved without AC, since there is no obvious way to define a well-ordering on any familiar uncountable set such as $\mathbb{R}$. But a nice result (of ZF) known as **Hartogs's theorem** states that $\forall x\, \exists \alpha (\alpha \not\preceq x)$. This implies that for any ordinal there's another one of larger cardinality. It follows (using replacement) that there's a one-to-one correspondence between the infinite von Neumann cardinals and all ordinals. The first uncountable von Neumann cardinal is denoted ω_1, the next one is ω_2, etc.; and if $\text{Lim}(\lambda)$, ω_λ is just $\bigcup_{\alpha<\lambda} \omega_\alpha$. Under this definition, there's an ω_α for every α. (Technically, $\omega_0 = \omega$, but this subscript is usually dropped.)

Outside of foundations, Cantor's notation $\aleph_\alpha$ is more common than ω_α. In ZFC, these notations may be used interchangeably. It's often clearer to use alephs when doing cardinal arithmetic, since cardinal arithmetic is different from ordinal arithmetic. For example, CH is usually written $2^{\aleph_0} = \aleph_1$. It can't be written $2^\omega = \omega_1$, since $2^\omega = \omega$ as ordinals.

The cumulative hierarchy

Here is possibly the most important transfinite inductive definition in axiomatic set theory:

Definition.

(a) $V_0 = \emptyset$.

(b) For every α, $V_{\alpha+1} = \mathcal{P}(V_\alpha)$.

(c) If Lim(λ), then $V_\lambda = \bigcup_{\alpha<\lambda} V_\alpha$.

Intuitively, we think of V_α as the set of all sets that are formed *before* stage α.

Proposition 2.21.

(a) *V_α is transitive, for every α.*

(b) *If $\alpha < \beta$, then $V_\alpha \subset V_\beta$.*

Proof.

(a) By transfinite induction: certainly, $\emptyset$ is transitive. If V_α is transitive, then so is $V_{\alpha+1}$, by Proposition 2.5(a). Finally, if Lim(λ) and V_α is transitive for all $\alpha < \lambda$, then $V(\lambda)$ is a union of transitive sets, which must be transitive by Proposition 2.5(b). ■

Exercise 22. Prove part (b) of this proposition.

Lemma 2.22. *For any set, there is a transitive set that contains it (as a subset).*

Proof. Given x, let $y_0 = x$ and, inductively, $y_{n+1} = y_n \cup (\bigcup y_n)$, that is, all the members of y_n together with all of their members. Then let $z = \bigcup_{n<\omega} y_n$. Clearly, $x \subseteq z$, and it is easy to show that z is transitive. ■

Clearly, the set z defined in this proof is the smallest transitive set containing x. This set is called the **transitive closure** of x, denoted $TC(x)$. By the way, the word "contains" is often ambiguous in set theory, as in the statement of this lemma. If we wanted the smallest transitive set containing x as a *member*, we would simply use $TC(\{x\})$.

Here is the most important property of the sets V_α:

Theorem 2.23. *Every set is in some V_α.*

Proof. Assuming that x is in no V_α, let $y = TC(\{x\})$. Then $\{u \in y \mid u$ is in no $V_\alpha\}$ is nonempty, and we may choose an $\in$-minimal element v

of this set, by regularity. Since y is transitive, every element of v is in some V_α. For $w \in v$, let $g(w)$ be the least α such that $w \in V_\alpha$. Then we can form $\{g(w) \mid w \in v\}$ by replacement, and let β be the LUB (union) of this set of ordinals. So $v \subseteq V_\beta$. But then $v \in V_{\beta+1}$, a contradiction. ■

Definition. For every x, the least α such that $x \in V_{\alpha+1}$ is called the **rank** of x.

This definition is set up so that the rank of any ordinal is itself.

Set theorists use a picture to describe the content of this theorem. Think of the class of all ordinals *Ord* as a *very* long, vertical "spine" starting with $\emptyset$ and proceeding upward. For each α, think of the collection of sets of rank α (that is, $V_{\alpha+1} - V_\alpha$) as a horizontal layer at α. Remember, α represents a level or stage of construction. Since the V_α's get bigger as α increases, the width increases as you go up. Thus the entire picture looks like a letter V (see Figure 2.3), and the class of all sets is denoted V. This categorization of sets is called the **cumulative hierarchy**. As we will see in Chapter 6, several variations of this idea have been extremely fruitful in modern set theory.

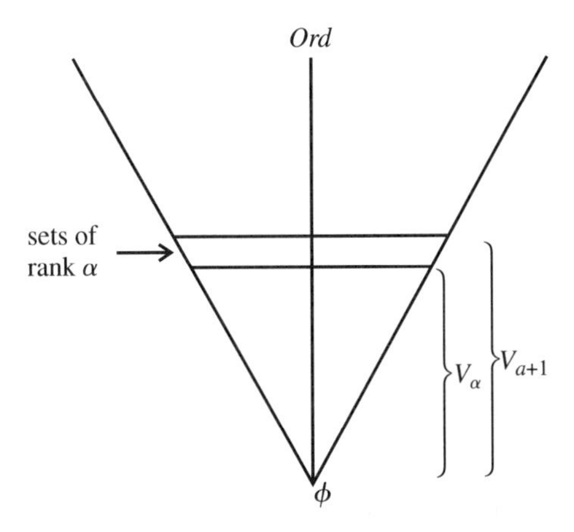

Figure 2.3. The cumulative hierarchy of sets

A set is called **hereditarily finite** if its transitive closure is finite. This means that the entire "membership tree" of the set is finite. For example, $\{\omega\}$ is obviously finite but it is not hereditarily finite. The following characterization of these sets is sometimes useful:

Proposition 2.24. *The set of all hereditarily finite sets exists and is precisely* V_ω.

Proof. We first show that every set in V_ω is hereditarily finite. By Proposition 2.21(a), each V_α is transitive. Also, it is well known (and easy to prove) that the power set of a finite set is finite. Therefore, V_n is finite for every $n \in \omega$. If $x \in V_\omega$, then $x \in V_n$ for some $n \in \omega$, and therefore $TC(x)$ is a subset of the finite set V_n. Therefore, $TC(x)$ is finite.

For the other direction, assume that x is a hereditarily finite set that is not in V_ω. An argument similar to the proof of Theorem 2.23 leads to a contradiction—see the following exercise. ∎

Exercise 23. Complete the second part of this proof. You may use the fact that a finite set of finite ordinals has a finite supremum.

Here, as promised, is the standard way to define cardinality rigorously in ZF. This clever adaptation of Frege's definition is due to Dana Scott:

Definition. The **cardinal** of any set x is the set of all sets *of least rank* that are the same size as x.

In other words, given x, let α be the least ordinal such that $\exists y \in V_{\alpha+1} (x \sim y)$. Then the cardinal of x is $\{y \in V_{\alpha+1} \mid x \sim y\}$.

Under this definition, a set is usually not a member of its own cardinal. But, trivially, x and y have the same cardinal(ity) under this definition if and only if $x \sim y$ in the sense of Section 2.2.

We have not finished our study of set theory. Having covered the basic concepts of the subject, we are almost ready to discuss the brilliant work of Gödel, Paul Cohen, and others that led to the enormous advances in set theory in the second half of the twentieth century. We will take up this discussion in Chapter 6, after we cover some more prerequisite topics.

CHAPTER 3

Recursion Theory and Computability

3.1 Introduction

Why should students of mathematics want to know something about recursion theory and computability? Here is one answer that applies to many other people beside mathematicians: a rudimentary knowledge of these subjects helps one to understand what tasks can and can not be done by digital computers. To be sure, today's powerful computers, with their slick graphical interfaces and "continuous" input and output devices (the mouse, speech, etc.), are very different from the simple computing devices such as Turing machines that are studied in elementary recursion theory. Yet, in an important sense, the most powerful supercomputer is still equivalent to a "universal" Turing machine with some extra bells and whistles thrown in. Many intelligent people are surprised that computers can perform calculations at lightning speed and play championship-level chess, but cannot process a natural language such as English (even in writing) at a four-year-old's level. But this situation is easier to understand for someone who knows a bit about the theory of computation. Clearly, this knowledge is more useful to computer scientists and programmers than to mathematicians, but it can also be helpful in a variety of mathematical studies.

Another reason to know something about recursion theory is that it provides a foundation for understanding Gödel's incompleteness theorems, possibly the most philosophically significant results in the history of mathematics. These theorems, which show that the most important mathematical systems have to be incomplete, are the subject of the next chapter.

A third reason to understand the basics of computability theory is that it is a prerequisite for the study of complexity theory and feasible computation. This subject, which is outlined in Section 3.5, is one of the most fruitful applications stemming from any part of the foundations of mathematics. Many important "real world" problems hinge upon theoretical questions from complexity theory.

The development of the subject of recursion theory was more or less an outgrowth of Hilbert's formalist program. Hilbert wanted there to be an "effective" or "mechanical" procedure that would verify the consistency of the formal axiomatic systems corresponding to ordinary mathematics. Phrases like "finitary algorithm" were also used in this regard. So it was natural to ask exactly what these terms should mean. In around 1936, several mathematicians (Alonzo Church, Stephen Kleene, Emil Post, and Alan Turing) independently proposed precise definitions of the notion of an effective procedure. Even though their definitions were very different from each other conceptually, it was proved that these definitions are all equivalent, in the sense that every function that can be computed under one definition can be computed using the others as well. The equivalence of these definitions, as well as our substantial experience with computer languages and programs, provide strong empirical evidence that these definitions do in fact correctly represent the intuitive notion of an effective procedure or mechanical computation. There is no way to prove this, but it is standard to view it as a sort of informal axiom, called **Church's thesis**.

For a more detailed treatment of the material in Sections 3.1 through 3.4, see [Rog], [End], or [Odi]. Most of the important early papers in the subject can be found in their original form in [Dav]. To learn more about complexity theory, see [DSW] or [LP].

Emil Post (1897–1954) was born in Poland but his family moved to New York in 1904. As a child, he lost an arm in an accident, but this was a minor irritant compared to the manic-depressive illness that plagued him throughout his adult life. Post's mental illness severely hampered his productivity at times, but he was helped enormously by his hard-working and loving wife Gertrude.

Post began his university career studying astronomy, but soon switched to mathematics. As an undergraduate, he wrote an unusual and original paper in which he attempted to generalize the differentiation operator D^n to nonintegral values of n. In his PhD thesis, he proved the consistency and completeness of propositional logic, generalized the method of truth tables to many-valued logic, and developed some important ideas for the analysis of formal systems—seminal ideas in the field of proof theory. In 1936, he introduced Post machines, which led to one of the equivalent formulations of the notion of a recursive function. Post is also known for his work on polyadic groups, recursively enumerable sets, and degrees of unsolvability, as well as his proof that the word problem for semigroups (Thue's problem) is recursively unsolvable.

In the 1920s, Post proved some results that were similar to Gödel's incompleteness theorems and the later ones of Church and Turing. But he didn't publish this work because he didn't consider it sufficiently thorough and polished. Later, he regretted this decision. In 1938, he wrote to Gödel, "As for any claims I might make perhaps the best I can say is that I would have proved Gödel's Theorem in 1921—had I been Gödel . . . after all it is not ideas but the execution of ideas that constitute a mark of greatness."

There were no drugs for manic depression when Post lived, and instead he underwent the trauma of electric shock treatments. He died of a heart attack soon after one of these sessions.

3.2 Primitive recursive functions

In this section we define a simple but surprisingly rich subset of the recursive (computable) functions. For the rest of this chapter, unless stated otherwise, "number" means natural number, and a **partial function** of k variables is a function from a *subset* of $\mathbb{N}^k$ to $\mathbb{N}$. A **total function** is a partial function whose domain is all of $\mathbb{N}^k$. (So every total function is also a partial function.) In this section, the unmodified word "function" always means a total function, but later in the chapter the context will determine the meaning of this word.

If f is a partial function of k variables, the **graph** of f is, as usual,

$$\{(x_1, \dots, x_k, x_{k+1}) \mid f(x_1, \dots, x_k) = x_{k+1}\}.$$

In pure set theory, the graph of a relation *is* the relation. But we will occasionally want to view them as distinct objects.

We will use the notation $\vec{x}$ to abbreviate a finite sequence of objects $(x_1, \dots, x_k)$.

Definition. The set of **primitive recursive (PR)** functions is the smallest set $\mathcal{F}$ of functions such that:

1. The *successor* function $f(x) = x + 1$ is in $\mathcal{F}$.
2. For every k and n in $\mathbb{N}$, the *constant* function $f(\vec{x}) = n$ of k variables is in $\mathcal{F}$.
3. For every k and j in $\mathbb{N}$ with $1 \leq j \leq k$, the *projection* function $f(\vec{x}) = x_j$ of k variables is in $\mathcal{F}$.
4. $\mathcal{F}$ is closed under *composition*. That is, if $g_1, \dots, g_m$ are functions of k variables in $\mathcal{F}$, and h is a function of m variables in $\mathcal{F}$, then the function $h[g_1(\vec{x}), g_2(\vec{x}), \dots, g_m(\vec{x})]$ of k variables is in $\mathcal{F}$.
5. $\mathcal{F}$ is closed under definition by *induction* or *recursion*. That is, if h is a function of k variables in $\mathcal{F}$ and g is a function of $k+2$ variables in $\mathcal{F}$, then the function f of $k+1$ variables defined inductively by

$$f(\vec{x}, 0) = h(\vec{x}),$$

$$f(\vec{x}, y + 1) = g(\vec{x}, y, f(\vec{x}, y))$$

is also in $\mathcal{F}$. In this clause, k could be 0, in which case h would be a constant.

As usual, saying that $\mathcal{F}$ is the smallest set satisfying (1) through (5) is the same as saying that it is the intersection of all sets satisfying (1) through (5). More concretely, it is clear that $f \in \mathcal{F}$ if and only if there is a sequence $(f_1, f_2, \dots, f_n)$ such that $f_n = f$, and each f_j is a function of type (1), (2), or (3), or is defined from earlier functions in the sequence by rule (4) or (5). Informally, we will call such a sequence a **generating sequence** for f. More precisely, a generating sequence must also specify, for each j, which of the five clauses in the definition of PR functions, and which earlier functions in the sequence (for the last two clauses) have been used to define f_j. In other words, a generating sequence for a PR function also includes complete instructions for obtaining the sequence of functions $(f_1, f_2, \dots, f_n)$.

By examining the clauses in the definition, it is easy to see that every PR function is total. Primitive recursive functions were first defined precisely by Gödel in [Go31], where they were called recursive functions.

Example 1. The ordinary addition function of two variables is PR. To see this, first recall the definition of addition by induction, as in Section 2.4: $m + 0 = m$, and $m + (n + 1) = (m + n) + 1$. So we can define a generating sequence for addition as follows:

Let f_1 be the projection function with $j = k = 1$. In other words, f_1 is the identity function $f_1(x) = x$.

Let f_2 be the successor function.

Let f_3 be the projection function with $j, k = 3$. So $f_3(x, y, z) = z$.

Let f_4 be the composition $f_2 \circ f_3$. So $f_4(x, y, z) = z + 1$.

Finally, let f_5 be defined by induction, with $h = f_1$ and $g = f_4$. It is easy to verify that f_5 is addition.

Exercise 1. Prove that multiplication and exponentiation are PR functions of two variables. The standard inductive definition of multiplication is also given in Section 2.4.

Every "elementary" function from some $\mathbb{N}^k$ to $\mathbb{N}$ is PR. For example, in addition to the functions just discussed, the greatest common divisor and least common multiple functions are PR. The function f such that $f(n)$ is the nth prime is PR. The factorial function is PR. The bijections B_k between $\mathbb{N}^k$ and $\mathbb{N}$ defined in Appendix C are PR. (The bijection B defined there is not PR, only because of the technicality that its domain is not of the right form.) All "elementary" combinations of PR functions, such as sums and compositions, are PR.

Definition. A k-ary relation on $\mathbb{N}$ is called primitive recursive if its characteristic function (with domain $\mathbb{N}^k$) is primitive recursive.

Every "elementary" subset of $\mathbb{N}$, such as the set of all even numbers or the set of all primes, is PR. The binary relations $=$, $<$, $>$, $\leq$, and $\geq$ are PR. Unions, intersections, and complements of PR relations are PR. (Here, a union or intersection means a finite union or intersection of subsets of the same $\mathbb{N}^k$.)

Exercise 2. Assuming that the basic arithmetical operations on $\mathbb{N}$ are PR (addition, multiplication, and a modification of subtraction that outputs zero instead of negative numbers), show that intersections, complements, and unions of PR relations are PR.

Suppose A is a denumerable set, and g is a "nice" bijection between A and $\mathbb{N}$. Such a bijection allows us to "identify" A with $\mathbb{N}$, which means we can talk about PR functions with inputs and/or outputs from A. So a function f from A to A, for instance, would be called PR if and only if the function $g \circ f \circ g^{-1}$ from $\mathbb{N}$ to $\mathbb{N}$ is literally PR. The same convention applies to the other types of functions we will define later, such as partial recursive functions.

But what do we mean by a "nice" bijection? If the set A has no structure at all, then this term is meaningless and any bijection will do as well any other. In practice, however, the set A usually has some kind of structure, and a "nice" bijection is one that is clearly computable with respect to that structure; often, this can be obtained simply by listing the members of A in a systematic way. For example, if A is either $\mathbb{N}^j$ for some j or $\mathbb{N}^{<\omega}$ (the set of all finite sequences of natural numbers), then the bijections B_j and B would certainly be considered "nice." In particular, this allows us to consider PR functions from $\mathbb{N}^k$ to $\mathbb{N}^j$, not just from $\mathbb{N}^k$ to $\mathbb{N}$. Other sets that can be "nicely" matched with $\mathbb{N}$ are $\mathbb{Z}$, $\mathbb{Q}$, and $S^{<\omega}$ for any countable set S (provided that S has at least two members and can itself be "nicely" matched with a subset of $\mathbb{N}$). The crucial process of Gödel numbering that we will soon define is based on identifying sets of the form $S^{<\omega}$ with $\mathbb{N}$.

Example 2. We can define a "nice" bijection g between $\mathbb{Q}$ and $\mathbb{N}$ if we can list the rationals systematically. One standard way to do that is to list all fractions that are in lowest terms, in groups based on the sum of the numerator and the denominator:

$$\frac{0}{1}, \frac{1}{1}, -\frac{1}{1}, \frac{1}{2}, -\frac{1}{2}, \frac{2}{1}, -\frac{2}{1}, \frac{1}{3}, -\frac{1}{3}, \frac{3}{1}, -\frac{3}{1}, \frac{1}{4}, -\frac{1}{4}, \frac{2}{3}, -\frac{2}{3}, \frac{3}{2}, \ldots.$$

So we can let $g(0) = 0$, $g(1) = 1$, $g(-1) = 2$, $g(\frac{1}{2}) = 3$, etc. Note that $\frac{2}{2}$ and $-\frac{2}{2}$ are not included among the fractions whose numerator and denominator add to 4, because these two fractions are not in lowest terms. Figure 3.1 illustrates this idea, and more directly provides a bijection between $\mathbb{N} \times \mathbb{N}$ and $\mathbb{N}$.

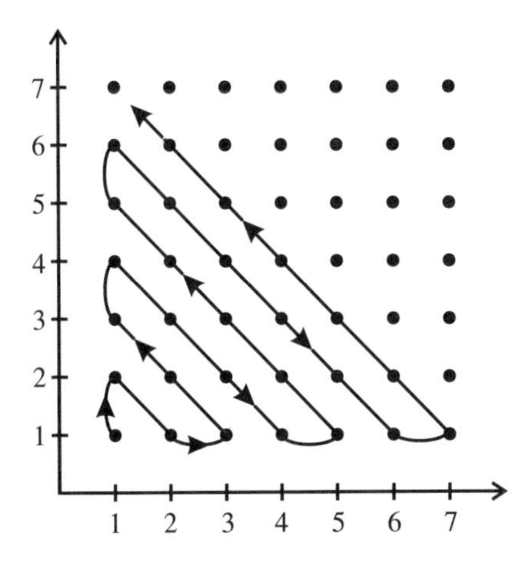

Figure 3.1. One approach to defining a bijection between $\mathbb{Q}$ and $\mathbb{N}$

Exercise 3.

(a) Define a bijection between $\mathbb{Z}$ and $\mathbb{N}$ that you would consider "nice."

(b) Let S be a finite set, say with m elements ($m \geq 2$). Explain how to set up a "nice" bijection between $S^{<\omega}$ and $\mathbb{N}$. The simplest approach is to first list the empty sequence, then all sequences from S of length 1, then all sequences of length 2, etc.

Now, consider the "elementary" functions encountered in basic algebra, trigonometry and calculus. These include polynomial, rational, trigonometric, exponential, and logarithmic functions, and simple combinations thereof (including compositions and inverses). Are these functions PR? This question is not clearly posed, since almost all of these functions have uncountable domains (some $\mathbb{R}^k$ or a substantial subset of it). The simplest way to make the question meaningful is to restrict the domain of these functions to a denumerable set like $\mathbb{Q}^k$. In the case of polynomials with rational coefficients, rational inputs always yield rational outputs, so we can use a "nice" bijection between $\mathbb{Q}$ and $\mathbb{N}$ to view any polynomial with rational coefficients and inputs as a function from $\mathbb{N}^k$ to $\mathbb{N}$. In this sense, every such polynomial is PR. So is every rational function with rational coefficients and inputs, provided that we add a special output value that represents an undefined computation. (We must do this since true PR functions are never undefined.)

With other elementary functions such as transcendental ones and those involving radicals, things get trickier because rational inputs often produce irrational outputs. For these functions, we might allow rational inputs and round the answers off to a fixed number of significant figures. (This is how ordinary calculators and most computer languages handle such functions.) With this type of restriction on the inputs and outputs, and some careful way of dealing with undefined values, every elementary function (and so, in particular, every function that can be evaluated on an ordinatory calculator) becomes PR. In fact, it is far from obvious how to define a function on a denumerable domain that cannot be considered PR.

Based on this, we might conjecture that the PR functions could be an appropriate choice for the category of computable functions. Our next task is to show that this is not reasonable. Certainly, every PR function is computable, intuitively; and in fact every PR function is recursive. But we will show that not every computable function is PR.

To prove this, we must carry out what is called the **arithmetization** of the PR functions. To do this, we map each PR function (more precisely, each generating sequence) to a number, called its **Gödel number** or its **index**, in a systematic way. Via this mapping, PR functions may be identified with numbers, and therefore they can in essence be inputs and outputs of computable functions. We will define this process in some detail, because Gödel numberings are needed at several other points in this chapter and the next, and we will not provide many details after this. I urge you to try to get a good "feel" for this process now in order not to be confused by it later.,

Given a PR function f, let $(f_1, f_2, \dots, f_n)$ be a generating sequence for f. Each PR function has many generating sequences and therefore will have many Gödel numbers, but that is not a problem. Furthermore, it is not important to describe an effective procedure for finding a generating sequence of a given PR function. What is important (and true) is that this mapping is effective in the other direction: the instructions contained in a generating sequence for f provide all the information needed to compute f on any input.

The set of all possible generating sequences is denumerable, and in principle all we need to Gödel number the PR functions is a "nice" bijection between this set and $\mathbb{N}$, which can be obtained by systematically listing all the generating sequences. For now, we will do something a bit easier, namely define a bijection between the set of all generating sequences and a proper subset of $\mathbb{N}$:

So, let $(f_1, f_2, \dots, f_n)$ be a generating sequence. Starting with f_1, we construct a finite sequence s_j of numbers corresponding to each f_j in this generating sequence, as follows:

1. If f_j is the successor function, let $s_j = (1)$.
2. If f_j is a constant function, let $s_j = (2, k, n)$, where k and n are as in clause 2 of the definition of PR functions.

3. If f_j is a projection function, let $s_j = (3, k, j)$, where k and j are as in clause 3 of that definition.
4. If f_j is defined by composition from functions h and g_1 through g_m occurring earlier in the generating sequence, let $s_j = (4, m, a, b_1, \ldots, b_m)$, where a and b_1 through b_m are the *positions* of h and g_1 through g_m in the generating sequence $(f_1, f_2, \ldots, f_n)$.
5. Finally, if f_j is defined by induction from functions g and h occurring earlier in the generating sequence, let $s_j = (5, a, b)$, where a and b are the positions of g and h in the generating sequence.

Then, having defined these sequences $s_1, s_2, \ldots, s_n$, concatenate them in order to form a single finite sequence s.

Example 3. Consider the following generating sequence for the function $f(x) = x + 2$:

$$\begin{aligned} f_1(x) &= x + 1, && \text{by clause 1,} \\ f_2(x) &= f_1(f_1(x)), && \text{by clause 4.} \end{aligned}$$

The sequences of numbers corresponding to f_1 and f_2 are $s_1 = (1)$ and $s_2 = (4, 1, 1, 1)$. Therefore, the sequence s that encodes this generating sequence is $(1, 4, 1, 1, 1)$.

Example 4. Consider the generating sequence given for the addition function in Example 1. The sequences of numbers corresponding to f_1 through f_5 are $s_1 = (3, 1, 1)$, $s_2 = (1)$, $s_3 = (3, 3, 3)$, $s_4 = (4, 1, 2, 3)$, and $s_5 = (5, 4, 1)$. By putting these sequences together, we say that the single sequence

$$s = (3, 1, 1, 1, 3, 3, 3, 4, 1, 2, 3, 5, 4, 1)$$

encodes the generating sequence of Example 1, and therefore encodes the addition function.

Thus we have, for each PR function f, a finite sequence s of numbers that encodes a complete set of instructions for computing f on any input. Using the bijection B of Appendix C, we can say that each PR function is actually encoded by the single number $B(s)$, its (nonunique) **Gödel number**. We can also talk about the unique Gödel number of any generating sequence.

Exercise 4. Calculate the Gödel number of the generating sequence in Example 3. You might want to use a calculator. This Gödel number is not huge, but you can see that the Gödel number of a long generating sequence (even the one in Example 4) would be enormous.

As we mentioned earlier, this Gödel numbering is not onto $\mathbb{N}$. For example, (4, 7) does not encode a legal generating sequence, and therefore $B(4, 7) = 2^5 \cdot 3^8 - 1$ is not the Gödel number of any PR function. This is not a major difficulty, but it can be inconvenient. If we want to fix this, we can use a bijection between the original set of Gödel numbers and $\mathbb{N}$, as in the next example. Henceforth, we will assume that the Gödel numbering we are using is onto $\mathbb{N}$, so that every natural number encodes a PR function. We will also make the same assumption for most other Gödel numberings that we use.

Example 5. Suppose that we have a bijection f between some set A and the set of all prime numbers, but we really want a bijection between A and $\mathbb{N}$. Simply let g be the function that lists all the primes, so that $g(0) = 2$, $g(1) = 3$, $g(2) = 5$, etc. Then $g^{-1} \circ f : A \to \mathbb{N}$ is a bijection. Furthermore, g is PR and therefore, if f is "nice," so is $g^{-1} \circ f$.

We now define a special function $\Psi(w, x)$ of two variables. To compute this function, begin with w and, using our Gödel numbering in reverse, determine the sequence of numbers and then the generating sequence corresponding to w. That means we have a complete description of the PR function f encoded by w. Next, determine the number k of input variables that f requires, and then use the bijection B_k to find the k-tuple corresponding to x. Finally, evaluate f on this k-tuple. In other words, we have $\Psi(w, x) = f({B_k}^{-1}(x))$ if w is the Gödel number of a PR function f of k variables, and x is any natural number. We call Ψ a **universal function** or an **enumeration function** for PR functions. (We refer to *a*, not *the*, universal function for PR functions because the function Ψ depends on the choice of the Gödel numbering and the B_k's.)

We can now prove our claim, provided that we are comfortable with computability as a nonrigorous notion:

Theorem 3.1. *Not every computable function is primitive recursive.*

Proof. It is intuitively clear that the function Ψ defined above is computable. Also, Ψ is total, since each PR function is defined on all of the appropriate $\mathbb{N}^k$. Now we use a simple diagonalization similar to the one used in the proof of Cantor's theorem: define a function of one variable by $\psi(x) = \Psi(x, x) + 1$. Clearly, ψ is computable. But if f is any PR function of one variable, let w be a Gödel number for f. Since B_1 is the identity on $\mathbb{N}$, we have $\psi(w) = \Psi(w, w)+1 = f(w)+1 \neq f(w)$. So $\psi \neq f$. Therefore, ψ is not PR. ∎

You might wonder why this diagonalization argument would not apply to any proposed definition of algorithms for computable functions. As long as such algorithms can be listed "reasonably" in a single infinite sequence (which would be true if, for example, the algorithms were programs in any standard programming language), it would seem that we could define ψ as above to get a new computable function. The answer to this apparent paradox is that any adequate definition of algorithms for computable functions must include algorithms for *nontotal* functions. This will become more clear in the next section.

Exercise 5. Prove that if a function $f(x, y)$ is PR, then so is the function $g(x) = f(x, x) + 1$. Conclude, as a corollary to the previous theorem, that the function Ψ is also not PR.

Exercise 6. Show how to define a set of natural numbers that is intuitively computable (meaning that the characteristic function of its graph is computable) but not PR.

Example 6. Here is a more concrete example of a computable function that is not PR. The following clauses define a function A of two variables, a modification of a remarkable function of three variables devised by Wilhelm Ackermann:

1. $A(0, n) = n + 1$, for every n.
2. $A(1, 0) = 2$.
3. $A(2, 0) = 0$.
4. $A(m + 3, 0) = 1$, for every m.
5. $A(m + 1, n + 1) = A(m, A(m + 1, n))$, for every m and n.

A bit of computation using these clauses should convince you that they really define a function, and that this function is computable. In fact, it is quite easy to write a program to compute A in any programming language that allows the type of recursion in clause (5). But it can be shown that A is not PR, and neither is the function $A(n, n)$ of one variable. More specifically, $A(n, n)$ "grows faster" than any PR function of one variable.

Exercise 7.

(a) Show that $A(1, n) = n + 2$, $A(2, n) = 2n$, and $A(3, n) = 2^n$.

(b) Describe the function $A(4, n)$ of one variable, preferably with a clear notation for it.

(c) Show that the number $A(4, 5)$ has more than 10,000 digits.

(d) Show that $A(5, 3) = 2^{16}$.

(e) Try to calculate $A(5, 4)$, and then explain why this number is incomprehensibly huge, but is dwarfed by $A(5, 5)$.

Exercise 8. Prove that, for each fixed m, the function $A_m(n) = A(m, n)$ is PR. Use induction on m.

Exercise 9.

(a) Write a computer program to calculate the function A for arbitrary inputs.

(b) Use your program to try to compute $A(4, 5)$ and then $A(5, 4)$. Don't be surprised if the program crashes!

3.3 Turing machines and recursive functions

We will now define Turing's version of the notion of "mechanical procedure." I have chosen to use Turing machines as the basis for computability because there is a certain elegant simplicity, perhaps even charm, to them.

Definition. A **Turing machine** is a device that operates as follows:

1. The machine has a positive finite number of **internal states**. We may assume these are numbered from 1 to some n, and it may be helpful to imagine that they are arranged on a wheel.
2. For input, output, and "memory," the machine uses a paper tape that stretches infinitely in both directions, which are called "left" and "right." The tape is divided into square cells, each of which has 0 or 1 written in it at any time.
3. Before the machine starts, and after every "step," one internal state is the **current state**, and one cell is the **current cell**. The current state and the number in the current cell constitute the **configuration** of the machine at any point. We may imagine that one "wedge" of the wheel of internal states is lined up with one cell of the tape. (See Figure 3.2.) Note that there are only $2n$ possible configurations.
4. The machine starts in internal state 1 and with some cell chosen as the current cell, and then proceeds in discrete steps (perhaps one per second). The action of the machine at any step is determined *solely by its configuration right before the step begins*, and consists of two parts. The first part of the action is to possibly change the internal state to any of the $n - 1$ other ones. The second part of the action consists of exactly one of these four possibilities: change the number in the current cell; move the tape one cell to the left; move the tape one cell to the right; or halt the computation.

This definition is more mechanical than mathematical. More rigorously, a Turing machine is defined by specifying the number of internal states and the action corresponding to each configuration. The initial placement of numbers on the tape (including the choice of the

Alan Turing (1912–1954) was an eccentric genius and one of the most tragic figures in the history of mathematics. Alan was raised in England while his parents were living in India. He showed brilliance at school but his individuality prevented him from obtaining exceptional grades. Turing was a superb athlete who, at age fourteen, commuted sixty miles daily by bicycle. Later, he rode a defective bicycle that required a precisely syncopated pedaling rhythm in order not to throw the chain. He chose not to fix it because he enjoyed owning a bike that no one else could ride.

Turing's dissertation, completed in 1935 at King's College, Cambridge, was in probability theory and contained an independent proof of the central limit theorem. In the following year he finished his seminal paper in which he defined Turing machines and recursive functions, proved the unsolvability of the halting problem, and used this result to solve the Entscheidungsproblem. During World War II, he did brilliant work for British intelligence, leading the team that broke the Germans' "Enigma" machine codes and contributed to the development of "real" computers later on. After the war Turing did important work in logic, physics, biology, and computer science, including a 1950 paper that laid the foundations of artificial intelligence.

Turing was a homosexual who, in 1952, naively reported an affair to the police because he had been blackmailed. He was tried and convicted for this activity, and forced to take estrogen injections for a year in order to avoid prison. Worse, he was judged to be a security risk, lost his clearance, and found himself under constant police scrutiny. Turing became depressed and died from poisoning by a cyanide-laden apple. An inquest concluded that it was suicide, even though his mother always maintained that it was an accident. In the 1980s, a play about his life entitled *Breaking the Code* had successful runs in several countries.

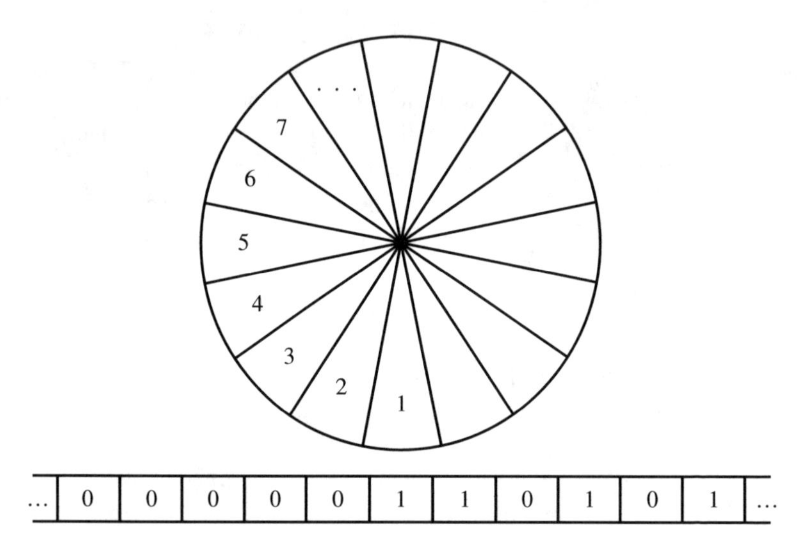

Figure 3.2. Diagram of a Turing machine

initial current cell) is the input for the computation, which is not part of the definition of the machine. Thus, a Turing machine may be represented formally as a function $M(u, v) = (y, z)$, where u and y are numbers between 1 and some fixed n (representing internal states), $v = 0$ or 1, and z is a number between 1 and 4, corresponding to the four possibilities for the second part of the action. Thus (u, v) is any configuration and (y, z) is the action specified by that configuration. Note that such a function may be viewed as a finite set of 4-tuples.

Exercise 10. How many Turing machines with n internal states are there?

Perhaps you find it hard to believe that a Turing machine could do much of anything. It can only "read" and "write" one bit at a time. It has no way of storing or retrieving a significant amount of information. And yet Turing machines can in theory compute the same class of partial functions as any digital computer. One enlightening and painless way to get a feel for Turing machines and their capabilities is to experiment with one of the Turing machine simulators that are available on the internet.

Specifically, let M be a Turing machine and k a natural number. Then M can be used to compute a partial function f of k variables, as follows: start the machine in internal state 1. Given an input $(x_1, x_2, \ldots, x_k)$, set up the tape with 0's in every cell except that, starting at the initial current cell, there are (as we move toward the right, say) $(x_1 + 1)$ 1's, then a 0, then $(x_2 + 1)$ 1's, then a 0, and so on until there are $(x_k + 1)$ 1's. If the machine ever halts, $f(x_1, x_2, \ldots, x_k)$ is the total number of 1's on the tape at that point. If the machine never halts, $f(x_1, x_2, \ldots, x_k)$ is undefined. For example, a Turing machine that computes the sum of two numbers must take a tape that reads 1111101111 (and 0's elsewhere), starting at the initial current cell, and halt with a total of seven 1's on the tape, since $4 + 3 = 7$.

Note that any single Turing machine can technically accept all finite sequences of natural numbers as inputs. But normally, when we define a Turing machine, we are interested only in the partial function of k variables that it computes, for some fixed k.

Example 7. Let's define a Turing machine to compute the constant function $f(x) = 2$. This Turing machine must take any input tape that consists of all 0's except for one nonempty sequence of 1's (starting at the initial current cell and moving to the right), and halt with exactly two 1's on the tape.

The first task of this machine must be to change all the 1's on the tape to 0's. This requires pairs of steps, consisting of changing a 1 to 0 and then moving one cell to the right. So, we'll say that whenever the machine is in state 1 and the current cell is 1, the machine goes to state 2 and changes the 1 to 0. (In the notation introduced earlier, $M(1, 1) = (2, 1)$.) Then, whenever the machine is in state 2, the machine goes back to state 1 and moves the tape one cell to the right. Thus $M(2, 0) = M(2, 1) = (1, 3)$.

Eventually, the machine will erase and get past all the 1's that were on the input tape. Then it just needs to write two 1's on the tape, and halt. So when it's in state 1 and there's a 0 in the current cell, it goes to state 3 and changes the 0 to 1. From state 3, it goes to state 4 and moves one to the right. From state 4, it goes to state 5 and changes the value in the current cell. Finally, from state 5, it halts. It is clear that this machine always halts with two 1's on the tape, provided that the input tape is of the correct form.

The complete instructions for this Turing machine are encoded by the following list of 4-tuples: $(1, 1, 2, 1)$, $(1, 0, 3, 1)$, $(2, 0, 1, 3)$, $(2, 1, 1, 3)$, $(3, 0, 4, 3)$, $(3, 1, 4, 3)$, $(4, 0, 5, 1)$, $(4, 1, 5, 1)$, $(5, 0, 5, 4)$, $(5, 1, 5, 4)$. Actually, we don't even need all of these 4-tuples, since certain configurations can't occur, such as (2, 1) and (3, 0).

Exercise 11. Define Turing machines to compute the following functions:

(a) $f(x, y) = x + y + 2$.

(b) $f(x, y) = x + y$.

(c) $f(x) = x + 3$.

Exercise 12. Consider the following nine-state Turing machine. (Since there are fewer than ten states, we specify the machine by giving a list of four-digit numbers. For example, the number 1014 signifies that $M(1, 0) = (1, 4)$, in the notation defined above):

1014, 1121, 2033, 2122, 3034, 3141, 4053, 4144, 5063, 5153, 6071, 6163, 7081, 7173, 8092, 8182, 9094, 9122.

Let f be the partial function of one variable computed by this Turing machine. Evaluate $f(0)$, $f(1)$, and $f(2)$. Then conjecture the general rule for $f(x)$, and show that it is correct.

Definition. A partial function of k variables that can be computed by some Turing machine is called **partial recursive**. If the function is total, it is called **total recursive** or simply **recursive**. (Remember that "PR" stands for "primitive recursive," not "partial recursive"!)

This definition bears out the point made at the end of the previous section: any adequate definition of the notion of an effective procedure must include procedures that give rise to nontotal functions as well as total ones. It might seem more logical to refer to "recursive partial" functions rather than "partial recursive" ones, but the latter term is the standard one.

Definition. Let $A \subseteq \mathbb{N}^k$. A is called **recursive** if its characteristic function is recursive.

So this definition is analogous to the definition of a primitive recursive relation. Note that there is no meaningful concept of "partial recursive relation."

As indicated in the introduction to this chapter, there have been many independent definitions of computability for functions, including several definitions from the 1930s as well as later ones based on more modern concepts of what a computer program is. They all turn out to be equivalent to the above definition of partial recursive and recursive functions. In this context, Church's thesis becomes the succinct claim that "recursive" correctly captures the concept of "computable." Almost everyone believes this, but of course there is no way to prove it mathematically because it involves a term that is not mathematically precise. Church's thesis is similar to the claim that some degree of differentiability of a function corresponds to "smoothness" of its graph. (An exception regarding vertical tangents must be made for this claim; for example, the cube root function is not differentiable at 0, but its graph is just as smooth as the graph of the cubing function.)

Church's thesis is an extremely useful tool for proving *nonrigorously* that a function is partial recursive. We describe a finite algorithm of some sort that, given any sequence of inputs for the function in question, fails to halt whenever the function is undefined, and computes the correct output in a finite number of steps whenever the function is defined. Then we say that we are done "by Church's thesis." This is usually much easier than defining a Turing machine for the given function! We will often use this shortcut, but in all such cases, rigorous proofs can also be given.

Proposition 3.2. *Every primitive recursive function is recursive (but not conversely). The functions Ψ and ψ used in Theorem 3.1 are also recursive.*

Proof. In the previous section we reasoned informally that these functions are computable. Hence, by Church's thesis, we conclude that they are recursive. The words in parentheses are, in the presence of Church's thesis, a restatement of Theorem 3.1. ∎

It follows immediately from this proposition that every primitive recursive subset of $\mathbb{N}^k$ is recursive. The converse of this also fails, by Exercise 6 of the previous section.

We will need to assume that we have at our disposal a Gödel numbering of all Turing machines. We have mentioned that any Turing machine can be described by a finite set of 4-tuples of numbers, which can be turned into a single finite sequence of numbers. Again using something like the bijection B of Appendix C, as we did for the Gödel numbering of PR functions, we can encode all the instructions for the machine into a single number. Also, we will again assume for convenience that our Gödel numbering of Turing machines is *onto* $\mathbb{N}$.

Notation. We let M_w denote the Turing machine encoded by the number w, while $\varphi_w^{(k)}$ denotes the partial recursive function of k variables computed by M_w. We may drop the superscript k when it is unimportant or when $k = 1$.

Definition. Let $\Phi^{(k)}$ be the partial function of $k + 1$ variables defined by

$$\Phi^{(k)}(w, x_1, \ldots, x_k) = \varphi_w^{(k)}(x_1, \ldots, x_k).$$

We call $\Phi^{(k)}$ a **universal function** or an **enumeration function** for partial recursive functions of k variables. (When we define a partial function in this way or equate two partial functions, it is always understood that both functions are undefined at exactly the same values.)

Theorem 3.3 (Kleene's Enumeration Theorem). *The functions $\Phi^{(k)}$ defined above are partial recursive.*

Proof. The complete description of the Turing machine M_w can be reconstructed from its Gödel number w. It is thus intuitively clear that the functions $\Phi^{(k)}$ are computable. So by Church's thesis, we conclude that they are partial recursive. ■

Definition. A Turing machine that computes $\Phi^{(k)}$ is called a **universal Turing machine** for partial recursive functions of k variables.

The functions $\Phi^{(k)}$ do the same thing for partial recursive functions as the function Ψ used in Theorem 3.1 does for PR functions. If we wanted to, we could define each $\Phi^{(k)}$ as a function of two variables or even one variable by using the bijections B_k. Using these bijections also allows us to say that each $\Phi^{(k)}$ contains all the "information" in all of them. Furthermore, we could use the bijection B to define a single partial recursive universal function for all partial recursive functions. Thus, there exists a single universal Turing machine that computes all partial recursive functions, of any number of variables.

Suppose that a certain Turing machine, with a certain k-tuple as input, halts after a finite number of steps. Then a complete record of that computation can also be encoded into a sequence of numbers, and from there into a single number. Let us assume that this **computation record** includes the definition of the Turing machine itself, the k-tuple being used as input, and the sequence of configurations that the machine goes through. The output value of the computation is not directly included in the computation record but can be computed from it.

Notation (Kleene's T-Predicate). We write $\mathrm{T}(w, x, z)$ to mean that the Turing machine M_w with input $B^{-1}(x)$ halts, with computation record z.

Also, we let $U(z)$ be the output value corresponding to the computation record z. Let's once again assume that the coding of computation records is arranged to be onto $\mathbb{N}$, making U a total function. U is known as the **upshot function**.

Proposition 3.4. *The relation* T *and the function U are PR.*

We omit the proof. It is intuitively clear that T and U are recursive. It takes quite a bit of work to show that they are PR, but because they require only routine arithmetic to evaluate and are clearly total, there should be no doubt that they are PR.

The mu operator

Our definition of recursive functions does not indicate the close connection between them and PR functions. To demonstrate that connection, we now give another (equivalent) characterization of recursiveness. The material in the remainder of this section is not essential for understanding the rest of the book.

Notation. The symbol μ stands for "the least." In other words, if $f(\vec{x}, y)$ is a partial function, then $\mu y[f(\vec{x}, y) = z]$ denotes the partial function of $\vec{x}$ and z defined to be the least y in $\mathbb{N}$ such that $f(\vec{x}, y) = z$. If there is no such y, this new function is undefined at $(\vec{x}, z)$.

If $f(\vec{x}, y)$ is a recursive function and $g(\vec{x}, z)$ is defined to be $\mu y[f(\vec{x}, y) = z]$, then it is intuitively clear that g is *partial* recursive. For, to evaluate $g(\vec{x}, z)$, it is only necessary to evaluate $f(\vec{x}, 0)$, $f(\vec{x}, 1)$, $f(\vec{x}, 2)$, etc. until we reach the first number y for which $f(\vec{x}, y) = z$. If there is such a y, then $g(\vec{x}, z)$ is found after a finite amount of computation. But if there is no such y, then $g(\vec{x}, z)$ is undefined. Thus, if we were to use the μ-operator to expand the class of PR functions, we would introduce partial functions.

It would be nice if we could define the set of partial recursive functions by simply taking the definition of the PR functions and adding the μ-operator as clause 6. But this doesn't work. We have just seen that when the μ-operator is applied once to a PR function, the new function may be partial. But if we then apply the μ-operator to this partial function, the resulting function might not be computable in any sense. For example, suppose that $f(\vec{x}, 1) = 3$ for every $\vec{x}$, but $f(\vec{x}, 0)$ requires a computation that may or may not halt, depending on $\vec{x}$. Then $\mu y[f(\vec{x}, y) = 3]$ is defined for every $\vec{x}$ (it always equals 0 or 1). But there is no apparent algorithm for this new function, and it can be shown that not every function of this type is recursive.

To avoid this problem, the μ-operator may be applied only once in the formation of partial recursive functions; it cannot be iterated unless we somehow know each time that the function to which it is being applied is total. Specifically, we have the following:

Theorem 3.5 (Kleene's Normal Form Theorem). *Let* $k \in \mathbb{N}^+$. *Then there are PR functions g and h such that, for every partial recursive function* $f(\vec{x})$ *of k variables, there is a number w such that f is the function* $g(\mu y[h(\vec{x}, w, y) = 1])$.

Proof. Let f be a partial recursive function of k variables. Then $f = \varphi_w^{(k)}$ for some w. Then we see that $f(\vec{x}) = U(\mu z[\mathrm{T}(w, B_k(\vec{x}), z)])$, for every $\vec{x}$. So we define g to be U, and h to be the characteristic function of $\mathrm{T}(w, B_k(\vec{x}), z)$. ■

Corollary 3.6. *A function* $f(\vec{x})$ *is partial recursive if and only if it is of the form* $g(\mu y[h(\vec{x}, y) = c])$, *where g and h are PR functions and c is a natural number.*

Proof. The forward direction of this is immediate from the theorem, while the reverse direction follows (by Church's thesis) from the discussion right after the definition of the μ-operator, and the fact that compositions of partial recursive functions are partial recursive. ■

The normal form theorem makes it clear that Kleene's T-predicate contains all the information contained in the universal functions $\Phi^{(k)}$ and Φ. This relationship can be used to prove Kleene's enumeration theorem (Theorem 3.3) rigorously.

There is also a bounded μ-operator, which does not create non-total functions from total ones. If $f(\vec{x}, y)$ is a function, then $\mu y < n[f(\vec{x}, y) = z]$ denotes the function of $\vec{x}$, z, and n defined to be the least y such that $f(\vec{x}, y) = z$ or $y = n$. If f is recursive, then so is this new function, by Church's thesis. It is also true (but less obvious) that this implication still holds if "recursive" is replaced by "PR."

Exercise 13. Given that (i) the usual multiplication function is PR, (ii) the graph of every PR function is PR, and (iii) the bounded

μ-operator preserves "PR-ness," as mentioned above, prove:

(a) the relation $m \mid n$ (m divides n, or m is a factor of n) is PR.

(b) the set of prime numbers is PR.

3.4 Undecidability and recursive enumerability

We have not yet given any examples of nonrecursive functions. Their existence is clear, since there are uncountably many functions from $\mathbb{N}$ to $\mathbb{N}$ but only a denumerable number of Turing machines. The purpose of this section is to demonstrate that many simple functions are not recursive. Some of these "negative" results, proved in the 1930s, were quite surprising to the pioneers of the subject.

Relations that are recursive are also called **decidable** relations. There is a variety of terminology related to decidability, especially when the relation under discussion consists of more sophisticated objects than k-tuples of numbers, such as Turing machines. The question of whether a relation A is decidable is called the **decision problem** for A. If A is decidable, we also say that A is **recursively solvable** or that there is a **decision procedure** for A. Here is Turing's classic result known as the recursive unsolvability of the halting problem:

Theorem 3.7. *The set of ordered pairs* $\{(w, x) \mid \varphi_w(x) \text{ is defined}\}$ *is undecidable.*

Proof. Let H denote this set of ordered pairs, and let G be its characteristic function. Using G, we construct another function by diagonalization, as in Theorem 3.1: let

$$f(x) = \begin{cases} 0, & \text{if } G(x, x) = 0, \\ \varphi_x(x) + 1, & \text{if } G(x, x) = 1. \end{cases}$$

The function f is defined in a simple way from G and a universal Turing machine. So if G is recursive, then it follows by Church's thesis that

f is also recursive. But f cannot be recursive because, for any number w, $f(w) \neq \varphi_w(w)$, and so $f \neq \varphi_w$. Therefore, G is not recursive. ■

In other words, there is no computer program that can examine a given Turing machine and a given input tape and always determine whether the resulting computation will halt. Note that without using G as an auxiliary function, there is no way to duplicate the diagonalization of Theorem 3.1 to construct a function that is not partial recursive.

Corollary 3.8. *The set of numbers* $\{x \mid \varphi_x(x)$ *is defined*$\}$ *is undecidable.*

The following powerful result shows that there is a great abundance of undecidable problems.

Theorem 3.9 (H. G. Rice). *Let* $\mathcal{C}$ *be any nonempty proper subset of the set of all partial recursive functions of one variable. Then the set* $\{w \mid \varphi_w \in \mathcal{C}\}$ *is not decidable.*

Proof. Assume $\mathcal{C}$ is such a set and $\{w \mid \varphi_w \in \mathcal{C}\}$ is decidable, with characteristic function f. We may assume $\emptyset \in \mathcal{C}$. Let g be any fixed partial recursive function that is not in $\mathcal{C}$. (The case $\emptyset \notin \mathcal{C}$ is handled by letting $g \in \mathcal{C}$.)

Using the recursiveness of f, we describe a decision procedure for the halting problem, contradicting Theorem 3.7. Given x and y, construct a Turing machine that, given an input u, first tries to compute $\varphi_x(y)$. If and when this halts, it then computes $g(u)$. The Gödel number w of this Turing machine is certainly a recursive function of x and y. It is also clear that $\varphi_w = g$ if $\varphi_x(y)$ is defined, and $\varphi_w = \emptyset$ if $\varphi_x(y)$ is undefined. Thus we can determine whether $\varphi_x(y)$ is defined by evaluating $f(w)$. Since this is impossible, f must not be recursive. ■

There is also, for each k, a version of Rice's theorem for functions of k variables.

Think about what Rice's theorem says. Among all the nontrivial questions (ones that have both yes and no answers) that can be asked about the partial function generated by a Turing machine, not a single one of them is decidable! In my opinion, this is truly remarkable. For example, there is no decision procedure that determines whether any given Turing machine computes a total function, or halts for at least one input, or computes a function with an infinite range, etc. Similarly, there is no decision procedure that determines whether any two given Turing machines compute the same partial function of one variable.

Rice's theorem is relevant to computer science, because it lets software designers know that most types of "program verification" software cannot be theoretically perfect. In particular, it would be useful to be able to examine the code of a computer program and determine, before running the program, whether it will go into an "infinite loop." Turing's result and Rice's theorem tells us that this cannot be done perfectly, so one needs to be content with tests that only find some infinite loops.

Recursive enumerability

We now turn our attention to an important category of sets that includes many undecidable ones. Consider again the set H of Theorem 3.7, which we know is undecidable. In spite of this, we can create an effective procedure that will *list* all the ordered pairs in H, and no others: using a universal Turing machine, first carry out one step of the computation of $\varphi_w(x)$ for each pair (w, x) with $w, x \leq 1$. Then carry out *two* steps of the computation of $\varphi_w(x)$ for each pair (w, x) with $w, x \leq 2$. Continue in this way. Whenever a computation halts, add the appropriate pair (w, x) to the list (and then don't waste time by repeating that computation). This effective procedure eventually lists every ordered pair that it should.

Note that this procedure never tries to compute any particular $\varphi_w(x)$ for an unlimited number of steps, because such a computation might not halt and then the list would not get completed. This technique of switching among computations is called **dovetailing** and is the idea behind the extremely important notion of timesharing on a computer.

Also note that the set H is the domain of the universal function $\Phi^{(1)}$. We have the following:

Theorem 3.10. *For any $A \subseteq \mathbb{N}$, the following are equivalent:*

(a) *A is the domain of some partial recursive function.*

(b) *A is the range of some partial recursive function.*

(c) *$A = \emptyset$ or A is the range of some (total) recursive function.*

(d) *$A = \{y \mid \exists x[(x, y) \in R]\}$, for some PR relation R.*

(e) *$A = \{y \mid \exists x[(x, y) \in R]\}$, for some recursive relation R.*

Proof. (a) implies (b): Assume that A is the domain of a partial recursive function φ_w. We then modify M_w as follows: given an input x, perform the same computation that M_w does, but also keep a "copy" of x. If and when the computation is about to halt, change the output from $\varphi_w(x)$ to x. This machine computes a partial recursive function g such that $g(x) = x$ when defined, and $Rng(g) = Dom(g) = A$.

(b) implies (c): Assume that $A \neq \emptyset$ and A is the range of a partial recursive function φ_w. We describe a Turing machine that computes a total recursive function with the same range. Let c be some fixed number in A. Given an input x, first compute the numbers x_1 and x_2 such that $B_2(x_1, x_2) = x$. Then carry out up to x_2 steps in the computation of $\varphi_w(x_1)$ by M_w. If this computation halts, output $\varphi_w(x_1)$. If this computation does not halt in x_2 steps, output c. Clearly, this machine computes a total recursive function g with $Rng(g) \subseteq A$. Conversely, assume that $y \in A$. Then there is a number x_1 such that $\varphi_w(x_1) = y$. Suppose that M_w takes x_2 steps to compute $\varphi_w(x_1)$. Then if $x = B_2(x_1, x_2)$, we have $g(x) = y$, so $y \in Rng(g)$.

(c) implies (d): If $A = \emptyset$, just let $R = \emptyset$. Otherwise, assume A is the range of φ_w. (We don't need to assume that φ_w is total.) Then

$$A = \{y \mid \exists w, x, z[T(w, x, z) \wedge U(z) = y]\},$$

where T and U are defined before Proposition 3.4. By using the bijection B_3, the three quantified variables in this description of A can be merged into one. So A has the desired form, since T and U are PR.

(d) implies (e): Trivial.

(e) implies (a): See the next exercise. ∎

Exercise 14. Prove that (e) implies (a) in the previous theorem.

Definition. A subset of $\mathbb{N}$ is called **recursively enumerable (RE)** if it satisfies the conditions of Theorem 3.10.

Since the functions B_k provide "nice" bijections between $\mathbb{N}^k$ and $\mathbb{N}$, Theorem 3.10 and the concept of recursive enumerability also apply to subsets of $\mathbb{N}^k$, with $k > 1$. Or one can simply define a subset of $\mathbb{N}^k$ to be RE if it is the domain of some partial recursive function.

The word "enumerable" means "listable," so this definition is meant to describe sets that can be listed by an effective procedure. One way to make this rigorous would be in terms of a Turing machine that starts with an empty tape and performs a single computation that never halts, creating a tape that contains all the members of A as sequences of 1's, separated by 0's.

Here is another nice way to characterize RE sets: by definition, a set A is recursive if and only if there is a Turing machine that always correctly "answers" 1 or 0, depending on whether the input is or is not in A. By Theorem 3.10, A is RE if and only if there is a Turing machine that always halts (and answers 1, if desired) for inputs that are in A, and fails to halt for inputs that are not in A.

Proposition 3.11. *Every recursive set is RE.*

Exercise 15. Prove this proposition.

Theorem 3.12. *For any $A \subseteq \mathbb{N}$, the following are equivalent:*

(a) *A is recursive.*

(b) *A and $\mathbb{N} - A$ are both RE.*

(c) *A is finite or there is an effective procedure that lists the members of A in increasing order (that is, a strictly increasing recursive function whose range is A.)*

(d) *$A = \emptyset$ or there is an effective procedure that lists the members of A in nondecreasing order (that is, a nondecreasing recursive function whose range is A.)*

Proof. (a) implies (b): Immediate from the previous proposition, since if A is recursive, then so is $\mathbb{N} - A$.

(b) implies (c): Assume A and $\mathbb{N}-A$ are both RE, and A is infinite. We may also assume that $A \neq \mathbb{N}$, since the conclusion is trivial when $A = \mathbb{N}$. Here is an effective procedure for listing the members of A in increasing order: start listing the members of both A and $\mathbb{N} - A$, using two recursive functions. (This again requires dovetailing.) Eventually, 0 appears in one of the lists. If 0 appears in the list for A, it goes in the increasing list for A; otherwise, it doesn't. Then follow the same procedure for 1, 2, 3, etc.

(c) implies (d): See the next exercise.

(d) implies (a): Case 1: Assume A is finite. Then A is certainly recursive.

Case 2: Assume A is infinite, and g is a nondecreasing recursive function whose range is A. Here is an algorithm to decide membership in A: given n, compute successive values of g, starting with $g(0)$, until an output $\geq n$ is obtained. Then n is in A if and only if one of the outputs computed for g so far is n. (Note that this algorithm involves the μ operator. So this is an example in which the function we define might be recursive but not PR, even if g is PR.) ■

Exercise 16. Prove that (c) implies (d) in this theorem.

Exercise 17.

(a) Show that if A and B are recursive subsets of $\mathbb{N}$, then so are $A \cup B$, $A \cap B$, and $\mathbb{N} - A$. Feel free to use Church's thesis.

(b) Which parts of (a) hold if "recursive" is replaced by "RE"?

Exercise 18. Put each the following sets into one of these categories, with some justification: (1) recursive; (2) RE, but not necessarily recursive; (3) "co-RE" (that is, the complement of an RE set), but not necessarily recursive; or (4) not necessarily RE or co-RE.

(a) $\{n \in \mathbb{N} \mid$ there is a sequence of at least n 3's in the decimal expansion of $\pi\}$.

(b) $\{n \in \mathbb{N} \mid$ there is a sequence of exactly n 3's in the decimal expansion of $\pi\}$.

(c) the set of counterexamples to Goldbach's conjecture, namely, even numbers greater than 2 that cannot be written as the sum of two primes.

(d) the set of counterexamples to de Polignac's conjecture, namely, even numbers that cannot be written as the *difference* of two primes.

Exercise 19. Let f be a partial function. Prove that f is partial recursive if and only if the graph of f is RE. (Therefore, the domain, range, and graph of any partial recursive function must be RE.)

Exercise 20. Prove that the graph of a (*total*) recursive function must be recursive. Therefore, from the previous exercise, the following are equivalent, for any total function f:

(a) f is recursive.

(b) The graph of f is RE.

(c) The graph of f is recursive.

Exercise 21. Show that the graph of a partial recursive function is not always recursive. (Hint: for example, consider the halting problem.)

The undecidable problems that we have presented so far are all from within the subject of recursion theory itself. After these early results were obtained, people began to identify undecidable problems in "mainstream" branches of mathematics such as group theory and topology. One example that is striking because of the simplicity and importance of the problem is the negative answer to **Hilbert's tenth problem**. As part of his famous list of twenty-three problems posed at the Second International Congress of Mathematicians in 1900, Hilbert asked this natural question about Diophantine equations: Is there an effective procedure that, given a polynomial $f(x_1, x_2, \ldots, x_n)$ with integer coefficients, determines whether there is an n-tuple of natural numbers that makes the polynomial equal 0?

After several decades of serious effort, this question was finally answered in the negative in 1970. The main lemma for this result is that for every RE set A, there is a polynomial $f(x_1, x_2, \ldots, x_n, y)$ with integer coefficients such that any given natural number m is in A if and only if there is an solution in natural numbers to the equation $f(x_1, x_2, \ldots, x_n, m) = 0$. If there were a decision procedure for the existence of solutions to Diophantine equations, this lemma would imply the existence of a decision procedure for membership in an arbitrary RE set, which is impossible by Theorem 3.7. Important progress toward the main lemma was made by Martin Davis, Hilary Putnam, and Julia Robinson, and the final piece was provided by Yu Matijacevič at the age of twenty-two. For a thorough discussion of all of Hilbert's problems, see [Bro].

Here is a "sharper" version of this result that is reminiscent of Rice's theorem. Note that the solution set (in natural numbers) of a polynomial equation can have any finite cardinality, or it can be denumerable. In other words, its Von Neumann cardinal can be any member of $\omega + 1$.

Theorem 3.13 (Davis). *Let A be any fixed nonempty, proper subset of $\omega + 1$. Then there is no decision procedure that determines, for any polynomial equation $f(x_1, x_2, \ldots x_n) = 0$ with integer coefficients, whether the cardinality of its solution set (in natural numbers) is in A.*

So the negative solution to Hilbert's tenth problem is this theorem for the special case $A = \{\emptyset\}$.

3.5 Complexity theory

Much of this chapter has dealt with the question of what is computable in a strict mathematical sense: a computable function is one that can be evaluated by a Turing machine or other mechanical procedure. Soon after digital computers came into use in the 1950s, people began to consider the more practical question of which functions are *feasibly* computable. In the "real world" of science, business, etc., it is of little value to write a computer program to perform some task if the program might take absurdly long to complete that task. Thus, a different type of theory of computation called **complexity theory**, much more applicable than classical recursion theory, came into being.

Example 8. For any $n > 1$, let $f(n)$ be the smallest prime factor of n. By Church's thesis, f is certainly recursive. (Technically, the domain of a recursive function should be all of $\mathbb{N}$, so we could let $f(0) = f(1) = 0$.) In fact, based on properties of the bounded μ-operator mentioned at the end of Section 3.3, it follows that f is primitive recursive. But let's consider how to compute $f(n)$. For any nonprime $n > 1$, we know that $f(n) \leq \sqrt{n}$, and the obvious algorithm for determining $f(n)$ is to divide n by 2, 3, etc. until we find a factor of n.

Now suppose that n is a hundred digit number. If n is not prime, then $f(n)$ might still have as many as fifty digits, and therefore almost 10^{50} divisions could be required to compute $f(n)$. Computers are amazingly fast and getting faster all the time. As of 2003, the fastest computers can perform about ten billion (10^{10}) divisions per second. On such a computer, this program could still require over 10^{32} *years* to compute $f(n)$—a duration that dwarfs the age of the universe!

You might respond that it's not surprising that $f(n)$ would take a huge number of steps to compute, since n itself is huge. But note that the "data" required to input n into a computer is a string of only a hundred characters. A program that requires eons to run on an input string of only a hundred characters is of limited practical value.

Definitions. A Turing machine or other mechanical procedure is said to run in **polynomial time** if there is a polynomial p such that the procedure halts in fewer than $p(m)$ steps whenever the input consists of m characters.

A function is called **polynomially computable** if there is a Turing machine that computes it and runs in polynomial time. Similarly, we can talk about **polynomially decidable** sets, relations, and problems. The class of all polynomially decidable problems is denoted **P**.

In practical terms, programs are considered "efficient" or "feasible" if and only if they run in polynomial time. You might object that a program of this type would not be particularly efficient if all the polynomials p that fit the definition for it were of very large degree. This point has some validity. However, the polynomially computable problems that occur in practice almost always have bounding polynomials of low degree, typically no more than four.

If, in the above definition, we change "$p(m)$" to "$2^{p(m)}$," we get the definitions of programs that run in **exponential time**, **exponentially computable** functions, and **exponentially decidable** sets, relations, and problems. Since $x < 2^x$ for all real numbers, every program that runs in exponential time also runs in polynomial time, and every exponentially definable object is also polynomially definable. Moreover, recall that exponential functions "grow much faster" than polynomials. In other words, if p and q are any polynomials such that q is nonconstant and has a positive leading coefficient, then $\lim_{m\to\infty}[p(m)/2^{q(m)}] = 0$.

Example 9. Consider the function f defined in the previous example. If n is an m-digit number, then n is approximately equal to 10^m, and so it takes up to $\sqrt{10^m}$ steps to compute $f(n)$, which is fewer than 2^{2m} steps. Therefore, f is exponentially computable.

It is widely believed, although not proven, that f is not polynomially computable. The most important contemporary theory of cryptography (secret codes) is based on this assumption. Specifically, the so-called **RSA public key encryption** method is based on the notion that there is no feasible algorithm for factoring the product of two large primes, which is a special case of computing f. If someone were to find a way of computing f in polynomial time, it would have enormous practical consequences.

Example 10. Most easily definable subsets of $\mathbb{N}$, such as the set of all odd numbers and the set of all squares, are polynomially decidable. What about the set of all primes? This set is clearly decidable, but the most obvious algorithm for deciding primality is similar to the obvious algorithm for computing the function f—it can take up to $\sqrt{n}$ divisions to determine whether n is prime by "trial division." So one might expect the set of primes not to be in **P**. But in 2002, after years of intense study and various deep partial results, a polynomial time algorithm (of degree about 8, which is relatively high) for testing primality was found. The proof that this algorithm works is quite difficult, but the algorithm itself is surprisingly simple. It remains to be seen whether this breakthrough will lead to any changes in cryptography.

It might appear that the question of whether some procedure runs in polynomial time would depend on what type of computer it is run on. In the above example, the program for f can require about 3^m steps when the input contains m characters, so it appears to run in exponential time but not necessarily in polynomial time. But what if we were to run this program for f on a Turing machine? Then the string of characters required for the input n has length n, so if the output could be computed in no more than $\sqrt{n}$ steps, f would become polynomially computable. However, this turns out not to be the case. In this situation, the Turing machine would not be able to compute $f(n)$ in $\sqrt{n}$ steps. Instead, the number of steps required would become an exponential function of n, and so the program would still not run in polynomial time. It can be shown that the broad categories of polynomial and exponential time for algorithms do not depend on the type of computer or programming language that is used to implement the algorithm.

Almost all of the decidable problems that arise in practical situations are exponentially decidable. However, not all computable objects are exponentially computable. In fact, the exponentially computable functions are a proper subset of the primitive recursive functions. For instance, the simple function 2^{2^n} is primitive recursive but not exponentially computable.

Nondeterministic Turing machines, and P vs. NP

Polynomial time computability and exponential time computability are a couple of very general criteria of complexity. There are many ways of measuring complexity that are more sophisticated and precise. We will discuss just one other important measure, which is based on the concept of a **nondeterministic Turing machine**. A standard Turing machine is deterministic, in the sense that the current configuration determines the next one uniquely. In a nondeterministic Turing machine, there can be more than one possible successor configuration for a given configuration, and so there can be more than one possible computation for a given input. We will not give a more rigorous definition, but roughly this means that the program is allowed to take "guesses." In an actual computer, this effect could be obtained with a "random number generator."

Definition. Let A be a subset of $\mathbb{N}$ (or some other "nice" countable domain such as $\mathbb{Z}$, $\mathbb{N} \times \mathbb{N}$, etc.). We say that A is in **NP** if there is a nondeterministic Turing machine and a polynomial p such that, for any natural number n, $n \in A$ if and only if there is *at least one* computation with input n that halts in fewer than $p(m)$ steps. As before, m is the number of characters required to input n.

NP stands for "nondeterministic polynomial-time." Note that the classes **P** and **NP** are collections of sets (or relations or "problems"), not functions.

The class **P** is obviously closed under complementation: if A is in **P**, so is $\mathbb{N} - A$. The same holds for exponentially decidable sets, primitive recursive sets, and decidable sets. This follows from the symmetry of their definitions with respect to the outputs 0 and 1 (or "No" and "Yes"). What about the class **NP**? Its definition is deliberately asymmetric, so we might expect that this class, like the class of recursively enumerable sets, is not closed under complementation. As we will soon see, it is not known whether this is so, but it is considered quite likely.

Example 11. Let A be the set of composite natural numbers. We have mentioned that it is now known that the set of all primes is in **P**, and therefore A is also in **P**. But what can we say about these sets without using sophisticated methods? It is easy to see that A is in **NP**, because we can always show quickly that a large number is composite if we are lucky enough to guess a factor! More precisely, we can define a nondeterministic Turing machine that, given an m digit input n, guesses a sequence of up to about $m/2$ digits, and then tests whether the guessed sequence of digits is a factor of n. If n is composite, at least one such guessed number (sequence of digits) will verify that fact.

On the other hand, no lucky guessing of a number can provide a quick verification that n is prime. So there is no obvious proof that the set of primes is in **NP**.

Exercise 22. Our definition of **NP** looks very different from our earlier definition of **P**. But we can unify them if we wish: show that if we simply replace "nondeterministic" by "ordinary" in the definition of **NP**, we get a correct definition of **P**. (Of course, the italicized quantifier "at least one" becomes silly in this case, because there is only one computation for each input.) You will need to use the fact that every polynomial is recursive.

Proposition 3.14. $\mathbf{P} \subseteq \mathbf{NP}$, *and every member of* **NP** *is exponentially decidable.*

Proof. Since every ordinary Turing machine is a nondeterministic Turing machine, the previous exercise establishes the first claim. Here is an outline of a proof of the second claim: assume $A \in \mathbf{NP}$. Then there exist a nondeterministic Turing machine and a polynomial p satisfying the definition of that. If k is the total number of configurations possible in this Turing machine, then there are at most $k^{p(m)}$ possible computations for an input n with m characters. Therefore, we can design an ordinary Turing machine that will carry out all of these computations, thereby determining whether n is in A in something like $p(m) \cdot k^{p(m)}$ (or fewer) steps. And, given k and p, it is not hard to find another polynomial q such that $p(m) \cdot k^{p(m)} \leq 2^{q(m)}$ for every m in $\mathbb{N}$. ∎

What about the converses of these results? A relatively straightforward diagonalization argument can be used to construct an exponentially decidable problem that is not in **NP**. But the converse of the first assertion is open. That is, no one knows whether $\mathbf{P} = \mathbf{NP}$. This is certainly the most important open problem in complexity theory, and one of the most important in all of mathematics. Much of its importance stems from the fact that a great many problems that don't appear to be in **P** are known to be in **NP**. Therefore, knowing whether these classes are equal would help to determine the computational feasibility of many practical problems.

To make this more clear, we need one more concept: it is possible and in fact quite normal for one computer program to "query" another program for information. It is also useful in the theory of computation to define this idea mathematically. Let B be a fixed subset of $\mathbb{N}$, and imagine a computer program that can, whenever it chooses, ask whether any particular number is in B, and receive the correct answer in what is considered one step. In the literature, this is called a program that consults an **oracle**. (A bit more generally, the oracle could provide the value of some fixed function on any number provided by the program.) Without going into details, let us assume that we have defined this concept of a Turing machine that can consult an oracle.

Note that if B is not decidable, then the class of problems decidable by Turing machines that can consult B will include all decidable problems and others as well, for instance B. But if B is decidable, then no new problems are decidable (nor any new functions computable) by using B as an oracle. All that is gained by using a decidable oracle is a decrease in the number of steps required in computations.

Definition. For any $A, B \subseteq \mathbb{N}$, we say that A is **polynomially reducible to** B if there is a Turing machine with oracle B that decides A in polynomial time.

Definition. A problem is called **NP-complete** if it is in **NP**, and every problem in **NP** is polynomially reducible to it.

Think of an **NP**-complete problem as a "maximally hard" **NP** problem. It's in **NP**, but it's complex enough so that every **NP** problem can be solved efficiently in terms of it. The next proposition follows easily from the definition:

Proposition 3.15. *If any* **NP**-*complete problem is in* **P**, *then* **P** = **NP**.

Earlier, we mentioned that many practical problems that don't appear to be in **P** are in **NP**. Furthermore, a large number of these problems are actually **NP**-complete. Here are a few examples intended to convey a sense of the surprising number and variety of **NP**-complete problems:

Example 12. Given a network of cities and a set of roads between pairs of them, the **Hamiltonian circuit problem** asks whether there is a route that starts and ends in a given city and visits every other city exactly once. This problem is **NP**-complete.

Example 13. Given a network of cities and the distance between each pair of them, one version of the **traveling salesman problem** asks for the shortest route that starts in a given city and visits every city. However, this version is not a "problem" in our strict sense, since it doesn't ask for a yes or no answer. But this equivalent version is **NP**-complete: given such a network of cities and distances and an integer b (the "budget"), is there a route of length b or less that starts in a given city and visits every city?

Example 14. Perhaps you know that the famous **four-color theorem**, which states that every conceivable map can be "properly" colored with no more than four colors, was finally proved in 1976 after being open for 124 years. To say that a map is properly colored means that countries that border each other must get different colors. For this theorem, it is also assumed that no country consists of two or more disconnected pieces. It turns out that the **three-color problem**, which asks whether a given map can be properly colored with no more than three colors, is **NP**-complete. (The question of how to input a map into a computer is answered in Section 5.3.)

Example 15. The problem of determining whether a given statement of propositional logic (built up from propositional variables using connectives only) is consistent is **NP**-complete. A more useful variant of this problem involves the fact that every statement of propositional logic is equivalent to a statement in conjunctive normal form (CNF). The problem of deciding whether a given CNF statement is consistent, called the **satisfiability problem**, is also **NP**-complete. This result is known as **Cook's theorem**. Furthermore, this problem remains **NP**-complete even if we allow no more than three disjuncts in each disjunction of the CNF statement.

Example 16. Given integers a, b, and c, does the equation $ax^2 + by = c$ have a solution in integers? This is called the **quadratic Diophantine equation problem**, and it is also **NP**-complete.

It would be easy to fill several more pages with interesting **NP**-complete problems that have no apparent connection to each other. And yet, we know that all of these problems are "equally hard." An efficient (polynomial-time) program for any one of them would show that all of them, and indeed all **NP** problems, are efficiently solvable. Conversely, a proof that any one of them is not in **P** would show that none of them is in **P**. Either of these results would be a breakthrough of tremendous importance.

CHAPTER 4

Gödel's Incompleteness Theorems

4.1 Introduction

Why should students of mathematics want to know something about Gödel's incompleteness theorems? Here is one answer: as we mentioned in Chapter 3, they are possibly the most philosophically significant results in the history of mathematics. These theorems expose substantial and unavoidable limitations on the power and perhaps even the reliability of mathematics. They are remarkably similar in flavor to the uncertainty principle in physics, another amazing discovery of the early part of the twentieth century. The undecidable statements in Gödel's results are unnatural and contrived, but in the last section of this chapter we will discuss some more recent results that provide rather concrete, ordinary statements about natural numbers that are also true but unprovable in Peano arithmetic. Gödel's work and these extensions of it deserve to be of interest to the general mathematical community.

It is also worthwhile to understand the powerful methods that Gödel developed in order to prove his incompleteness theorems. In some ways, this "machinery" was almost as revolutionary as the incompleteness theorems themselves, and various parts of it have been extremely useful tools ever since. Specifically, these methods played a vital role in the creation of recursion theory a few years later, and as a result they were also instrumental in the development of computers and the theory of computation.

Yet another reason to become familiar with Gödel's incompleteness theorems is that they are closely related to the fascinating, classic concepts of self-reference and circularity. Several ancient Greek philosophers proposed paradoxes of various types, and they were aware that logical paradoxes (as opposed to time and motion paradoxes, for example) are generally based on circular definitions or self-referential statements. The paradoxes of set theory, such as Russell's paradox and the Burali–Forti paradox, are also based on these ideas. Even though Gödel's proof of the incompleteness theorems was ingeniously innovative, it was also clearly motivated by these traditional motifs. This chapter will place the incompleteness theorems in the context of several fascinating results that involve self-reference and fixed points.

For a more detailed treatment of the incompleteness theorems, see [Smu]. The Pulitzer Prize winning [Hof] is a brilliant, fascinating, eclectic romp through a variety of disciplines, highlighting the theme of self-reference. To read the seminal papers in this subject in their original form, see [Dav] (all authors) or [Go86] (Gödel only).

4.2 The arithmetization of formal theories

In the last chapter, we discussed the arithmetization of primitive recursive functions and Turing machines. We now describe the arithmetization of Peano arithmetic (PA) and other first-order theories. When this was first carried out by Gödel, it was a radically new concept, but in our computer age, it is a familiar process. Letters, digits, and other characters (including non-printing characters such as carriage returns) are stored in a computer as numbers. The standard "ASCII" coding for this assigns a number between 0 and 255 (eight bits) to every common character. A document created in a word-processing program is basically a finite sequence of characters, so it is stored in the computer as a finite sequence of numbers, which can then be encoded into a single number. In the same way, the terms, formulas, and finite sequences of formulas of a typical first-order language can be represented as finite sequences of standard characters, and thus they can be encoded as numbers. So we will assume henceforth that we have (separate) Gödel

If someone were to survey all logicians or even all mathematicians, asking them to name the most brilliant and/or productive logician of all time, it is safe to say that **Kurt Gödel** (1906–1978) would win this contest easily. Starting when he was barely in his twenties, he made substantial technical breakthroughs in every branch of foundations. His incompleteness theorems have certainly captured the fancy of the nonmathematical public more than any other results within foundations.

Gödel was born in Brno, Moravia, and in 1924 went to the University of Vienna to study physics. Soon his love of precision caused his main interest to change to mathematics, and then to foundations. He proved the completeness theorem (see Chapter 5), his Ph.D. dissertation in 1929, and the incompleteness theorems just one year later. His relative consistency results for set theory (Chapter 6) were obtained in 1938. In other words, Gödel produced three of the most fundamental and powerful theorems in foundations, by age 32. His collected works, including notes and correspondence, comprise almost a thousand pages.

When he was 21, Gödel fell in love with a 27-year old divorced nightclub dancer. His parents disapproved, but ten years later he and Adele were married and stayed together for life. In 1933, Gödel spent a year at the brand new Institute for Advanced Study in Princeton, and settled there permanently in 1940 after the Nazi takeover of Austria. There his closest companion was Einstein, but Gödel was for the most part reclusive and antisocial. He had recurrent bouts of depression, hypochondria, and paranoia, which became more severe as he aged. He ultimately refused to eat because he believed that people wanted to poison him, and he died of malnutrition. Unfortunately, Adele was hospitalized at the time and could not care for him as was her habit.

For more about Gödel's life and work, see [Daw] or [Go86].

numberings of the terms, formulas, and finite sequences of formulas of whatever first-order language we are working with. As in Chapter 3, we will make the convenient and harmless assumption that these Gödel numberings are onto $\mathbb{N}$.

One subtle point about this process is that a first-order language always has an infinite number of variables and therefore, in a sense, an infinite number of symbols. But it only requires a finite number of *characters* to express these variables.

Example 1. In the system TEX, the variable v_{274} is created by the sequence "v _ 2 7 4". So twelve characters (v, _, and the ten digits) suffice to represent the denumerable number of variables of a first-order language. The same reasoning applies to a first-order language with a denumerable number of relation, constant, and/or function symbols.

So the only first-order languages that cannot be encoded into $\mathbb{N}$ are those with an uncountable number of symbols, which are used only in specialized situations. Note that a first-order language with a denumerable number of symbols has only a denumerable number of formulas.

Notation. Let $\mathcal{L}$ be a countable first-order language. For any $\mathcal{L}$-formula P, #P will denote its Gödel number. For any number n, A_n will denote the formula it encodes.

The two operations defined by this notation are by definition inverses of each other. So $\#(\mathrm{A}_n) = n$ for every number n, and $\mathrm{A}_{\#\mathrm{P}} = \mathrm{P}$ for every $\mathcal{L}$-formula P.

Assuming that the Gödel numbering of formulas is done sensibly, every important function or relation that is based on the syntax of the language will be PR. Much of the "nitty-gritty" of [Go31] consisted of showing this fact. For example, if $f(m, n)$ is defined to be $\#(\mathrm{A}_m \wedge \mathrm{A}_n)$, then f is PR. Similarly, the function g defined by $g(m, n) = \#(\forall v_m \mathrm{A}_n)$ is PR.

When we call a theory (in a countable language) PR, recursive, or RE, we are of course speaking of the set of Gödel numbers of the formulas that comprise the theory. Also, a theory is called **recursively axiomatizable**, or simply **axiomatizable**, if it is equivalent to a recursive set of formulas. So a recursive set of formulas is automatically axiomatizable. One related bit of terminology bears mentioning because it is potentially confusing: when a theory T is called **decidable**, this does not mean that T itself is recursive. Rather, it means that $Thm(T)$ is recursive, which is stronger than saying that T is axiomatizable.

The set of axioms of first-order logic in $\mathcal{L}$ is PR, as is the set of axioms of every common formal theory used in mathematics.

Example 2. What would it mean if someone were to ask whether first-order group theory is decidable? The question would certainly not be referring to the usual list of *axioms* for this theory (as in Appendix D). It has only a finite number of proper axioms, so the set of all the axioms for this theory is clearly PR. The question presumably refers to the set of *theorems* of this theory. In Chapter 5, we will see that this is precisely the set of formulas of the appropriate language that are true in every group. This is a much harder question, whose answer is negative.

If T is a PR set of axioms, any function or relation involving the syntax of proofs from T will also be PR. Here is Gödel's [Go31] main result of this type:

Lemma 4.1. *Let T be a fixed (primitive) recursive theory. Then the binary relation*

$$\{(m, n) \mid m \text{ is the Gödel number of a proof (from } T\text{) of the formula } \mathrm{A}_n\}$$

is (primitive) recursive.

This lemma makes sense, because the set of logical axioms is PR and the rule of inference modus ponens is syntactically very simple. At least, it is clear that this binary relation is recursive when T is recursive.

Theorem 4.2. *For any theory T, the following are equivalent:*

(a) *$Thm(T)$ is RE.*

(b) *T is axiomatizable.*

(c) *T is equivalent to an RE set of formulas.*

Proof. (a) implies (b): Assume $Thm(T)$ is RE. It is automatically nonempty (since it includes all the logical axioms), so by Theorem 3.10(c) there is a recursive function f such that $Thm(T) = \{\mathrm{Q}_n \mid n \in \mathbb{N}\}$, where $f(n) = \#(\mathrm{Q}_n)$ for every $n \in \mathbb{N}$. Define the new theory $T' = \{\mathrm{P}_n \mid n \in \mathbb{N}\}$, where $\mathrm{P}_n = (\mathrm{Q}_0 \wedge \mathrm{Q}_1 \wedge \cdots \wedge \mathrm{Q}_n)$. Clearly, T' is equivalent to $Thm(T)$. So $Thm(T') = Thm(Thm(T)) = Thm(T)$. Thus we will be done if we can show that T' is recursive. But, using f, we can define a recursive function g such that $g(n) = \#(\mathrm{P}_n)$ for every n. By the way the P_n's were defined, g is increasing. Therefore, by Theorem 3.12, the range of g is recursive. In other words, T' is recursive. (This proof is known as **Craig's trick**.)

(b) implies (c): Immediate, since every recursive set is RE.

(c) implies (a): See the next exercise. ■

Exercise 1.

(a) Using Lemma 4.1, complete the proof of this theorem. That is, show that if T is RE, then so is $Thm(T)$. This can be done informally, using Church's thesis.

(b) Show that if T is an axiomatizable and complete set of sentences, then T is decidable. (Hint: From these givens, it is easy to show that the set of sentences in $Thm(T)$ is recursive. But remember that $Thm(T)$ also includes non-sentences. So you will probably want to use the generalization theorem. As in part (a), feel free to use Church's thesis as well.)

Convention. For the rest of this chapter, K always denotes an axiomatizable first-order theory that is an extension of PA. For simplicity, we will generally talk as if K is a theory in the same language as PA. But K could be a theory in a different countable language, such as the language of set theory, in which PA can be nicely "embedded."

Because the language of PA has the symbols $\overline{0}$ and S (for the successor function), every natural number has a term that denotes it, in an obvious way. For example, the term $S(S(S(\overline{0})))$ denotes the number 3. These terms for numbers are called **numerals**.

Notation. We write $\overline{n}$ as an abbreviation for the numeral denoting n.

Using numerals, Gödel defined a particularly useful syntactic function, the **substitution function**. We will work with the following slightly simplified version of it:

Proposition 4.3. *Let $Sub(m, n)$ be the Gödel number of the sentence obtained by replacing all occurrences of free variables in the formula* A_m *by the numeral $\overline{n}$. Then the function Sub is PR.*

We have been using the word "arithmetization" to refer to the process of assigning Gödel numbers to objects such as Turing machines and first-order formulas. But there is another major component of the arithmetization process, namely showing that the material we have been discussing in this chapter can be developed in PA. Almost every result in this chapter, except for later ones that refer to "true" statements, can be stated in the language of PA, using our various Gödel numberings, and then proved in PA. Our next task is to describe this formalization process.

Definition. Let R be a k-ary relation on $\mathbb{N}$, and let $\mathrm{P}(x_1, \ldots, x_k)$ be a formula of PA. We say that P **represents** R in K if, for any $a_1, \ldots, a_k \in \mathbb{N}$:

$$(a_1, \ldots, a_k) \in R \text{ implies } K \vdash \mathrm{P}(\overline{a_1}, \ldots, \overline{a_k}), \quad \text{and}$$

$$(a_1, \ldots, a_k) \notin R \text{ implies } K \vdash \sim \mathrm{P}(\overline{a_1}, \ldots, \overline{a_k}).$$

We say that R is **representable** in K if there is a formula that represents it in K. A partial function is called representable if its graph is representable.

Note that if K is consistent, the two implications in this definition become biconditionals. Here, without proof, is one of the major technical results in [Go31]:

Theorem 4.4. *Every PR relation and function is representable in K.*

Corollary 4.5. *Assume K is consistent. Then a relation is representable in K if and only if it is recursive.*

Proof. Assume that K is consistent and R is a representable relation of k variables. Then there is a formula P such that, for any $a_1, a_2, \ldots, a_k \in \mathbb{N}$, $(a_1, \ldots, a_k)$ is in R if and only if

$$K \vdash \mathrm{P}(\overline{a_1}, \ldots, \overline{a_k}),$$

which in turn says

$$\exists m[m \text{ is the Gödel number of a proof (in } K) \text{ of } P(\overline{a_1}, \ldots, \overline{a_k})].$$

This has the form $\exists m[\mathrm{Q}(m, \overline{a_1}, \ldots, \overline{a_k})]$, where Q is PR by Lemma 4.1. Therefore, R is RE, by Theorem 3.10(e). Similar reasoning shows that the complement of R is RE. Therefore, by Theorem 3.12, R is recursive.

For the reverse direction, assume that R is any recursive relation of k variables. Choose w such that φ_w is the characteristic function of R. Now assume that $a_1, \ldots, a_k \in \mathbb{N}$ and $R(a_1, \ldots, a_k)$ holds. That means the Turing machine M_w, with input $\vec{a}$, halts with output 1 and with some computation record z. So $\mathrm{T}(w, b, z)$ holds, where $b = B_k(\vec{a})$. Also, we have $U(z) = 1$. Here, T and U are Kleene's T-predicate and the upshot function defined in Section 3.3.

By the previous theorem, there are formulas $\mathrm{P_T}$, P_U, and P_B that represent T, U, and B_k, respectively. Therefore, the sentences $\mathrm{P_T}(\overline{w}, \overline{b}, \overline{z})$, $\mathrm{P}_B(\overline{a_1}, \ldots, \overline{a_k}, \overline{b})$, and $\mathrm{P}_U(\overline{z}, \overline{1})$ are all provable in K. Now let $\mathrm{Q}(\vec{x})$ be the formula $\exists b, z[\mathrm{P_T}(\overline{w}, b, z) \wedge \mathrm{P}_B(\vec{x}, b) \wedge \mathrm{P}_U(z, \overline{1})]$. We will show that Q represents R in K. By elementary logic (technically "existential generalization," discussed in Section 1.3), if a formula is provable for the particular numerals $\overline{b}$ and $\overline{z}$, then the same formula is provable with existentially quantified variables in place of those numerals. Therefore, $\mathrm{Q}(\overline{a_1}, \ldots, \overline{a_k})$ is provable in K.

Similarly, if $R(a_1, \ldots, a_k)$ is false, then the sentence $\mathrm{Q}(\overline{a_1}, \ldots, \overline{a_k})$ with $\mathrm{P}_U(z, \overline{1})$ replaced by $\mathrm{P}_U(z, \overline{0})$ is provable in K. From this it follows easily that $\sim \mathrm{Q}(\overline{a_1}, \ldots, \overline{a_k})$ is provable in K. This completes the proof that R is representable in K. ∎

Corollary 4.6. *Every (total) recursive function is representable in K.*

Proof. By Exercise 20 of Section 3.4, the graph of a recursive function must be recursive. ■

The next result, while not explicitly stated in [Go31], may be viewed as the main lemma for all of the incompleteness theorems. The construction Gödel used to prove this lemma is one of the most original and ingenious in the history of mathematics.

Theorem 4.7 (Fixed-Point Lemma, or Diagonalization Lemma). *Let* P *be a formula with one free variable. Then there is a sentence* Q *such that* $K \vdash [\mathrm{Q} \leftrightarrow \mathrm{P}(\overline{\#\mathrm{Q}})]$.

Proof. Let P be given with only v_k as a free variable. First define the formula $\mathrm{R}(v_k)$ to be $\mathrm{P}[Sub(v_k, v_k)]$. Then let $m = \#\mathrm{R}$, and let Q be $\mathrm{R}(\overline{m})$.

It is clear (in fact, trivial!) that the following chain of equivalences is provable in K: $\mathrm{Q} \leftrightarrow \mathrm{R}(\overline{m}) \leftrightarrow \mathrm{P}[Sub(\overline{m}, \overline{m})] \leftrightarrow \mathrm{P}[Sub(\#\mathrm{R}, \overline{m})] \leftrightarrow \mathrm{P}[\#(\mathrm{R}(\overline{m}))] \leftrightarrow \mathrm{P}(\overline{\#\mathrm{Q}})$. ■

Example 3. Suppose $\mathrm{P}(v_3)$ is the formula that expresses that v_3 is a multiple of 19. Then $\mathrm{R}(v_3)$ is the formula expressing, "If you take the formula with Gödel number v_3, and replace all free variables in it with the numeral for v_3, the Gödel number of the resulting formula is a multiple of 19." This formula $\mathrm{R}(v_3)$ has a Gödel number; let's suppose it's 1234567890. Now let Q be the formula $\mathrm{R}(\overline{1234567890})$. It is clear that Q and $\mathrm{P}(\overline{1234567890})$ "say the same thing"; more precisely, they are equivalent in PA.

It should be clear why this is called the fixed-point lemma: any P with one free variable defines a function on sentences, with each sentence Q mapped to $\mathrm{P}(\overline{\#\mathrm{Q}})$. The lemma says that this function must have a fixed point, "up to equivalence in K." It is also called the diagonalization lemma because the formula $Sub(v_k, v_k)$ represents the diagonal of the binary predicate $Sub(v_i, v_j)$. So this lemma traces its lineage back to Cantor's original diagonalization argument.

There are several ways to strengthen the fixed-point lemma. In particular, we will make use of the fact that the Gödel number of Q can be found effectively from the Gödel number of P and the Gödel number of a Turing machine that decides membership in some recursive set of axioms for K. Also, P can be allowed to have any nonzero number of free variables. Then, in the conclusion, Q has one fewer free variable than P does, and the numeral $\overline{\#\mathrm{Q}}$ just replaces one particular free variable of P.

The recursion theorem

We have now developed all of the machinery needed to prove the incompleteness theorems. But first, in order to gain more insight into the fixed-point lemma, let's take a look at a different fixed-point result that is one of the most surprising and useful theorems of recursion theory. Suppose that, for some reason, we want to know whether there is a w such that the range of φ_w consists precisely of all multiples of w. In the words of [Rog], "At first glance, the existence of such a w might appear to be an arbitrary and accidental feature" of the Gödel numbering that we use. "Indeed, it might seem likely that, in our indexing, no such w exists." But the recursion theorem, which was proved by Kleene (years after Gödel's [Go31]), shows that a w with this property, or any other similar property, must exist.

Theorem 4.8 (Recursion Theorem). *For any total recursive function f of one variable, there is a w such that $\varphi_{f(w)} = \varphi_w$.*

Proof. We first define a recursive function g. We would like to define $g(u)$ to be $\varphi_u(u)$. The problem is that this g would not be total, so we have to define g in a more roundabout way: given any number u, construct a Turing machine that, for any input x, first tries to compute $\varphi_u(u)$. If and when that computation halts with some output v, the machine then tries to compute (and give as output) the number $\varphi_v(x)$. Naturally, if $\varphi_u(u)$ is not defined, then this Turing machine will not halt for any input x. We define $g(u)$ to be the Gödel number of this Turing machine.

Thus g is a total recursive function such that, whenever $\varphi_u(u)$ is defined, $\varphi_{g(u)} = \varphi_{\varphi_u(u)}$, and whenever $\varphi_u(u)$ is undefined, $\varphi_{g(u)}$ is the empty function. Now let any recursive function f be given. Then the composite function $f \circ g$ is also (total) recursive. Let z be a Gödel number of $f \circ g$. Then we see that $\varphi_{f \circ g(z)} = \varphi_{\varphi_z(z)} = \varphi_{g(z)}$. Let $w = g(z)$. ■

The number w in this theorem is called a **fixed-point value** for f. Of course, w is not a true fixed point of f, since we do not expect that $f(w) = w$. Rather, $f(w)$ and w are the Gödel numbers of "equivalent" Turing machines, in terms of the functions of one variable that they compute.

Example 4. By the recursion theorem, there must be a w such that $\varphi_w = \varphi_{w+1}$, a w such that $\varphi_w = \varphi_{w+3}$, a w such that $\varphi_w = \varphi_{15w^3+7}$, etc. The necessity of these facts is not at all obvious.

Example 5. Let's use the recursion theorem to answer the question we posed before stating the theorem. Let u be given. Then we can construct a Turing machine that, for any input x, determines whether x is a multiple of u, and then halts (with output x) if and only if it is. Let $f(u)$ be the Gödel number of such a Turing machine. Then f is a recursive function, as long as we construct the Turing machine in terms of u in a systematic way. By applying the recursion theorem to this f, we get the desired number w.

As another example of the power of the recursion theorem, here is a short, cute proof of Rice's theorem: let $\mathcal{C}$ be any nonempty proper subset of the set of all partial recursive functions of one variable. Choose fixed numbers a and b such that $\varphi_a \in \mathcal{C}$ and $\varphi_b \notin \mathcal{C}$, and then define a total function f of one variable by

$$f(x) = \begin{cases} b, & \text{if } \varphi_x \in \mathcal{C}, \\ a, & \text{if } \varphi_x \notin \mathcal{C}. \end{cases}$$

If f were recursive, there would be a w such that $\varphi_{f(w)} = \varphi_w$, by the recursion theorem. But this is impossible, because if $\varphi_w \in \mathcal{C}$, then $\varphi_w = \varphi_{f(w)} = \varphi_b \notin \mathcal{C}$, and if $\varphi_w \notin \mathcal{C}$, then $\varphi_w = \varphi_{f(w)} = \varphi_a \in \mathcal{C}$. Thus f is not recursive, and so neither is the set $\{x \mid \varphi_x \in \mathcal{C}\}$, because f is essentially the characteristic function of that set.

We mentioned some ways to strengthen the fixed-point lemma. The recursion theorem can also be strengthened in several ways. One way that is obvious from the proof is that w can be computed as a recursive function of the Gödel number of f. More important, if f is a function of more than one variable, then the result still holds and w can be effectively computed from the extra variables of f as well as from the Gödel number of f.

The recursion theorem is a close cousin of the fixed-point lemma, and their proofs are based on the same idea. I have chosen to include separate proofs of these two results, in order to give you two perspectives on this brilliant, almost magical, construction. I suggest that you compare these proofs. In the proof of the recursion theorem, f, $f \circ g$, z, and w correspond to P, R, m, and Q in the earlier proof, respectively.

4.3 A potpourri of incompleteness theorems

We are finally "there." All of the results in this section are in the general category of incompleteness theorems. In order to give a smooth and efficient exposition, I will first present these results in nearly antichronological order. A brief history follows.

Theorem 4.9 (Church's Theorem). *There is no consistent, decidable extension of Peano arithmetic.*

Proof. Assume that K is a decidable extension of PA. That means $Thm(K)$ is recursive, which by Corollary 4.5 implies that $Thm(K)$ is representable in K. From this, we need to show that K is inconsistent.

So assume that $\mathrm{R}(x)$ represents $Thm(K)$ in K. Then apply the fixed-point lemma with P being $\sim$ R, to obtain a sentence Q. So $K \vdash (\mathrm{Q} \leftrightarrow \sim \mathrm{R}(\overline{\#\mathrm{Q}}))$. If $K \vdash \mathrm{Q}$, then $K \vdash \sim \mathrm{R}(\overline{\#\mathrm{Q}})$. But because R represents $Thm(K)$ in K, we also have that $K \vdash \mathrm{R}(\overline{\#\mathrm{Q}})$, and thus K is inconsistent. On the other hand, if $K \nvdash \mathrm{Q}$, then $K \nvdash \sim \mathrm{R}(\overline{\#\mathrm{Q}})$, which means that it's not the case that $\mathrm{Q} \notin Thm(K)$. But that says $K \vdash \mathrm{Q}$, a contradiction. ■

Theorem 4.10 (Gödel–Rosser Incompleteness Theorem). *There is no complete, axiomatizable extension of PA.*

Proof. Such a theory would be consistent (by the definition of completeness) and decidable (by Exercise 1(b)). This would violate Church's theorem. ∎

Corollary 4.11. *If PA is consistent, then it is undecidable, and therefore incomplete.*

The rest of this chapter will occasionally refer to "true" statements in the language of PA, a concept that will not be defined rigorously until Chapter 5. But in essence it means just what it should mean in ordinary mathematics: "true in the real world," or, more precisely, "true in the **standard model of arithmetic**" (the set $\mathbb{N}$ together with the usual arithmetical operations on it, to be defined more rigorously in Section 5.2). Let us accept without proof that all the axioms, and therefore all the theorems, of PA are true. The statements that we know to be true are, generally, those that we can prove in "ordinary mathematics" (say, ZFC set theory). However, definitions and theorems that mention truth cannot be formalized in PA.

Corollary 4.12 (Church). *The set of all true sentences in the language of PA is not recursive and, hence, not representable in K.*

Proof. This set of sentences is automatically complete. So it's not recursive, since that would violate the Gödel–Rosser theorem. Then Corollary 4.5 implies that it cannot be representable in K. ∎

Another way to state the first part of this corollary is that if T is any consistent, axiomatizable theory in the language of PA, then there are true sentences that are not provable in T.

Exercise 2. Show that neither the set of true sentences of the language of PA, nor the set of false sentences, is RE.

Surprisingly perhaps, if the multiplication operator is removed from the language of PA, then the set of all true sentences in this reduced language is recursive. However, this is not a theory in which very much serious mathematics can be carried out.

Another way to prove this result of Church is by using the unsolvability of the halting problem: Kleene's T-predicate is PR and therefore representable in PA, and so the rules under which Turing machines operate can be accurately formalized in PA. Therefore, if there were an effective procedure for identifying true sentences of arithmetic, we could use it to solve the halting problem.

Furthermore, it can be shown that only a finite number of the proper axioms of PA are needed to formalize the operation of Turing machines; that is, only a finite number of instances of induction are needed. Let P be the conjunction of this entire finite list of proper axioms. Then, given a Turing machine M_w and an input x, let Q be a sentence of PA that expresses that M_w with input x halts. Then M_w with input x halts if and only if P $\rightarrow$ Q is a law of logic. Therefore, if there were an effective procedure for identifying laws of logic, we could use it to solve the halting problem. We have just outlined a proof of the negative answer to the "Entscheidungsproblem" (decision problem), another one of the problems on Hilbert's famous 1900 list:

Theorem 4.13 (Turing, Church). *The set of laws of logic in the language of PA (and many other first-order languages) is undecidable.*

A more common way to express this result is that "pure predicate calculus" is undecidable. More specifically, it holds for any first-order language with at least one relation or function symbol of two or more variables. It fails for simpler languages such as the language of **pure identity**, in which there are no function or constant symbols and the only atomic formulas are equations.

We have mentioned that the notion of truth for arithmetical statements cannot be formalized correctly in Peano arithmetic. Corollary 4.12 makes this point, and in Section 5.7 we will present a stronger theorem of this type, due to Alfred Tarski. Here is a syntactic result in the same vein, which doesn't mention the standard model or any infinite structure and therefore can be formalized in PA:

Definition. A **truth predicate** for K is a formula Tr with one free variable such that, for any sentence Q of the language of K, $K \vdash [\mathrm{Q} \leftrightarrow \mathrm{Tr}(\overline{\#\mathrm{Q}})]$.

Theorem 4.14 (Tarski's Truth Theorem). *If K is consistent, then there is no truth predicate for K.*

Exercise 3. Prove this theorem. You can use an argument that is very similar to the proof of Church's theorem, but even simpler.

Tarski's truth theorem and the other incompleteness theorems are related to, and to some extent based on, the ancient **liar's paradox**: the simple sentence "I am lying" or "This statement is a lie" cannot be given a truth value consistently. Of course, these modern results are much more technically sophisticated.

Exercise 4. Show that if K is complete, then any formula that represents $Thm(K)$ in K must also be a truth predicate for K. Using this, derive the Gödel–Rosser incompleteness theorem from Tarski's truth theorem and Corollary 4.5, without directly using the fixed-point lemma or Church's theorem.

The contradiction that results from assuming that K has a truth predicate is called **Tarski's truth paradox**. It is interesting to note that there do exist truth predicates for sentences of bounded quantifier complexity. Specifically, for each n, there is a truth predicate for all Σ_n (respectively, Π_n) sentences, and this truth predicate can itself be made to be Σ_n (respectively, Π_n). Indeed, one way to interpret Tarski's truth theorem is that it would be unlikely that a single predicate, of some fixed complexity, could accurately "reflect" the truth of all sentences of arbitrarily large complexity.

An historical perspective on Gödel's work

This seems like an appropriate place to say a bit about who did what, and when. None of the incompleteness results given so far in this section appeared explicitly in Gödel's revolutionary 1931 article. So exactly what did Gödel do in [Go31]? As we have already mentioned, he described the PR functions and showed that all important syntactic functions and predicates are PR (as in Lemma 4.1 and Proposition 4.3). He then proved that every PR function is representable (Theorem 4.4) and, implicitly, proved the fixed-point lemma.

With this machinery, Gödel then proved his two surprising incompleteness theorems. The first theorem is essentially the Gödel–Rosser theorem with the assumption of consistency replaced by a slightly stronger, and somewhat awkward, condition. For the record, here is the definition of this concept, followed by the theorem itself:

Definition. A theory T in the language of PA is called **ω-consistent** ("omega-consistent") if there is no formula $\mathrm{P}(x)$ of one free variable for which $T \vdash \exists x \mathrm{P}(x)$ but $T \vdash \sim \mathrm{P}(\overline{n})$ for each number n.

Note that ω-consistency trivially implies consistency, since every formula is provable in an inconsistent theory.

Theorem 4.15 (Gödel's First Incompleteness Theorem). *Let K be an axiomatizable extension of PA. Then there is a sentence* Q *of the language of K such that:*

(a) *If K is consistent, then $K \nvdash \mathrm{Q}$, and*

(b) *If K is ω-consistent, then $K \nvdash \sim \mathrm{Q}$, so* Q *is independent of K.*

Proof. Given K, Lemma 4.1 and Theorem 4.4 guarantee the existence of a formula $\mathrm{Prf}_K(m, n)$ that represents "m is the Gödel number of a proof of A_n from K," in K. Next, let $\mathrm{Prov}_K(n)$ be the formula $\exists m\, \mathrm{Prf}_K(m, n)$. This formula is intended to express (but not necessarily represent, in our rigorous sense) "A_n is provable from K," that is to say $A_n \in \mathit{Thm}(K)$. Then, apply the fixed-point lemma with P being $\sim \mathrm{Prov}_K(n)$, to obtain a sentence Q.

Now, if $K \vdash \mathrm{Q}$, then there is an actual proof of Q from K, encoded by some number m. Then $K \vdash \mathrm{Prf}_K(\overline{m}, \#\mathrm{Q})$, which in turn means that $K \vdash \mathrm{Prov}_K(\#\mathrm{Q})$. But, by the definition of Q, we also have $K \vdash \sim \mathrm{Prov}_K(\#\mathrm{Q})$. Therefore, K is inconsistent, so part (a) is proved.

On the other hand, if $K \vdash \sim$ Q, a similar argument does not quite work to show that K is inconsistent. All we can conclude is that K is not ω-consistent, yielding (b). ■

Exercise 5. Prove part (b) of this theorem. Explain why you cannot derive that K is inconsistent from the assumption that $K \vdash \sim$ Q.

Note how similar Gödel's proof is to the proof of Church's theorem. The main difference is that the sentence Q used for Church's theorem is hypothetical, whereas Gödel's Q is directly definable from K. More precisely: Q can be found effectively from the Gödel number of a Turing machine that decides membership in some set of axioms for K. This sentence Q is remarkable in that it quite directly says "I am not provable in K." The fixed-point lemma enables us to define sentences that "talk about themselves"!

If we assume that everything provable from K is true (which is stronger than assuming ω-consistency), then the proof of part (a) of Gödel's theorem tells us that Q is in fact true, since Q is not provable and therefore $\sim \mathrm{Prov}_K(\#\mathrm{Q})$ is true. Thus Gödel's construction can be used to prove Corollary 4.12, but Gödel was more interested in proving results that could be formalized in arithmetic. Gödel could not quite obtain the "cleaner" Gödel–Rosser version of the incompleteness theorem, but what he did prove was still a shattering revelation.

A few years later, J. Barkley Rosser devised a very clever variant of Gödel's sentence Q, through which the assumption of ω-consistency can be weakened to ordinary consistency. That is, if K is consistent, then Rosser's Q is independent of K, and the Gödel–Rosser version of the theorem follows. Whereas Gödel's sentence expresses that "There is no proof of me," Rosser's sentence expresses the more intricate notion, "For any proof of me, there is a proof of my negation with a smaller Gödel number." This was a brilliant innovation on Rosser's part. However, Church's theorem, with its relatively simple proof, is a more straightforward route to the Gödel–Rosser theorem.

Exercise 6. Using the fixed-point lemma and the formula Prf_K, show how to construct Rosser's sentence, and use it to prove the Gödel–Rosser theorem.

By the way, almost all of the incompleteness results that we have mentioned, except those appearing in [Go31], were proved in about 1936. Since that year also marks the introduction of Turing machines and at least two other definitions of computability, the proof of the unsolvability of the halting problem, and Church's thesis, it was truly a remarkable year for recursion theory and the foundations of mathematics!

Gödel's second incompleteness theorem

There is one more important piece to the refutation of Hilbert's program. Let Con_K be the sentence $\sim \mathrm{Prov}_K[\#(\overline{0} = \overline{1})]$. Con_K is intended to formalize the statement that K is consistent; this works because we know that $K \vdash \overline{0} \neq \overline{1}$. The statement $\overline{0} = \overline{1}$ could be replaced by any contradiction. Here is the other major result of [Go31]:

Theorem 4.16 (Gödel's Second Incompleteness Theorem). *If K is a consistent, axiomatizable extension of PA, then $K \nvdash \mathrm{Con}_K$.*

Proof. We outline the proof, which is a brilliant example of how fruitful it can be to go back and forth between looking at a formal theory "from without" and "from within."

Assume that $K \vdash \mathrm{Con}_K$. Then, consider the proof of Gödel's first incompleteness theorem, in which we defined a sentence Q and showed that if K is consistent, then $K \nvdash$ Q. This entire argument can be formalized and proved in PA, and therefore in K. In other words, we have that $K \vdash [\mathrm{Con}_K \to \sim \mathrm{Prov}_K(\#\mathrm{Q})]$. So, by modus ponens, $K \vdash \sim \mathrm{Prov}_K(\#\mathrm{Q})$. But by the way Q was defined, that means that $K \vdash$ Q. As we have just noted, this implies that K is inconsistent. ■

Isn't this an amazing line of reasoning? We will say a bit more about this proof below. Surprisingly, however, Gödel's second incompleteness theorem is less "absolute" than one might expect. Solomon Feferman [Fef60] proved that if the axioms of PA are encoded in a more elaborate way, then it can occur that PA $\vdash \mathrm{Con}_{PA}$.

Corollary 4.17. *If PA is consistent, then PA+ $\sim \text{Con}_{PA}$ is consistent but not ω-consistent.*

Exercise 7. Prove this corollary.

After Gödel's work became known, efforts were made to distill exactly what properties of the provability predicate are needed to prove the second incompleteness theorem from the first. Bernays provided the first clear answer to this question; later, Martin Löb obtained this simpler list:

Proposition 4.18 (Derivability Conditions). *Let K be a consistent axiomatizable extension of PA, and let* P *and* Q *be any sentences of the language of K. Then:*

(a) *If* $K \vdash \text{P}$, *then* $K \vdash \text{Prov}_K(\#\text{P})$.

(b) $K \vdash [\text{Prov}_K(\#\text{P}) \wedge \text{Prov}_K(\#(\text{P} \to \text{Q})) \to \text{Prov}_K(\#\text{Q})]$.

(c) $K \vdash [\text{Prov}_K(\#\text{P}) \to \text{Prov}_K(\text{Prov}_K(\#\text{P}))]$.

The derivability conditions and their proofs are implicit in Gödel's [Go31]. They are just three instances of the general fact that K is powerful enough to reason about its own syntax. Also note that condition (c) is simply the formalization of condition (a) in K.

Exercise 8. Assuming Gödel's first incompleteness theorem and the derivability conditions, prove Gödel's second incompleteness theorem.

Now, here is an interesting feature of the derivability conditions. If you refer to the discussion of modal logic in Section 1.7, you will see that the derivability conditions are essentially axioms 2 through 4 of the system S4, stated within K and with $\Box$ replaced by Prov_K. This makes sense in that it is not unreasonable to identify provable statements with "necessarily true" statements. It turns out that modal logic is a very fruitful viewpoint from which to analyze incompleteness phenomena—see [Smo].

Exercise 9. Does axiom 1 of S4 also hold in K with $\Box$ replaced by Prov_K? In other words, do we have $K \vdash [\text{Prov}_K(\#\text{P}) \to \text{P}]$, for each sentence P? Why or why not?

Hilbert's formalist program, revisited

Let us now take a closer look at the formalist program and what Gödel's results did to it. Hilbert proposed working with two formal systems, S and T. S was to be the "finite" (or "finitistic"), "meaningful" part of mathematics, while T would be "transfinite," "idealized" mathematics. T should be an extension of S. For simplicity, we could take S to be PA (although Hilbert seems to have envisioned a weaker theory known as **primitive recursive arithmetic**), and T could be some theory of analysis or even all of ZFC.

To counter the objections of the intuitionists and others to "idealized" (abstract) methods in mathematics, Hilbert wanted to prove, in S, that T doesn't prove any *meaningful* statements that are not already provable in S. This would be comforting since it would show that idealized methods help to prove idealized statements only, not meaningful statements. A related goal proposed by Hilbert was to prove, in S, that T is consistent. He also expected S to be complete (with respect to meaningful statements), and hoped that this fact could be proved in S. So all the relevant metamathematics was to be carried out in S.

Every one of these goals was refuted by Gödel's work. The statement Q constructed in Gödel's first incompleteness theorem demonstrates the impossibility of the first goal. Q is certainly a meaningful statement in the sense that Hilbert intended. (Structurally, Q is quite simple, a Π_1 sentence of arithmetic, a category that Hilbert explicitly considered meaningful.) Q is true, and provable in any reasonable candidate for T. But Q (which depends on S) cannot be provable in S, assuming that S is consistent and is powerful enough to express basic arithmetical notions.

Clearly, Gödel's first incompleteness theorem also shows that no reasonable candidate for S (or T) can be complete. Finally, his second incompleteness theorem shows that no reasonable formal system can prove even its own consistency. So there is no way that the system S could prove the consistency of the stronger system T. By the way, Gödel's results really did come as a complete surprise, even a shock, to the mathematical community. Apparently, neither Hilbert nor any of his colleagues had an inkling that the formalist program might be doomed to failure.

Hilbert's Entscheidungsproblem is usually not considered to be part of the formalist program, but it is certainly related. Hilbert asked whether the set of laws of predicate logic is decidable, even though the notion of decidability had not been precisely defined at that time (1928). Hilbert probably expected an affirmative answer (recall Exercise 1(b)), which would imply that all finitely axiomatizable first-order theories are decidable. But, as we have seen, Church and Turing provided a strong negative answer to this question as well: pure predicate calculus (in all but the simplest languages) and most interesting first-order theories are undecidable.

On the other hand, let's keep in mind that a large part of Hilbert's program was extremely successful: the formalization of just about all of contemporary mathematics, including both the language and the axiomatization of every important branch of the subject. This formalization has played a vital role in the subsequent progress in all areas of the foundations of mathematics, and it has also has had a strong influence on the current tenor of mathematics in general. Indeed, none of the "negative" results described in this section could have been obtained without the achievement of formalization.

By the way, the word "formalism" has taken on a somewhat different meaning in recent times, as more or less the opposite of "Platonism" or "realism." A Platonist mathematician is one who believes that abstract objects like uncountable sets really exist, and therefore that statements about such objects are meaningful, and either true or false. A Platonist would believe that GCH is true or false, even though we have no idea now which one it is, and we may never know. A formalist takes a more cautious (or perhaps even negative) position regarding the existence of abstract objects, but believes that it's fine to study them mathematically (by working from axioms about them), without worrying about their existence or whether statements involving them are meaningful, let alone true or false.

4.4 Strengths and limitations of PA

Gödel provided specific statements that are expressible in the language of Peano arithmetic and are true, but cannot be proved in Peano arithmetic. The sentence Q used to prove the first incompleteness theorem is in this category. So is Con_{PA}, which is perhaps easier to understand.

However, these are not the sort of statements that arise in ordinary mathematics. Ever since Gödel's surprising results, mathematicians have been interested in finding more "natural" mathematical statements that are expressible in PA but independent of PA. In this section we will consider this quest. But let's begin with the positive side, by describing what sorts of things *can* be proved in PA. As we will see, this theory is actually quite powerful.

Before we concern ourselves with what can be *proved* in PA, we should first have some understanding of what can be *stated* in PA, since you can't prove something unless you can state it. Because PA is intended to be a theory that describes $\mathbb{N}$ and the usual operations on it, we might expect that all of the standard theorems of number theory can be stated in PA. This is indeed the case, provided that we allow some leeway about precisely what it means to state something in a formal language. Recall that the language of PA has symbols for addition, multiplication, and the successor operator, as well as the constant 0. From these, it is not difficult to define subtraction, division (with remainder), and the ordering on natural numbers within this language.

Exponents present more of a challenge. One important result of number theory states that every natural number can be written as the sum of four squares of integers. There is no problem stating this in PA, because the term n^2 can be replaced by $n \cdot n$. But what about exponents that are variables? Fermat's famous "last theorem" states that there are no positive integers a, b, c, and n such that $n > 2$ and $a^n + b^n = c^n$. It is less obvious how to express this in PA, but we can be sure that it is possible because we know that the exponentiation function is recursive (by Church's thesis or, more concretely, by writing an explicit primitive recursive definition for it). Then Corollary 4.5 guarantees that this function is representable in PA. In order to give a direct proof of the expressibility of exponentiation in PA, the standard argument uses the so-called **Chinese remainder theorem**, a classic fact of number theory. (See [End].) In fact, Gödel used this theorem in [Go31] to prove the incompleteness theorems.

Statements that involve quantification over all subsets of $\mathbb{N}$ with a particular number of elements, or over all finite subsets of $\mathbb{N}$, can be expressed in PA using the bijections B_k and B defined in Appendix 3. So can statements that involve quantification over all recursive functions, or all recursive sets, or all RE sets. But statements that involve quantification over all subsets of $\mathbb{N}$ cannot be expressed in PA, since such statements are second-order. However, statements of this type rarely occur in number theory.

There are theorems of number theory that seem especially difficult to state in PA. Consider the **prime number theorem**, proved independently by Jacques Hadamard and Charles de la Vallée-Poussin in 1896. Let $\pi(n)$ denote the number of prime numbers that are less than or equal to n. Then the prime number theorem states that, as $n \to \infty$, $\pi(n)$ is close to $n/\ln(n)$. More precisely,

$$\lim_{n\to\infty} \frac{\pi(n)\cdot\ln(n)}{n} = 1.$$

In other words: for large n, the probability that a natural number less than n is prime is approximately $1/\ln(n)$.

At first glance, it might seem unlikely that this theorem can be stated in PA. But in fact it can. The theorem has the form $\lim_{n\to\infty} f(n) = 1$, where f is a function from $\mathbb{N}$ to $\mathbb{R}$. As we indicated in Section 3.2, even a function as complex as this f is PR, except for the problem of rounding off outputs. Specifically, it can be shown that there are PR functions g and h of two variables such that, whenever $g(n,k) = a$ and $h(n,k) = b$, the fraction a/b matches $f(n)$ to k decimal places. The theorem then has the equivalent formulation

$$\lim_{n,k\to\infty} \frac{g(n,k)}{h(n,k)} = 1.$$

Now, since g and h are PR, their graphs are representable in PA by some formulas P and Q, with three free variables each. Using these formulas, it is not hard to state the prime number theorem in PA, in the form above that is based on g and h.

Exercise 10. Using this P and Q, write out a statement of PA that expresses the prime number theorem in the form suggested. For simplicity, you may assume that inequalities are allowed in the language of PA. But your solution should not include fractions or absolute values.

Thus we see that Peano arithmetic provides a rich enough language and theory to express some very complex mathematical statements, even ones that do not seem to be strictly about whole numbers. PA is also powerful enough to prove almost all mathematical theorems that can be stated in its language. For instance, the prime number theorem can be proved in PA. So can the above-mentioned theorem that every natural number can be written as the sum of four squares. Fermat's last theorem is an interesting case. As of 2003, logicians were not sure whether Wiles's difficult proof could be translated into PA. But most specialists believe that some proof of the theorem will be found that can be carried out in PA. It would be rather remarkable if this turned out not to be the case.

By the way, it can be shown that PA is equivalent to ZF set theory with the axiom of infinity replaced by its negation. (Here, "equivalent" means that there is a "nice" bijection between formulas of these two theories that maps theorems to theorems, in both directions.) In essence, this means that PA may be viewed as a theory about natural numbers and/or hereditarily finite sets.

Ramsey's theorems and the Paris–Harrington results

We will devote most of this subsection to describing what is considered by many to be the first example of a reasonably "natural" number-theoretic theorem that cannot be proved in PA. In order to understand this result, we must first understand the **finite Ramsey's theorem**.

Frank Ramsey (1903–1930) showed great promise in many fields at an early age. When he was just twenty-one, he became only the second person ever to be elected to a fellowship at King's College, Cambridge, who had not previously studied at King's. He later became a Director of Studies in Mathematics at King's College. He was an excellent lecturer, and he published in many fields: foundations of mathematics, combinatorics, probability, economics, and philosophy. Among mathematicians, he is remembered primarily for his work in combinatorics, but his great love was actually philosophy. His tragic death at age twenty-six, from an attack of jaundice, deprived the twentieth century of one of its finest minds. In any case, fame did come to his family: his brother Michael became the Archbishop of Canterbury.

Ramsey's mathematical research was primarily within the logicist school of Russell and Whitehead. He accepted the notion that mathematics is a part of logic, and made several improvements to the axiomatization found in *Principia Mathematica*. He attacked intuitionism, calling it the "Bolshevik menace of Brouwer and Weyl," and said that "Brouwer would refuse to agree that either it was raining or it was not raining, unless he had looked to see." He also criticized Hilbert's formalism as an attempt to reduce mathematics to "a meaningless game with marks on paper." (Hilbert had in fact used this phrase, but only to emphasize a point about formalism and the axiomatic method. As the leading mathematician of the time, Hilbert certainly did not believe this literally.) Ramsey's combinatorial theorem also came about via his work in foundations: he was trying to solve a special case of the decision problem for first-order logic. This theorem and its consequences have become more important than any of Ramsey's direct work in foundations.

A cute classic problem is to prove that in any set of six people, there must be a subset of three of them who all know each other, or a subset of three none of whom know each other. For this to work, we must assume that the relation "x knows y" is defined (true or false) for all pairs of people, and symmetric: x knows y if and only if y knows x. We can state this problem more formally as follows: for any set A and natural number n, let $A^{(n)}$ denote the collection of all subsets of A with exactly n elements. (Note how $A^{(n)}$ differs from A^n, which is a set of *ordered* n-tuples.) Then what this problem says is that if A is any set with six elements and $A^{(2)}$ is partitioned into two sets K and U (K consisting of pairs who know each other, in the "people version" of the problem), then there is a $B \subseteq A$ such that B has three elements and B is **homogeneous** with respect to this partition—meaning that $B^{(2)} \subseteq K$ or $B^{(2)} \subseteq U$. This is one of the simplest cases of a theorem proved by Frank P. Ramsey.

Exercise 11. Prove this case of Ramsey's theorem. (Hint: Consider any one person out of the six. This person must either know three others in the group, or be unacquainted with three others.)

Exercise 12. Show that this theorem fails for sets of five people instead of six.

Technically, a partition of $A^{(n)}$ into m subsets is simply a function $f : A^{(n)} \to m$, and a subset B of A is homogeneous with respect to f if f is constant on $B^{(n)}$. To state the full (finite) Ramsey's theorem, it is helpful to use a special notation. For natural numbers j, k, m, and n, we write

$$j \to (k)^n_m$$

to mean that for any partition of $j^{(n)}$ into m subsets, there is a k-element subset of j that is homogeneous with respect to this partition. In this notation, our problem involving six people says that $6 \to (3)^2_2$, and the last exercise states that $5 \not\to (3)^2_2$. (It makes no difference whether this notation is defined to be about the set j only, or more generally about all j-element sets A.)

Theorem 4.19 (Ramsey). *For any k, m, and n, there is a j such that $j \to (k)^n_m$.*

The proof of this result, which we will not provide, can be carried out in PA. The theorem asserts the existence of a total function R such that $R(k, m, n)$ is the least j such that $j \to (k)^n_m$. For instance, $R(3, 2, 2) = 6$. This function is certainly recursive, but it grows very rapidly and there is no simple formula for it. In a few special cases, R does not grow too fast. For example, it can be shown that $4^{k-1} \to (k)^2_2$ for every k, so $R(k, 2, 2) \leq 4^{k-1}$.

There are also important infinite versions of Ramsey's theorem, in which j, k, m, and n are allowed to be arbitrary cardinals, not just finite ones. Here is the primary one of these, usually referred to simply as Ramsey's theorem. The proof is involved and requires the axiom of choice.

Theorem 4.20. *If m and n are any natural numbers, $\aleph_0 \to (\aleph_0)^n_m$.*

For instance, $\aleph_0 \to (\aleph_0)^2_2$ says that if we "color" every (unordered) pair of natural numbers red or blue, then there must be an infinite subset of $\mathbb{N}$ whose pairs are all the same color. The investigation of infinite Ramsey's theorems has turned out to be an extremely fruitful area of research in set theory, as we will see in Chapter 6.

Exercise 13. Show that given any infinite $A \subseteq \mathbb{N}$, there is an infinite $B \subseteq A$ such that either every number in B is a factor of every larger one, or no number in B is a factor of any other one.

In the 1970s, Jeff Paris and Leo Harrington [PH] concocted a very clever modification of the finite Ramsey's theorem that is still true and can be stated in PA, but whose proof requires the infinitary Ramsey's theorem. They were able to show that their modified version cannot be proved in PA, assuming that PA is consistent. Here is their innovation:

Definition. A nonempty finite subset B of $\mathbb{N}$ is called **relatively large** if its cardinality is equal to or greater than its smallest element.

Example 6. The set {3, 7, 16} is relatively large. The set of all integers between two million and three million is not.

Theorem 4.21 (Paris–Harrington). *Let* P *be the finite Ramsey's theorem, with the stronger requirement that the homogeneous subset that is asserted to exist must be a relatively large set with at least k elements. Then* P *is true (in other words, it can be proved in "ordinary mathematics"). But* P *is not provable in* PA, *if* PA *is consistent.*

Interestingly, the words "at least" cannot be omitted here. We omit the proof, except that we will show how to derive P from the infinite Ramsey's theorem in Section 5.3, and this establishes that P is true. Of course, this also gives the finite Ramsey's theorem as a corollary of the infinite one. The second claim of Theorem 4.21 is the more substantial one. To prove it, Paris and Harrington showed that $\mathrm{P} \to \mathrm{Con}_{\mathrm{PA}}$ is provable in PA. By the way, for each fixed n, in the notation introduced earlier, the Paris–Harrington refinement is provable in PA.

You might object that the idea of a relatively large set is quite artificial, and therefore the Paris–Harrington modification of Ramsey's theorem is not a very "natural" statement. This is a valid point, but the combinatorial statement P is still much more in the realm of ordinary mathematics than Gödel's independent statements, which involve metamathematical techniques such as coding formal languages into arithmetic and are based on the unusual concept of self-reference.

Since the Paris–Harrington result, there have been others along the same lines. In 1982 Paris and Laurence Kirby proved that **Goodstein's theorem**, a surprising number-theoretic fact, is independent of PA. Goodstein's theorem pertains to a certain class of functions on $\mathbb{N}$. More precisely, there is a denumerable collection of Goodstein's functions G_n. By definition, $G_n(0) = n$, and the values of $G_n(k)$ for $k > 0$ are determined by the same inductive rule for all the G_n's. The theorem asserts that these functions, most of which grow extremely rapidly for a long time as k increases, actually converge to zero as $k \to \infty$. For example, $G_{19}(0) = 19$, $G_{19}(1) = 7625597484990$, and $G_{19}(6)$ has over fifteen million digits! The function G_4 does not increase very rapidly at first, but the smallest k for which $G_4(k) = 0$ is still greater than 10^{121}. The proof of Goodstein's theorem is an ingenious argument using countable ordinals and the fact that they are well ordered. What is particularly appealing about the Paris–Kirby result is that the original Goodstein's theorem, with no artificial changes, is shown to be independent of PA. An elementary and very readable treatment of Goodstein's theorem (but not the independence result) is given in [Hen].

CHAPTER 5
Model Theory

5.1 Introduction

Why should students of mathematics want to know something about model theory? Here is one answer: model theory helps one to understand what it takes to specify a mathematical structure uniquely.

Let us consider what is undoubtedly the most well-known problem of this sort, the problem of defining or characterizing the field of real numbers. (These are really two separate problems; we are more concerned with characterization.) This was a major concern in the eighteenth and nineteenth centuries. One of the main reasons it took nearly two centuries after the invention of calculus to put the subject on a firm theoretical foundation is that there was no satisfactory characterization of the reals.

So, how do we characterize the reals? The first part of the answer is that we expect the reals, unlike more "meager" structures such as the integers, to allow subtraction and division as well as addition and multiplication. That is, we expect $\mathbb{R}$ to be a field. We also believe that the reals correspond to points on a line; this requires $\mathbb{R}$ to be an ordered field.

The property of being an ordered field does not specify the structure of $\mathbb{R}$ uniquely, because $\mathbb{Q}$ is also an ordered field and doesn't even have the same cardinality as $\mathbb{R}$. What are some additional properties that we might require of $\mathbb{R}$ to set it apart from $\mathbb{Q}$? In terms of algebraic properties, we also expect $\mathbb{R}$ to be a **real-closed** ordered field, that is, an ordered field in which every nonnegative number has a square root and every polynomial of odd degree has a zero. Another nice way of characterizing real-closed ordered fields is as ordered fields satisfying the intermediate value theorem for polynomials.

The defining properties of a real-closed ordered field are easy enough to state in the standard first-order language of an ordered ring with unity (with symbols $+$, $\cdot$, $-$, 0, 1, and $<$), although the part about polynomials requires an infinite list of axioms. In Section 5.5 we will see that this theory is complete, and therefore becomes inconsistent if any more sentences are added to it. Does this perhaps suggest that the property of being a real-closed ordered field characterizes $\mathbb{R}$ uniquely? It turns out that there is no way this could be so. One of the most important results in model theory, the **Löwenheim–Skolem–Tarski (LST) theorem**, guarantees that if a countable set of first-order sentences is satisfied by some infinite structure, then there are structures of every infinite cardinality that satisfy these sentences. So no countable list of first-order properties can "pinpoint" $\mathbb{R}$.

Yet, we know that it is possible to characterize $\mathbb{R}$ uniquely. The usual way is to add to the ordered field axioms the **completeness property** of Dedekind, that every nonempty set of reals with an upper bound has a least upper bound. But the completeness property is second-order: it refers to arbitrary *sets* of real numbers as well as individual numbers. Thus the LST theorem no longer applies. It is worthwhile to realize that the completeness property of the reals is a fundamentally more complex statement than the properties that define real-closed ordered fields, and cannot be replaced by any list of first-order properties. This situation, in which important properties cannot be stated accurately in first-order logic, is reminiscent of the first two examples in Section 1.5.

Another good reason to know something about model theory is that it has many applications outside of foundations, notably in algebra. Model theory is the study of **structures** for first-order languages, and all of the objects studied in abstract algebra—groups, rings, fields, modules, vector spaces, etc.—are structures in this technical sense. Furthermore, several of the main techniques within contemporary model theory, such as the analysis of the **definable** sets of structures, generalize methods that appear in many guises throughout algebra. Thus, model theory creates a means for illuminating and generalizing a variety of concepts from different parts of abstract algebra, and also provides some powerful tools for obtaining results in algebra as well as other branches of mathematics. We will see examples of algebraic applications of model theory in sections 5.4 through 5.7.

For a more thorough treatment of model theory, see [Hod] or [CK]. The "bible" of the subject is [CK], while [Hod] is a somewhat lighter introduction. [Mar] and [HPS] are good sources for model theory and its applications to algebra and other branches of mathematics.

5.2 Basic concepts of model theory

In this chapter, P and Q always denote formulas and T denotes a theory, in some first-order language $\mathcal{L}$ with equality. Here is the most basic concept of first-order model theory:

Definition. A **structure** $\mathfrak{A}$ for a language $\mathcal{L}$ (**$\mathcal{L}$-structure** for short) consists of:

(1) a nonempty set A or $|\mathfrak{A}|$, called the **universe** of $\mathfrak{A}$, to be used as the domain of the variables of $\mathcal{L}$.

(2) a subset of A^n assigned to each n-ary relation symbol in $\mathcal{L}$.

(3) a function from A^n to A assigned to each n-ary function symbol in $\mathcal{L}$.

(4) an element of A assigned to each constant symbol in $\mathcal{L}$.

In this chapter, $\mathfrak{A}$ and $\mathfrak{B}$ always denote structures, while $\mathcal{C}$ denotes a class of structures. We must often refer to a class of structures because many important collections of structures are not sets. For example, the collection of all groups is a proper class, as is the collection of all finite groups.

Example 1. The simplest first-order language is the language of pure identity, whose only atomic formulas are equations between variables. A structure for this language is just a nonempty set.

Example 2. Let $\mathcal{L}$ be the first-order language with no function or constant symbols and a single binary relation symbol R. This language is appropriate for many theories, including orderings (reflexive and irreflexive, partial and total), equivalence relations, and set theory. Of course, we would use a different abbreviation for atomic formulas $R(v_i, v_j)$ in each of these theories (probably $v_i \leq v_j$ or $v_i < v_j$ for an ordering, $v_i \sim v_j$ or $v_i \equiv v_j$ for an equivalence relation, and $v_i \in v_j$ for set theory). An $\mathcal{L}$-structure is simply any pair (A, S), where $A \neq \emptyset$ and $S \subseteq A \times A$.

An $\mathcal{L}$-structure provides a realization or interpretation of every relation symbol, function symbol, and constant symbol in $\mathcal{L}$, as well as for every bound variable appearing in an $\mathcal{L}$-formula. So it should be possible to define what it means for a given $\mathcal{L}$-formula to be true or false in a given $\mathcal{L}$-structure, as long as we also provide values for the free variables of the formula. Accordingly:

Definition. An **assignment** in $\mathfrak{A}$ is a function $g : V \to A$, where V is the set of variables of $\mathcal{L}$.

If g is an assignment, we write g^i_x to denote the assignment whose value on v_i is x, and which is otherwise identical to g.

Lemma 5.1. *If g is an assignment in an $\mathcal{L}$-structure $\mathfrak{A}$, there is a unique function $\hat{g} : \mathcal{T} \to A$, where $\mathcal{T}$ is the set of all terms of $\mathcal{L}$, satisfying:*

(a) *If t is a variable of $\mathcal{L}$, then $\hat{g}(t) = g(t)$.*

(b) *If t is a constant symbol of $\mathcal{L}$, then $\hat{g}(t)$ is the element of A corresponding to that constant symbol.*

(c) *If t is a term of the form $f(t_1, t_2, \ldots, t_n)$ then $\hat{g}(t)$ is*

$$F(\hat{g}(t_1), \hat{g}(t_2), \ldots, \hat{g}(t_n)),$$

where F is the function in $\mathfrak{A}$ that corresponds to f.

We will refer to $\hat{g}(t)$ as the **interpretation** of the term t based on the assignment g. We omit the proof of this lemma, which is a straightforward induction on the number of function symbols appearing in the term t. In fact, it would be reasonable just to *define* $\hat{g}$ without mentioning that anything needs to be proved. We will take this approach in the next definition.

Example 3. Let $\mathcal{L}$ be the language of a ring, $\mathfrak{A}$ the ring of real numbers, t the term $v_2 \cdot (v_5 + v_1)$, and g any assignment in which $g(v_1) = 5$, $g(v_2) = 3$, and $g(v_5) = 12$. Then $\hat{g}(t) = 3(12 + 5) = 51$.

Definition. Let g be an assignment in an $\mathcal{L}$-structure $\mathfrak{A}$, and let P be any formula of $\mathcal{L}$. Then we can define what it means for P to be **true** in $\mathfrak{A}$ (or **satisfied** by $\mathfrak{A}$) under the assignment g, using a straightforward inductive definition on the structure of P:

(1) If P is an equation $t_1 = t_2$, then P is true in $\mathfrak{A}$ under g iff $\hat{g}(t_1) = \hat{g}(t_2)$.

(2) If P is of the form $R(t_1, t_2, \ldots, t_n)$, where R is an n-ary relation symbol of $\mathcal{L}$, then P is true in $\mathfrak{A}$ under g iff

$$(\hat{g}(t_1), \hat{g}(t_2), \ldots, \hat{g}(t_n)) \in \hat{R},$$

where $\hat{R}$ is the subset of A^n that corresponds to R.

(3) If P is of the form $\sim$ Q, then P is true in $\mathfrak{A}$ under g iff Q is *not* true in $\mathfrak{A}$ under g.

(4) If P is of the form $\mathrm{Q}_1 \wedge \mathrm{Q}_2$, then P is true in $\mathfrak{A}$ under g iff Q_1 and Q_2 are both true in $\mathfrak{A}$ under g.

(5) If P is of the form $\forall v_i \mathrm{Q}$, then P is true in $\mathfrak{A}$ under g iff Q is true in $\mathfrak{A}$ under the assignment g^i_x, for every x in A.

Recall that we don't need to include clauses for the other connectives and $\exists$, because these are all definable from $\sim$, $\wedge$, and $\forall$.

Notation. We write $\mathfrak{A} \models \mathrm{P}[g]$ to mean that P is true in $\mathfrak{A}$ under the assignment g. We also write $\mathfrak{A} \models T[g]$ to mean that $\mathfrak{A} \models \mathrm{P}[g]$ for every $\mathrm{P} \in T$. In this notation, the symbol $\models$ is usually read "satisfies."

We say that P (or T) is **satisfiable** if it is satisfied by *some* $\mathfrak{A}$ and g, and **valid** if it is satisfied by *every* $\mathfrak{A}$ and g.

A straightforward induction on the structure of P shows that the truth of $\mathfrak{A} \models \mathrm{P}[g]$ depends not on all values of g, but only on the values of g on the free variables of P. In particular, if P is a sentence, then g is irrelevant and we simply write $\mathfrak{A} \models \mathrm{P}$, and say that $\mathfrak{A}$ is a **model** of P (or $\mathfrak{A}$ satisfies P, or P is true in $\mathfrak{A}$). In the same vein, if T is a set of sentences we write $\mathfrak{A} \models T$ instead of $\mathfrak{A} \models T[g]$, and we say that $\mathfrak{A}$ is a **model** of T.

Notation. We write $T \models \mathrm{P}$ to mean that whenever $\mathfrak{A} \models T[g]$, we also have $\mathfrak{A} \models \mathrm{P}[g]$.

Two ways to read this notation are to say that P is a **consequence** of T, or T **entails** P, since it says that whenever all the formulas in T are true, so is P. If P and all the formulas in T are sentences, $T \models \mathrm{P}$ says that every model of T is also a model of P. Even though the symbol $\models$ appears in both $\mathfrak{A} \models \mathrm{P}$ and $T \models \mathrm{P}$, these notations have very different meanings. $\mathfrak{A} \models \mathrm{P}$ simply says that P is true in the structure $\mathfrak{A}$. $T \models \mathrm{P}$ is a more complex statement involving all possible structures.

Example 4. Let $\mathcal{L}$ be the first-order language with no relation or constant symbols and a single binary function symbol $\cdot$. Then an $\mathcal{L}$-structure $\mathfrak{A}$ is simply a nonempty set A together with a function from A^2 to A.

Now let T be the first-order theory of a group, that is, the usual list of defining properties or axioms of a group (as in Appendix D). Then $\mathfrak{A} \models T$ says that every sentence of T is true in $\mathfrak{A}$. In other words, it says that $\mathfrak{A}$ is a group. Similarly (in a different language), a model of ring theory is simply a ring, a model of field theory is a field, etc. So there's nothing mysterious about the notion of a model of a theory.

Let P be the commutative law. Then it is clear that $T \not\models \mathrm{P}$, because not every group is abelian. Similarly, $T \not\models \sim \mathrm{P}$.

Finally, let $\mathrm{Q}(v_0)$ be the formula $\forall v_1 (v_0 \cdot v_1 = v_1 \cdot v_0)$, and let g be an assignment. Then $\mathfrak{A} \models \mathrm{Q}[g]$ if and only if $g(v_0)$ commutes with every element of A under the binary operation of $\mathfrak{A}$. If $\mathfrak{A}$ is a group, this says that $g(v_0)$ is in the **center** of $\mathfrak{A}$.

Example 5. The **standard model of arithmetic** is the structure $\mathfrak{N} = (\mathbb{N}, +, \cdot, S, 0)$. Here, a symbol like + denotes an actual function on $\mathbb{N}$; it is not a symbol of a formal language. (This sort of ambiguous usage is quite common, and at times one must be very careful to avoid difficulties that could occur from it.) $\mathfrak{N}$ is of course a structure for the language of Peano arithmetic. It is not trivial to prove that $\mathfrak{N} \models \mathrm{PA}$, but it is not very difficult.

The next example and the following two exercises examine whether certain axioms are true in structures that are not "intended" to be models of those axioms. This is good practice because it forces us to carefully examine the satisfaction relation without preconceived notions.

Example 6. Let $\mathcal{L}$ be the first-order language of set theory and let $\mathfrak{A}$ be the $\mathcal{L}$-structure $(\mathbb{R}, <)$. If P is the extensionality axiom of ZF, then the interpretation of P in $\mathfrak{A}$ is

$$\forall x, y \in \mathbb{R}[x = y \leftrightarrow \forall u \in \mathbb{R}(u < x \leftrightarrow u < y)].$$

This is clearly true, so $\mathfrak{A} \models \mathrm{P}$. If Q is the pairing axiom, the interpretation of Q in $\mathfrak{A}$ is

$$\forall x, y \in \mathbb{R} \exists z \in \mathbb{R} \forall u \in \mathbb{R}(u < z \leftrightarrow u = x \vee u = y).$$

This is clearly false, so $\mathfrak{A} \not\models \mathrm{Q}$. In fact, if the quantifiers on x and y were removed from Q, it would be false in $\mathfrak{A}$ under every assignment.

Exercise 1. Let $\mathfrak{A}$ be $(\mathbb{R}, <)$, as in Example 6. Show that the union axiom is true in $\mathfrak{A}$, but the empty set and power set axioms are not.

Exercise 2. Which proper axioms of ZFC are true in the structure $(\mathbb{N}, <)$?

Exercise 3. Show that all the proper axioms of ZFC except the axiom of infinity are true in the structure $(V_\omega, \in)$. Don't try to do this very rigorously, as it could get quite tedious. You may use Proposition 2.24 and other results from the last part of Chapter 2.

5.3 The main theorems of model theory

In this section we present the three theorems that form the foundation of first-order model theory: the completeness theorem, the compactness theorem, and the Löwenheim–Skolem–Tarski theorem.

In Chapter 1 we defined $T \vdash \mathrm{P}$ to mean that P can be deduced from T in first-order logic. This syntactic "single turnstile" relation, based on the notion of a formal proof, is conceptually very different from the semantic "double turnstile" relation $T \models \mathrm{P}$, which is based on the much more abstract concept of truth in structures. One of the most appealing features of first-order logic is that the two "turnstiles," which are the two reasonable notions of logical consequence, actually coincide:

Theorem 5.2 (Gödel's Completeness Theorem). *A set of formulas is satisfiable if and only if it is consistent.*

Proof. The forward direction is straightforward and is often stated separately as the **soundness theorem** for first-order logic. To prove it, assume that T is satisfiable, so that $\mathfrak{A} \models T[g]$ for some $\mathfrak{A}$ and g. Now consider any proof $\mathrm{Q}_1, \mathrm{Q}_2, \ldots, \mathrm{Q}_n$ from T. For any Q_i that is in T, we have $\mathfrak{A} \models \mathrm{Q}_i[g]$ by assumption. Next, it is straightforward to show that every axiom of first-order logic is true in every structure, under every assignment. So $\mathfrak{A} \models \mathrm{Q}_i[g]$ for any Q_i that is a logical axiom. Finally, whenever both P and $\mathrm{P} \to \mathrm{Q}$ are true, then Q must also be true, so modus ponens preserves truth in structures. Thus, by what is essentially induction on the length of a proof, every theorem of T is true in $\mathfrak{A}$ under g. But a contradiction can never be true in a structure, by the definition of $\models$, so T is consistent.

The reverse direction of the proof is complicated, so we will provide only a very bare outline of it. The full proof of both directions can be found in almost any text on mathematical logic or model theory, such as [End]. So assume that T is a consistent theory. For simplicity we first outline the proof under the additional assumption that the language $\mathcal{L}$ of T is denumerable:

(1) Add a denumerable set of new constant symbols $\{c_1, c_2, c_3, \dots\}$ to $\mathcal{L}$ to form a new language $\mathcal{L}'$. These constants are often described as "witnesses" for the elements of the model of T to be constructed. They are essentially Skolem constants, in the sense of Section 1.6.

(2) Create a list $\{\mathrm{P}_1, \mathrm{P}_2, \mathrm{P}_3, \dots\}$ of all $\mathcal{L}'$-formulas whose only free variable is v_0.

(3) Inductively define another list of formulas $\{\mathrm{Q}_1, \mathrm{Q}_2, \mathrm{Q}_3, \dots\}$, where Q_n is $\exists v_0 \mathrm{P}_n \rightarrow \mathrm{P}_n(c_k)$. Here, c_k is the first of the new constants that does does not appear in P_n or in any Q_m with $m < n$. Also, $\mathrm{P}_n(c_k)$ means the result of replacing every free occurrence of v_0 in P_n by c_k. The Q_n's are called **Henkin axioms**.

(4) Let T' consist of T and all the Q_n's. It is not hard to prove that T' is still consistent: each Q_n says that if a property holds for at least one object, it holds for the object determined by some new constant symbol. Intuitively, there is no way that statements of this form can lead to a contradiction.

(5) Extend T' to a complete $\mathcal{L}'$-theory T''. To do this, simply list all the sentences of $\mathcal{L}'$ and, starting with T', inductively add each sentence to the theory as long as its negation can't be proved from the theory constructed so far.

(6) Now let $\mathcal{T}$ be the set of all variable-free terms of $\mathcal{L}'$. We would like to use $\mathcal{T}$ as the universe of a model of T'' (and therefore a model of T). In a sense, this is easy, since the completeness of T'' means that T'' determines exactly how all the relation, function, and constant symbols of $\mathcal{L}'$ must be interepreted on elements of $\mathcal{T}$.

(7) However, there is one problem with the previous step: there might be distinct terms t_1 and t_2 in $\mathcal{T}$ such that the equation $t_1 = t_2$ is provable in T''. Then t_1 and t_2 must be interpreted as the same object in any model of T''. To rectify this, consider the equivalence relation on $\mathcal{T}$ defined by $T'' \vdash (t_1 = t_2)$. Then define A to be the set of all equivalence classes. Because of the "substitution of equals" axiom of predicate logic, it is clear that the relations, functions, and constants of $\mathcal{L}'$ are well-defined on A. Thus, we obtain

an $\mathcal{L}'$-structure with universe A, which can be shown to be a model of T''. Hence T'' is satisfiable, and so is T.

The proof when T is uncountable is similar, except that transfinite induction must be used in step (3), and some form of the axiom of choice is needed in step (5) (and perhaps step (2)). ■

For future reference, note that the cardinality of the model constructed in this proof is no greater than the cardinality of $\mathcal{L}$, since $\mathcal{T}$ has the same cardinality as $\mathcal{L}$. (When we refer to the cardinality of a structure, we mean the cardinality of its universe.)

Corollary 5.3. *A formula is valid if and only if it is a law of logic.*

Exercise 4. Prove this corollary.

Corollary 5.4. *For any T and* P, *$T \vdash$* P *if and only if $T \models$* P.

Exercise 5. Prove this corollary. Use Exercise 7(b) of Section 1.4.

This corollary is also referred to as Gödel's completeness theorem. The two versions are easily shown to be equivalent. Note that the completeness theorem does not say that first-order logic, or any particular first-order theory, is complete in the sense of Section 1.4. It is certainly not true that every sentence or its negation is provable in first-order logic. We have instead the weaker result that every valid sentence is provable.

Definition. Two $\mathcal{L}$-structures $\mathfrak{A}$ and $\mathfrak{B}$ are **elementarily equivalent**, denoted $\mathfrak{A} \equiv \mathfrak{B}$, if they satisfy the same sentences of $\mathcal{L}$.

Elementary equivalence is a relatively weak condition, as we will soon see.

Corollary 5.5. *A set of sentences is complete if and only if it has models and all of its models are elementarily equivalent.*

Exercise 6. Prove this corollary.

Theorem 5.6 (Compactness Theorem).

(a) *If every finite subset of a theory T is satisfiable, then T is satisfiable.*

(b) *If $T \models$ P, then there is a finite subset T_0 of T such that $T_0 \models$ P.*

Proof.

(a) Suppose that every finite subset of T is satisfiable. So every finite subset of T is consistent, by the completeness theorem ("soundness"). But then T must be consistent, because a proof only has a finite number of steps. Thus, by the completeness theorem, T is satisfiable.

The proof of (b) is similar. ∎

Convention. For the rest of this chapter it will be understood, unless stated otherwise, that a theory means a set of *sentences*.

Corollary 5.7. *If a first-order theory has arbitrarily large finite models, then it has an infinite model.*

Proof. Suppose that T has arbitrarily large finite models. Form a new theory T' by adding to T a denumerable set of new constant symbols $\{c_n \mid n \in \mathbb{N}\}$, and the axiom schema $\{c_m \neq c_n \mid m \neq n\}$. Clearly, every finite subset of T' has a model because we can interpret any finite set of the c_n's as distinct elements in a sufficiently large finite model of T. Therefore T' has a model by the compactness theorem, and any such model must be infinite because all the c_n's must be interpreted as distinct elements. ∎

This corollary establishes that in first-order logic, there is no way to state precisely, even with an infinite number of axioms, that there are a finite number of elements in the domain. It follows that there is no single axiom that expresses that there are an infinite number of elements. Note the similar point made in the next to last paragraph of Example 17 of Section 1.5. We will expand on this theme in the first part of Section 5.7.

Corollary 5.8. *If P is true in every infinite model of T, then P is also true in all sufficiently large finite models of T.*

Exercise 7. Prove this corollary from the previous one.

So now we know that every first-order sentence that is true in all infinite groups (or rings, or fields) is also true in all sufficiently large finite ones. By similar reasoning, we can show that every sentence that is true in all fields of characteristic 0 is also true in all fields of sufficiently large finite characteristic.

The compactness theorem is a very powerful tool. One of its most important applications, which will be the subject of Chapter 7, is the following: let T be some first-order theory in which one can carry out calculus or real analysis. T might be ZFC set theory, or it could be something more limited. Form a new theory T' by adding a new constant symbol h to the language of T, and then add to T the new axioms $h \in \mathbb{R}$, $h > 0$, and the schema $\{h < 1/n \mid n \in \mathbb{Z}^+\}$. If T is consistent, it is clear that every finite subset of T' is satisfiable, because we can always interpret h as a positive real number less than any given finite set of positive fractions. Therefore T' is satisfiable. But in a model of T', h must be interpreted as a positive real number that is less than every number of the form $1/n$. In other words, h must be a positive **infinitesimal** in such a model. This rather simple idea led to the development of **nonstandard analysis**.

To close our discussion of the compactness theorem, here are a couple of combinatorial applications of it. First, let's keep the promise that was made at the end of Chapter 4:

Proposition 5.9. *The infinite Ramsey's theorem (4.20) implies the statement* P *of Theorem 4.21 (and hence the finite Ramsey's theorem).*

Proof. Assume Theorem 4.20. Let natural numbers k, m, and n be given; we treat them as fixed. Recall that PA has numerals, terms that denote specific natural numbers. We will define a theory T, in the language of PA with an additional n-ary function symbol g. The proper axioms of T are:

(1) The usual proper axioms of PA.

(2) The statement that the value of g does not depend on the order of its arguments. For instance, if $n = 2$, this becomes $\forall x, y, [g(x, y) = g(y, x)]$. For $n > 2$, the number of equations increases. This statement implies that g defines a function on *unordered* n-tuples. (We don't care about g's values on n-tuples with repeated values.)

(3) The following axiom schema: for each natural number r, the statement that g's values on n-tuples of distinct numbers less than r are all less than m. For each r, this axiom implies that g defines a partition of $r^{(n)}$ into m subsets.

Now, let Q be the statement that says that there is a relatively large set B with at least k elements such that g maps all of $B^{(n)}$ to a fixed number less than m. Q can be expressed in the language of T using the bijection B of Appendix C. We claim that $T \models$ Q. To see this, first note that the universe of any model of PA has a subset that can be identified with ω within the model, namely the elements corresponding to the terms $\overline{0}, \overline{1}, \overline{2}, \ldots$. Then axioms (2) and (3) of T guarantee that g defines a partition of $\omega^{(n)}$ into m subsets. Therefore, by the infinite Ramsey's theorem, there must be an infinite homogeneous subset of ω with respect to g. And any infinite subset S of ω must contain a relatively large set with at least k elements: just take the first $k + s$ elements of S, where s is the smallest number in S.

By compactness, there is a finite subtheory T' of T such that $T' \models$ Q. Let j be the largest number r such that T' includes the statement of axiom schema (3) for r. To see that this j works to prove the desired result, suppose f is any partition of $j^{(n)}$ into m subsets. If we take the standard model of arithmetic, use f to interpret g on all n-tuples of distinct numbers less than j, and let g's value equal m on all other n-tuples, we obtain a model of T'. Since Q holds in this model, there is a relatively large homogeneous set with at least k elements, with respect to f. ∎

Exercise 8. Give a simpler proof than the one just given (still using compactness) of the regular finite Ramsey's theorem from the infinite one. Your theory T need not include any of the axioms of PA.

Definitions. A **(nondirected) graph** with domain G is a pair (G, R), where R is any subset of $G^{(2)}$. Think of $\{x, y\} \in R$ as meaning that the "vertices" x and y are "connected."

A **(finite) subgraph** of (G, R) is a pair of the form $(H, R \cap H^{(2)})$, where H is a (finite) subset of G.

A k-coloring of a graph (G, R), where $k \in \mathbb{Z}^+$, is a function $g : G \to \{1, 2, \ldots, k\}$ such that $g(x) \neq g(y)$ whenever $\{x, y\} \in R$.

Theorem 5.10 (de Bruijn). *If every finite subgraph of a graph has a k-coloring, then so does the whole graph.*

Exercise 9. Prove de Bruijn's theorem. You can use an argument similar to, but simpler than, the proof of the previous proposition. Note that here the compactness theorem allows us to go from a finite version of a result to an infinite one, whereas the previous argument works in the opposite direction.

When presented to nonmathematicians, the four-color theorem is usually stated in terms of maps, but it has an equivalent formulation in terms of graphs: every finite planar graph can be 4-colored. (A planar graph is one that can be drawn in a plane, with connections between vertices shown as nonintersecting continuous curves. For example, it's certainly possible for five points to be connected to each of the others in a graph, but this cannot occur in a planar graph.) By de Bruijn's theorem, it follows that every planar graph can be 4-colored, and so the four-color theorem also holds for maps with an infinite number of countries.

The Löwenheim–Skolem–Tarski theorem

We now present the final cornerstone of first-order model theory. Chronologically, it was actually the first—Löwenheim proved the simplest version of it in 1915, well over a decade before the completeness and compactness theorems were obtained.

Theorem 5.11 (Löwenheim–Skolem–Tarski (LST) Theorem). *Let T be a theory in a first-order language $\mathcal{L}$. If T has an infinite model,* *then it has a model of every cardinality equal to or greater than $Card(\mathcal{L})$.*

Proof. Once again, we just outline the proof. The first task is to prove that if T is any satisfiable theory in a first-order language $\mathcal{L}$, then T has a model of cardinality $Card(\mathcal{L})$ or smaller. This follows directly from the construction used to prove Gödel's completeness theorem—note the remark about cardinality following our outline of that construction. Löwenheim and Skolem's original argument was somewhat simpler, because they could construct a model of "Skolem terms" within a given model, instead of needing to build a model from scratch.

For clarity, this result is often called the *downward* Löwenheim–Skolem theorem. Löwenheim proved it for finite T, and Skolem extended it to denumerable theories a few years later. Uncountable theories were not considered at that time. In the next section we will state a stronger version of the downward theorem (Theorem 5.13).

The full LST theorem follows from the downward theorem by the following argument: assume that T is a theory in some language $\mathcal{L}$ and that T has an infinite model. Let κ be any cardinal with $\kappa \geq Card(\mathcal{L})$. We use a simple adaptation of the proof of Corollary 5.7: instead of a denumerable set of new constant symbols, introduce κ new constant symbols and the axiom schema that says they are all distinct. Call this expanded theory and language T' and $\mathcal{L}'$, respectively. Note that $Card(\mathcal{L}') = \kappa$. By the compactness theorem, T' is satisfiable. Therefore, by the downward Löwenheim–Skolem theorem, T' has a model of cardinality $\leq \kappa$. But any model of T' clearly has cardinality $\geq \kappa$. Therefore, T' has a model of cardinality κ. ∎

Independent of the downward theorem, the argument in the last paragraph of this proof shows that if a first-order theory has an infinite model, then it has models of arbitrarily large cardinality. This is sometimes called the *upward* Löwenheim–Skolem theorem, but it was actually proved by Tarski and Robert Vaught many years later. In fact, Skolem didn't even accept this result, since he did not believe in the existence of uncountable sets.

Corollary 5.12. *Let κ be an infinite cardinal and let $\mathfrak{A}$ be an infinite structure for a countable language. Then there is a structure of cardinality κ that is elementarily equivalent to $\mathfrak{A}$.*

Proof. Apply the LST theorem with T being the set of all sentences that are true in $\mathfrak{A}$. ∎

The completeness theorem shows the equivalence of the syntactic and semantic notions of consequence. By contrast, the compactness theorem and the LST theorem make no mention of $\vdash$, so they are purely semantic results. Our proof of compactness used completeness, but it is also possible to give a purely semantic proof.

As indicated in the introduction to this chapter, the LST theorem shows that first-order logic is quite limited in its ability to describe a specific structure. The above corollary makes this point more clear, and the next two examples illustrate it further:

Example 7. We would like Peano's axioms for arithmetic to define $\mathbb{N}$ (with the usual operations), and only that structure. But now we see that (first-order) PA must have uncountable models. What would an uncountable model $\mathfrak{A}$ of PA look like? In any model of PA, there must be distinct elements that are the interpretations of the terms $\overline{0}$, $S(\overline{0})$, $S(S(\overline{0}))$, etc. It is not hard to show that the addition and multiplication operations on these elements must correspond to the usual operations on $\mathbb{N}$. In other words, $\mathfrak{A}$ must have a "substructure" that looks exactly like the structure $\mathbb{N}$.

But $\mathfrak{A}$ must of course have many more elements, since it is uncountable. All of these elements must be greater than the elements corresponding to $\mathbb{N}$ in the ordering of $\mathfrak{A}$. Since it is provable in PA that every element except 0 has both an immediate sucessor and an immediate predecessor, the ordering on $\mathfrak{A}$ must look like "a copy of $\mathbb{N}$ followed by an uncountable number of copies of $\mathbb{Z}$." The addition and multiplication operations in $\mathfrak{A}$ must be defined in such a way that all the usual properties of arithmetic hold, including the Euclidean algorithm, the unboundedness of the set of prime numbers, etc., not to mention the entire first-order axiom schema of mathematical induction. It is difficult to picture such a model; in fact, it is difficult to picture any model of PA that does not look like the model $\mathbb{N}$.

Example 8. The LST theorem tells us that ZFC set theory must have countable models, if it's consistent. This seems impossible, since within any model of set theory there must be "uncountable" sets. This so-called **Skolem's paradox** disappears when one realizes that a set can satisfy the definition of being uncountable *within* a model without actually being uncountable.

More specifically, imagine a model of set theory in which the interpretation of $\mathbb{N}$ is itself (with its usual elements), and $\mathcal{P}(\mathbb{N})$ is interpreted as some set B. Within the model, B must be uncountable, but that merely means that there is no bijection between $\mathbb{N}$ and B *in the model.* B could really be countable; in fact, we know that the whole model could be countable. We will encounter several more examples of this type of phenomenon in this book. Because of this example and others like it, the downward Löwenheim–Skolem theorem was a very surprising development at the time.

5.4 Preservation theorems

In this section we present several results of first-order model theory that explain why various operations (unions, intersections, homomorphic images, etc.) on mathematical structures do or do not preserve various properties of those structures.

Recall that a function $f : A \to B$ automatically induces functions from A^n to B^n, for each $n \in \mathbb{N}$, as well as functions from $\mathcal{P}(A^n)$ to $\mathcal{P}(B^n)$. In the next three definitions, it is assumed that $\mathfrak{A}$ and $\mathfrak{B}$ are structures for the same first-order language $\mathcal{L}$.

Definition. An **isomorphism** f between $\mathfrak{A}$ and $\mathfrak{B}$ is, as usual, a bijection between A and B that preserves all structural components: for each constant symbol of $\mathcal{L}$, its interpretation in $\mathfrak{A}$ is mapped to its interpretation in $\mathfrak{B}$; and similarly for relation symbols and function symbols, using the functions induced by f.

We write $\mathfrak{A} \cong \mathfrak{B}$ to mean that $\mathfrak{A}$ and $\mathfrak{B}$ are isomorphic (that is, there is an isomorphism between them).

Example 9. It is very easy to show that isomorphic structures must be elementarily equivalent. Thus, for instance, the group of integers $(\mathbb{Z}, +)$ and the group of even integers $(2\mathbb{Z}, +)$, where $+$ denotes ordinary addition, are elementarily equivalent since they are isomorphic under the function $f(n) = 2n$.

On the other hand, the *rings* $(\mathbb{Z}, +, \cdot)$ and $(2\mathbb{Z}, +, \cdot)$ are not elementarily equivalent, since only the first one has a multiplicative identity, a property which can be stated in the first-order language of a ring. Therefore, these rings are not isomorphic.

Some standard examples of elementarily equivalent structures that are not isomorphic will be given in Example 12.

Example 10. Let $\mathfrak{A} = (\mathbb{Z} \times \mathbb{Z}, P)$, the usual direct product of the group of integers with itself. (So P is "componentwise addition.") Then $\mathfrak{A}$ is not isomorphic to the group of integers. One standard argument for this is that the group $(\mathbb{Z}, +)$ is cyclic, generated by a single element (1 or -1), while $\mathfrak{A}$ is not cyclic.

Exercise 10.

(a) Give a direct proof that there is no isomorphism between the groups $(\mathbb{Z}, +)$ and $\mathfrak{A}$ discussed in the previous example.

(b) Even though these groups are not isomorphic, the property that a group is cyclic cannot be stated in the first-order language of a group. So these groups might still be elementarily equivalent. Show that in fact they are not. (Hint: the fact that every integer is even or odd can be used to construct a first order sentence that is true in $(\mathbb{Z}, +)$ but not in $\mathfrak{A}$.)

Definition. We say that $\mathfrak{A}$ is a **submodel** or **substructure** of $\mathfrak{B}$, denoted $\mathfrak{A} \subseteq \mathfrak{B}$, if: (i) $A \subseteq B$; (ii) for each constant symbol of $\mathcal{L}$, its interpretations in $\mathfrak{A}$ and $\mathfrak{B}$ are the same; and (iii) for each relation symbol and function symbol of $\mathcal{L}$, its interpretation in $\mathfrak{A}$ is the restriction to the appropriate A^n of its interpretation in $\mathfrak{B}$. It is implicit in this definition that A must be closed under all the functions that are part of the structure $\mathfrak{B}$.

If $\mathfrak{A} \subseteq \mathfrak{B}$, we also say that $\mathfrak{B}$ is an **extension** of $\mathfrak{A}$.

Definitions. We say that $\mathfrak{A}$ is an **elementary submodel** of $\mathfrak{B}$ or $\mathfrak{B}$ is an **elementary extension** of $\mathfrak{A}$, denoted $\mathfrak{A} \preceq \mathfrak{B}$, if $\mathfrak{A} \subseteq \mathfrak{B}$ and, for every $\mathcal{L}$-formula P and every $\mathfrak{A}$-assignment g, $\mathfrak{A} \models \text{P}[g]$ if and only if $\mathfrak{B} \models \text{P}[g]$. (You may recall that we are also using the symbol $\preceq$ to compare the cardinality of sets. There is no connection between the two meanings of this symbol.)

An isomorphism between $\mathfrak{A}$ and an elementary submodel of $\mathfrak{B}$ is called an **elementary embedding** of $\mathfrak{A}$ in $\mathfrak{B}$. More concretely, an elementary embedding of $\mathfrak{A}$ in $\mathfrak{B}$ is a function f from A to B such that, for every $\mathcal{L}$-formula P and every $\mathfrak{A}$-assignment g, $\mathfrak{A} \models \text{P}[g]$ if and only if $\mathfrak{B} \models \text{P}[f \circ g]$.

The property $\mathfrak{A} \preceq \mathcal{B}$ is very strong, strictly stronger than the conjunction of $\mathfrak{A} \subseteq \mathfrak{B}$ and $\mathfrak{A} \equiv \mathfrak{B}$, because the definition of $\mathfrak{A} \preceq \mathcal{B}$ refers to formulas and assignments, whereas $\mathfrak{A} \equiv \mathfrak{B}$ pertains only to sentences. Even $\mathfrak{A} \subseteq \mathfrak{B}$ and $\mathfrak{A} \cong \mathfrak{B}$ together do not imply $\mathfrak{A} \preceq \mathfrak{B}$, as the next example shows.

Example 11. Let $\mathcal{L}$ be the first-order language of an ordering: besides equality, it has a single binary relation symbol. Consider the two $\mathcal{L}$-structures $\mathfrak{A} = (\mathbb{Z}, <)$ and $\mathfrak{B} = (2\mathbb{Z}, <)$, as defined in Appendix D. $\mathfrak{B} \subseteq \mathfrak{A}$ and $\mathfrak{B} \cong \mathfrak{A}$. But it is not the case that $\mathfrak{B} \preceq \mathfrak{A}$. For instance, the formula $\exists v_1 (v_0 < v_1 < v_2)$ is false in $\mathfrak{B}$ but true in $\mathfrak{A}$, if v_0 is assigned the value 0 and v_2 is assigned the value 2.

The following result of Tarski and Vaught is a strengthened version of the downward Löwenheim–Skolem theorem, provable by essentially the same construction.

Theorem 5.13. *Let $\mathfrak{A}$ be any structure for a first-order language $\mathcal{L}$. Then $\mathfrak{A}$ has an elementary submodel whose cardinality is no greater than that of $\mathcal{L}$.*

Example 12. It can be shown that the denumerable field $\mathbb{A}$ of complex algebraic numbers (see Theorem 8.2) is an elementary submodel of the uncountable field $\mathbb{C}$. Similarly, $\mathbb{R} \cap \mathbb{A} \preceq \mathbb{R}$.

Preservation under submodels and intersections

In the remainder of this section, we will present several **preservation properties**, all of which (with full proofs and much more detail) can be found in [CK].

Definition. Let T be a theory. T is said to be **preserved under submodels** if $\mathfrak{B} \models T$ and $\mathfrak{A} \subseteq \mathfrak{B}$ imply $\mathfrak{A} \models T$.

Similarly, we can define what it means for T to be preserved under finite intersections, preserved under arbitrary intersections, preserved under homomorphic images, etc.

Theorem 5.14 (Łoś–Tarski). *A theory is preserved under submodels if and only if it is equivalent to a set of* Π_1 *sentences.*

Exercise 11. Prove the reverse direction of this theorem. (This is the easy direction. The forward direction requires substantial proof.)

We mention the obvious "companion result":

Corollary 5.15. *A theory is preserved under extensions if and only if it is equivalent to a set of* Σ_1 *sentences.*

Now let's see some applications of this theorem. For example, is every submodel of a group also a group? It depends on what we mean by "submodel." If we express the axioms of a group in the first-order language with multiplication only, then the identity and inverse axioms require existential quantifiers, so the theorem tells us that this theory is not preserved under submodels. And, clearly, a subset of a group may be closed under multiplication without being a subgroup. But if we express the axioms of a group in the first-order language with symbols for the identity and inverses as well as multiplication, then the natural axiomatization consists of Π_1 sentences, and in this context every submodel is a subgroup. This situation applies to many types of algebraic structures:

Corollary 5.16.

(a) *Every subset of a group that contains* 1 *and is closed under multiplication and inverses is a subgroup.*

(b) *Every subset of a ring that contains* 0 *and is closed under* $+$, $-$, *and* $\cdot$ *is a subring.*

(c) *Every subset of a field that contains* 0 *and* 1 *and is closed under* $+$, $-$, $\cdot$, *and* $\div$ *is a subfield.*

Proof. Immediate from Theorem 5.14. ■

There are, of course, endless variations on this corollary. Also, note that the converses of the parts of this corollary are true by definition, so they could be stated as biconditionals: a subset of a group is a subgroup if and only if it contains 1 and is closed under multiplication and inverses, etc.

Corollary 5.17. *The intersection of any collection of subgroups of a group is also a subgroup (and similarly for rings and fields).*

Proof. It is clear that if each set in a collection is closed under a certain operation (e.g., multiplication), then so is the intersection of that collection. ■

So the well-known fact that most types of algebraic structures behave well under intersections may be viewed as a model-theoretic result. It is not hard to see why these same types of structures are not preserved under even finite unions of submodels: if two subsets of a set are closed under some *unary* function, then so is their union. But this property fails for functions of more than one variable, including binary operators such as + and $\cdot$.

Corollaries 5.16 and 5.17 also help clarify the notion of the subgroup of a given group (or subfield of a given field, etc.) **generated** by an arbitrary subset. See Appendix D for the various equivalent definitions of this concept.

Preservation under unions of chains

While unions of algebraic structures are generally not algebraic structures of the same type, unions of *chains* of algebraic structures usually are:

Definition. Let I be a well-ordered set. A collection of structures $\{\mathfrak{A}_i \mid i \in I\}$ indexed by I is called a **chain** (respectively, **elementary chain**) if $\mathfrak{A}_i \subseteq \mathfrak{A}_j$ (respectively, $\mathfrak{A}_i \preceq \mathfrak{A}_j$) whenever $i < j$.

For many applications, it suffices to consider chains in which $I = \mathbb{N}$.

If $\{\mathfrak{A}_i \mid i \in I\}$ is a chain, then we can define the new structure $\mathfrak{B} = \bigcup_{i \in I} \mathfrak{A}_i$ in the obvious way, and it is obvious that $\mathfrak{A}_i \subseteq \mathfrak{B}$ for every $i \in I$. The analogous result for elementary chains, due to Tarski, is not much more difficult to prove:

Exercise 12. Prove that if $\{\mathfrak{A}_i \mid i \in I\}$ is an elementary chain and $\mathfrak{B} = \bigcup_{i \in I} \mathfrak{A}_i$, then $\mathfrak{A}_i \preceq \mathfrak{B}$ for each $i \in I$. (Hint: The fact to be proved involves a formula P, so prove it by induction on the structure of P. The only nontrivial step is the one involving $\exists$—there is no need for a separate step involving $\forall$.) Consequently, every first-order sentence is preserved under unions of elementary chains.

Theorem 5.18 (Chang–Łoś–Suzsko). *For any theory T, the following are equivalent:*

(a) *T is equivalent to a set of Π_2 sentences. (Such a theory is called **inductive**.)*

(b) *T is preserved under unions of chains.*

(c) *T is preserved under unions of chains indexed by $\mathbb{N}$ (with the usual ordering).*

Proof. We prove the two easier parts of this theorem:

(a) implies (b): Assume that T is equivalent to a set T' of Π_2 sentences, and let $\{\mathfrak{A}_i \mid i \in I\}$ be a chain of models of T, with union $\mathfrak{B}$. We want to show $\mathfrak{B} \models T$, for which it suffices to prove that $\mathfrak{B} \models T'$.

So let P $\in T'$. P has the form $\forall x_1, x_2, \ldots, x_m \exists y_1, y_2, \ldots, y_n Q$, where Q is quantifier free. Let $a_1, a_2, \ldots, a_m \in B$. Then there must be an $i \in I$ such that $a_1, a_2, \ldots, a_m \in A_i$. Since $\mathfrak{A}_i \models T$, it follows that $\mathfrak{A}_i \models T'$, and thus $\mathfrak{A}_i \models$ P. So there exist $b_1, b_2, \ldots, b_n \in A_i$ such that $\mathfrak{A}_i \models Q$ when each x_j is assigned to a_j and each y_k is assigned to b_k. But then $\mathfrak{B} \models Q$ under the same assignment. Therefore $\mathfrak{B} \models$ P, as desired.

Trivially, (b) implies (c). We omit the proof that (c) implies (a). Interestingly, this proof requires the use of elementary chains, even though the theorem refers only to ordinary chains. ∎

It follows immediately from this theorem that the union of any chain of groups is a group, and similarly for rings, fields, and many other algebraic structures. However, there are some subtleties involved in this claim:

Exercise 13.

(a) Write out the usual axioms for a group in a first-order language with symbols for multiplication and the identity, both with and without a symbol for inverses. Note that these axioms are all Π_2, so the union of any chain of groups in this language is again a group.

(b) Now write out the usual axioms for a group in a language with a symbol for multiplication but no symbol for the identity or inverses. Note that these axioms are not all Π_2.

(c) Show that, in spite of part (b), the union of any chain of groups in this language is also a group. The core of the proof is to show that all the groups in the chain have the same identity element.

(d) Therefore, there must be a set of Π_2 axioms for a group in this language too. Find such a list of axioms.

Preservation under homomorphic images

Definition. A **homomorphism** from $\mathfrak{A}$ to $\mathfrak{B}$ is a function f from A to B such that for any atomic formula P and any $\mathfrak{A}$-assignment g, $\mathfrak{A} \models \mathrm{P}[g]$ implies $\mathfrak{B} \models \mathrm{P}[f \circ g]$.

We say that $\mathfrak{B}$ is a **homomorphic image** of $\mathfrak{A}$ if there is a homomorphism from $\mathfrak{A}$ to $\mathfrak{B}$ that is *onto* B.

Exercise 14. Consider a simple first-order language, such as the language with one binary function symbol $\cdot$ and one binary relation symbol $<$. This could be the language of an ordered group, without symbols for the identity and inverses. Now verify that the above definition coincides with the usual definition of homomorphism for structures of this language, namely: a function between their universes such that $f(x \cdot y) = f(x) \cdot f(y)$, and $x < y$ implies $f(x) < f(y)$, for every x and y in the domain of f.

Definition. A first-order formula is said to be **positive** if it does not contain the connectives $\sim$, $\rightarrow$, or $\leftrightarrow$. So it may contain $\wedge$ and $\vee$, as well as the quantifiers.

Clearly, a positive formula must be true whenever all of its atomic subformulas are true. (This fact still holds if the connectives $\rightarrow$ and $\leftrightarrow$ are also allowed.) Furthermore, if a positive formula is true under a certain interpretation, it must remain true if the values of one or more atomic subformulas change from false to true.

Theorem 5.19 (R. C. Lyndon). *Let T be a consistent theory. Then T is preserved under homomorphic images if and only if T is equivalent to a set of positive sentences.*

We omit the proof. The reverse direction is easy to prove—a simple induction on the structure of P shows that every true positive formula stays true in a homomorphic image. (In fact, the definition of a homomorphism says precisely that this holds for atomic formulas.) The proof of the forward direction is more difficult.

Example 13. The usual axioms for a group are positive, whether or not symbols for the identity and inverses are included in the language. Therefore, every homomorphic image of a group is a group. The same holds for rings.

Example 14. The usual axioms for a ring with unity are not all positive, since they include the statement $0 \neq 1$. A fortiori, the usual axioms for an integral domain or a field are not all positive. In fact, none of these theories is preserved under homomorphic images: the unique mapping from the ring $\mathbb{R}$ to the ring $\{0\}$ is a ring homomorphism, but the domain is a field while the range is not even a ring with unity.

The usual form of the multiplicative inverse axiom contains the connective $\rightarrow$, but it has a positive equivalent:

$$\forall x[x = 0 \vee \exists y(x \cdot y = 1)].$$

(It also has a positive equivalent if the language does not have the symbols 0 and 1.) So we get a theory that is preserved under homomorphic images if we omit $0 \neq 1$ from the field axioms. What results is the first-order theory of fields *and* one-element rings.

Preservation under direct products

One of the most important ways of combining mathematical structures is to form their direct product. One can form direct products of just about every type of algebraic structure (see Appendix D), as well as orderings, topological spaces, etc. If $\{\mathfrak{A}_i \mid i \in I\}$ is a collection of structures for the same first-order language $\mathcal{L}$, we use the usual notation $\prod_{i \in I} \mathfrak{A}_i$ for their direct product. Finite direct products such as $\mathfrak{A} \times \mathfrak{B}$ and $\mathfrak{A} \times \mathfrak{B} \times \mathfrak{C}$ may be viewed as special cases of the general notion.

It would be nice to have a clear preservation result about theories preserved under direct products, but unfortunately the situation here is not as clear as with submodels, chains or homomorphic images. For instance, it is well known that an arbitrary direct product of groups is a group, and similarly for rings. But the direct product of just two fields is never even an integral domain. This is reminiscent of the situation for homomorphic images, but it is much harder with direct products to identify a syntactic condition on sentences that is equivalent to preservation. (Such a syntactic condition exists, but it is very involved and we will not present it here.)

We begin with a fact that at least makes it easier to think about preservation under direct products:

Proposition 5.20. *If a sentence is preserved under direct products of two structures, then it is preserved under arbitrary direct products.*

But even with the simplification afforded by this fact, the preservation situation is quite complex. Here are the relevant syntactic notions:

Definitions. A **basic Horn formula** is a formula that is of one of the forms P_1, $\sim P_1$, or $(P_1 \wedge P_2 \wedge \cdots \wedge P_n) \rightarrow P_{n+1}$, where all of the P_i's are atomic. A **Horn formula** is a formula built up from basic Horn formulas using quantifiers and the connective $\wedge$. A **Horn sentence** is a Horn formula with no free variables.

Theorem 5.21.

(a) *If a sentence is equivalent to a Horn sentence, then it is preserved under direct products.*

(b) *If a Π_2 sentence is preserved under direct products, then it is equivalent to a Π_2 Horn sentence.*

The restriction to Π_2 sentences in (b) cannot be removed— there are Σ_2 counterexamples. The restriction can be removed if we talk about **reduced products** (a generalization of the notion of direct products) instead of direct products. In other words, a sentence is preserved under reduced products if and only if it is equivalent to a Horn sentence. However, we will not define reduced products or discuss them further because they are quite specialized.

Example 15. The axioms of a partial ordering (as in Appendix B) are Horn sentences, so any direct product of partial orderings is a partial ordering.

The extra axiom of a total ordering, trichotomy, is not a Horn sentence and there is no obvious equivalent Horn sentence. In fact there can't be one, because this property is not preserved under direct products. For instance, the product ordering on $\mathbb{R} \times \mathbb{R}$ is not total.

Example 16. The axioms of a group are all Horn sentences, so every direct product of groups is a group. This is true even if the language does not include symbols for the identity and inverses. The same holds for rings, commutative rings, and rings with unity.

Exercise 15. Write out the inverse axiom for groups in the language without symbols for the identity and inverses, and convince yourself that it's a Horn sentence.

Example 17. An integral domain is a commutative ring with unity that also has no **zero-divisors**: $\forall x, y(x \neq 0 \wedge y \neq 0 \rightarrow x \cdot y \neq 0)$. This is not a Horn sentence.

A field is a commutative ring with unity that also satisfies the multiplicative inverse axiom, $\forall x[x \neq 0 \rightarrow \exists y(x \cdot y = 1)]$. This is also not a Horn sentence.

In fact, neither of these two properties is equivalent to a Horn sentence, because they are not preserved under direct products. For instance, the ring $\mathbb{R} \times \mathbb{R}$ is the direct product of two fields but it is not even an integral domain.

5.5 Saturation and complete theories

In Chapter 4 we saw that many important first-order theories cannot be complete. We will now use model theory to identify some complete theories. It is valuable to know that a theory is complete, because then we know it cannot be strengthened without passing to a more powerful language. Also, if a theory T is axiomatizable and complete, then it is decidable, by Exercise 1(b) of Chapter 4. This means there is an effective procedure that determines whether or not any given sentence is provable in T, a useful thing to know.

In this section we will discuss a powerful and versatile method for proving that theories are complete. Its main drawback is that it is set-theoretic and rather abstract. In the next section we will describe another method that applies to fewer situations but provides more precise information when it does work. For most of the completeness results to be discussed, we will outline proofs using both methods.

Definition. Let κ be a cardinal. A theory T is said to be **categorical in power** κ or simply κ**-categorical** if it has, up to isomorphism, exactly one model of cardinality κ. In other words, it has models of cardinality κ, and they are all isomorphic. (The word "power" is sometimes used to mean "cardinality.")

Proposition 5.22 (Łoś–Vaught Test). *If T has only infinite models and is κ-categorical for some $\kappa \geq Card(T)$, then T is complete.*

Proof. Assume T is κ-categorical, with $\kappa \geq Card(T)$. Let $\mathfrak{A}_1$ and $\mathfrak{A}_2$ be any models of T. By Corollary 5.12, there is a structure $\mathfrak{A}'_i$ of cardinality κ such that $\mathfrak{A}'_i \equiv \mathfrak{A}_i$, for $i = 1, 2$. By categoricity, $\mathfrak{A}'_1 \cong \mathfrak{A}'_2$. Therefore, $\mathfrak{A}_1 \equiv \mathfrak{A}_2$, so we are done by Corollary 5.5. ■

The condition that T has only infinite models cannot be removed. As a trivial example, let T be the empty theory in the language of pure identity. Then a model of T is simply a nonempty set with no structure, so T is categorical in every nonzero power. But T is not complete; for instance, the statement that there are at least two elements is independent of T.

Notation. Suppose $(L, <)$ is a linear ordering, $y \in L$, and $C, D \subseteq L$. We write $C < D$ (respectively, $C < y < D$) to mean that $c < d$ (respectively, $c < y < d$) for every $c \in C$ and $d \in D$.

Lemma 5.23. *For any linear ordering $(L, <)$, the following are equivalent:*

(a) *$(L, <)$ is a dense unbounded ordering, that is, a dense ordering with no greatest or least element.*

(b) *Whenever C and D are finite subsets of L such that $C < D$, there is a $y \in L$ such that $C < y < D$.*

Exercise 16. Prove this lemma. Be sure to consider the possibility that C and/or D is empty.

We are now in a position to prove one of the simplest and oldest completeness results:

Theorem 5.24. *The theory of a dense unbounded ordering is* $\aleph_0$*-categorical, and therefore complete.*

Proof. Let T be this theory. Clearly, T is consistent (because $\mathbb{Q}$ is a model) and has no finite models. So once we show that T is $\aleph_0$-categorical, completeness follows by the Łoś–Vaught test.

Let $\mathfrak{A}$ and $\mathfrak{B}$ be any two countable models of T. Say $A = \{a_n \mid n \in \mathbb{N}\}$ and $B = \{b_n \mid n \in \mathbb{N}\}$. We will use the symbol $<$ to denote the ordering in both $\mathfrak{A}$ and $\mathfrak{B}$. We now use an inductive process to define an isomorphism f between $\mathfrak{A}$ and $\mathfrak{B}$:

At stage 1, let $f(a_1) = b_1$. (This is the only stage where we carry out one step instead of two.)

At stage $n+1$, up to $2n$ values of f will already have been defined, in such a way that f is order-preserving so far. Let E be the set of members of A at which f has already been defined. If $a_{n+1} \in E$, skip the rest of this paragraph. Otherwise, let $E_1 = \{x \in E \mid x < a_{n+1}\}$ and $E_2 = \{x \in E \mid x > a_{n+1}\}$. We need to define $f(a_{n+1})$ and keep f order-preserving. Let $C = f(E_1)$ and $D = f(E_2)$. Clearly $C < D$, and so by the previous lemma, there is a $y \in B$ such that $C < y < D$. Let $f(a_{n+1})$ be any such y.

The second step of stage $n+1$ is to include b_{n+1} in the range of f (if it's not already there), in such a way that f stays order-preserving. This process is nearly identical to the definition of $f(a_{n+1})$, and so we omit the details.

It is clear that the function f defined by this process is an order-preserving function from the whole set A onto the set B. ■

Note that the construction of f in this proof requires defining two values of f at each stage. This is the classic example, due to Cantor, of a **back-and-forth argument**, one of the most powerful techniques in model theory. For an introduction to the more sophisticated uses of this method, see [Bar73].

Exercise 17. Prove that the following theories are also complete: the theory of a dense ordering with greatest and least elements; the theory of a dense ordering with least element but no greatest element; and the theory of a dense ordering with greatest element but no least element.

These four theories are not categorical in any uncountable power. For example, $\mathbb{R}$, "a copy of $\mathbb{R}$ followed by a copy of $\mathbb{Q}$," and "a copy of $\mathbb{Q}$ followed by a copy of $\mathbb{R}$" are all dense unbounded orderings of the same uncountable cardinality, but no two of them are isomorphic.

Exercise 18. Prove everything stated in the previous sentence.

Why doesn't the construction of f in the proof of Theorem 5.24 work for uncountable dense unbounded orderings? The first obvious difference is that we must use transfinite induction instead of ordinary induction when the sets are uncountable. Then, within the induction process, Lemma 5.23 no longer suffices. Instead, we would need a generalization of this lemma that replaces "finite" by "of smaller cardinality than L." And this generalization is false.

These considerations led Felix Hausdorff and others to consider those uncountable orderings for which Lemma 5.23 can be generalized. Here are some relevant notions:

Definitions. A collection of sets is said to have the **finite intersection property** if the intersection of any finite number of those sets is nonempty. We will call a collection C of sets **resilient** if every subcollection of C with the finite intersection property has nonempty intersection.

Example 18.

(a) Every finite collection of sets is resilient, trivially.

(b) One way of defining compactness of a topological space is that the collection of all of its closed subsets is resilient. By the well-known Heine–Borel theorem, every closed, bounded subset of $\mathbb{R}$ is compact. Therefore, the collection of all closed, bounded subsets of $\mathbb{R}$ is resilient. In particular, the collection of all bounded closed intervals in $\mathbb{R}$ is resilient.

(c) The collection of closed rays $\{[n, \infty) \mid n \in \mathbb{N}\}$ in $\mathbb{R}$ has the finite intersection property but its intersection is empty. So this collection is not resilient.

Exercise 19. Find a countable collection of bounded open intervals in $\mathbb{R}$ that is not resilient.

We can now establish the correct generalization of Lemma 5.23 for uncountable orderings:

Proposition 5.25. *Let $(L, <)$ be a total ordering and let κ be an uncountable cardinal. Then the following are equivalent:*

(a) *$(L, <)$ is dense and unbounded, and every collection of fewer than κ open intervals and open rays in L is resilient.*

(b) *Whenever C and D are subsets of L of cardinality less than κ such that $C < D$, there is a $y \in L$ such that $C < y < D$.*

Proof. For the forward direction, assume the givens, and let C and D be subsets of L of cardinality less than κ such that $C < D$. We consider three cases: If $D = \emptyset$, consider the collection of all open rays $\{x \mid x > c\}$, where $c \in C$. Within any finite subset of this collection, there is a greatest element u among the left-hand endpoints of these rays, and since $(L <)$ is unbounded, there is an element of L that is greater than u. So this collection of rays has the finite intersection property, and therefore has nonempty intersection. And if y is in the intersection of all these rays, we clearly have $C < y < D$.

The proof for the case that $C = \emptyset$ is nearly identical.

Finally, assume C and D are both nonempty. Then consider the collection of all open intervals of the form (c, d), where $c \in C$ and $d \in D$. By Proposition C.1(c) in Appendix C, the cardinality of this collection is less than κ. Also, within any finite subset of this collection, there is a greatest element u among the left-hand endpoints of these open intervals, and a least element v among the right-hand endpoints of these open intervals. Since $C < D$, $u < v$. Because $(L <)$ is dense, there is an element of L between u and v. So this collection of intervals has the finite intersection property, and therefore has nonempty intersection. And if y is in the intersection of all these intervals, we clearly have $C < y < D$.

The proof of the other direction is left for the following exercise. ■

Exercise 20. Prove the reverse direction of this proposition.

Hausdorff defined an η_α-set to be a linear ordering that satisfies the conditions of Proposition 5.25 with $\kappa = \aleph_\alpha$. Over time, model theorists devised the notion of **saturation** to generalize this notion to first-order structures other than linear orderings. Before we can define saturation, we need to define a more basic and important concept that we have been mentioning informally since Chapter 1:

Definition. Let $\mathfrak{A}$ be an $\mathcal{L}$-structure. A subset of A is called **(first-order) definable** (in $\mathfrak{A}$) if it is of the form $\{x \in A : \mathfrak{A} \models \mathrm{P}[g_x^0]\}$, for some $\mathcal{L}$-formula P and some assignment g. If P has no free variables other than v_0, we call the set **definable without parameters** or **$\emptyset$-definable** ("zero-definable") in $\mathfrak{A}$.

Similarly, one can define the notion of a definable (with or without parameters) n-ary relation in $\mathfrak{A}$. A definable function is one whose graph is definable.

Example 19.

(a) In the structure $(\mathbb{R}, <)$, every interval is definable. For instance, $(e, 5\pi)$ is $\{x \in \mathbb{R} : e < x \wedge x < 5\pi\}$. The numbers e and 5π act as parameters here. More precisely, let P be the formula $(v_1 < v_0 \wedge v_0 < v_2)$, and let g be any assignment such that $g(v_1) = e$ and $g(v_2) = 5\pi$. Of course, not every interval in this structure is $\emptyset$-definable, because its language is denumerable while the number of intervals is uncountable.

(b) All intervals are also definable in the field of real numbers, with no inequality symbol. For instance, $(e, 5\pi)$ is

$$\{x \in \mathbb{R} \mid \exists y, z \in \mathbb{R}(y \neq 0 \wedge z \neq 0 \wedge e + y^2 = x \wedge x + z^2 = 5\pi)\}.$$

(c) Every finite subset of the domain of a structure is definable, because any finite set of elements can be used as parameters in a single formula. Similarly, every **cofinite** subset (a subset whose complement is finite) is definable. As in (a) and (b), there is no reason to expect these sets to be $\emptyset$-definable, in general.

(d) Any set consisting of a single algebraic number is $\emptyset$-definable in the ordered field $\mathbb{R}$. For example, the formula

$$(0 < x < 1) \wedge (x^3 - 3x + 1 = 0)$$

defines a unique algebraic number. It follows that any finite set of algebraic numbers is $\emptyset$-definable, as is the complement of such a set.

(e) Julia Robinson proved that $\mathbb{Z}$ and $\mathbb{N}$ are $\emptyset$-definable in the field of rationals. The defining formulas for these subsets are ingenious and sophisticated.

(f) An important theorem of number theory asserts that every natural number can be written as the sum of at most four squares of integers. Therefore, $\mathbb{N}$ is $\emptyset$-definable in the ring of integers.

Exercise 21. Show that the set of positive rationals is $\emptyset$-definable in the field of rationals. Conclude that every interval whose endpoints are rational numbers or $\pm\infty$ is $\emptyset$-definable in this field. What about an interval like $(0, \sqrt{2}) \bigcap \mathbb{Q}$? Is every interval of the form $(a, b) \bigcap \mathbb{Q}$, where a and b are reals, definable (with parameters) in the field of rationals?

Exercise 22. Show that, in contrast to Example 19(d), no nonempty set of transcendental numbers is $\emptyset$-definable in the field $\mathbb{R}$. You may use Example 12 from Section 5.4.

Exercise 23. Prove that the set $\{i\}$ is not $\emptyset$-definable in the field $\mathbb{C}$. Use the fact that the function $f(z) = \overline{z}$ is an automorphism of $\mathbb{C}$, that is, an isomorphism from $\mathbb{C}$ to itself. Here, $\overline{z}$ is the complex conjugate of z, as usual.

Julia Robinson (1919–1985) had more than her share of adversity in her childhood. Her mother died when she was two, and at age nine Julia had serious episodes of scarlet fever and then rheumatic fever that kept her out of school for over two years. The family's savings were wiped out in the great depression, and her father committed suicide in 1937. However, Julia was able to persist in her studies, and eventually she earned her bachelors and masters degrees from U. C. Berkeley, and her PhD from Princeton, all in mathematics. Along the way she married Raphael Robinson, one of her professors at Berkeley. Thus, for some time, there were three logicians named Robinson in the United States: Julia, Raphael, and Abraham.

Robinson made several significant contributions to foundations, mostly in model theory. In her dissertation, supervised by Tarski, she proved the $\emptyset$-definability of $\mathbb{Z}$ in $\mathbb{Q}$, mentioned above. In Section 3.4 we mentioned her crucial work on Hilbert's tenth problem. She also did important research in algebraic model theory and recursion theory. Julia Robinson received many honors. She was the first woman to be elected to the National Academy of Sciences in 1976, and became the first woman president of the American Mathematical Society in 1982. However, she made it clear that she wished to be remembered for her achievements as a mathematician, independently of her gender.

Exercise 24. Prove that the set of definable sets (with or without parameters) in any structure $\mathfrak{A}$ forms an **algebra** of subsets of A, meaning a collection of subsets of A that is closed under **Boolean combinations** (finite unions and intersections, and complementation).

By Dc Morgan's laws (translated from logic to set operations), a collection of subsets of a set A that is closed under finite unions and complementation, or closed under finite intersections and complementation, must be an algebra.

We are now ready to define saturation. To keep the definition simple, we will not give it in full generality, which would allow κ to equal $\aleph_0$ and/or the language of $\mathfrak{A}$ to be uncountable.

Definition. Let κ be an uncountable cardinal, and let $\mathfrak{A}$ be a structure for a countable language. Then $\mathfrak{A}$ is **κ-saturated** if every collection of fewer than κ definable subsets of A is resilient.

If $\mathfrak{A}$ is $Card(A)$-saturated, then we simply say that $\mathfrak{A}$ is **saturated**. It is implicit in this definition that a saturated structure must be uncountable.

Let $\mathfrak{A}$ be any infinite structure. Consider the collection of all sets of the form $A - \{b\}$, where $b \in A$. By Example 19(c), these subsets of A are all definable. Also, this collection has the finite intersection property, but the intersection of the entire collection is empty. In other words, this collection is not resilient, so $\mathfrak{A}$ cannot be $Card(A)^+$-saturated. Therefore, a saturated structure is as saturated as it can possibly be. For this reason, some authors call saturation "full saturation."

In a linear ordering, it is clear that every open interval and every open ray is definable. So if a dense unbounded ordering is κ-saturated, then the conditions of Proposition 5.25 hold. (Conversely, the conditions of Proposition 5.25 imply κ-saturation, but this is less obvious. See Proposition 5.4.2 of [CK] for the proof.) Therefore, from the discussion following Theorem 5.24, we can deduce that any two saturated dense unbounded orderings of the same cardinality are isomorphic. This result has an important generalization to all types of first-order structures:

Theorem 5.26. *Any two saturated, elementary equivalent structures of the same cardinality are isomorphic.*

This theorem can be rephrased as a uniqueness result: a complete theory has, up to isomorphism, at most one saturated model of any given cardinality. The proof is a sophisticated back-and-forth argument that traces its lineage all the way back to the proof of Theorem 5.24.

However, the usefulness of this theorem is diminished by the fact that saturated structures are quite hard to come by. For instance, Example 18(c) and Exercise 19 tell us that the ordering $(\mathbb{R}, <)$ is not even $\aleph_1$-saturated. More generally, the existence of saturated structures cannot be proved in ZFC. The main positive result is that if κ is an infinite cardinal and T is a consistent theory with $Card(T) \leq \kappa$, then T has a κ^+-saturated model of cardinality 2^κ [CK, Lemma 5.1.4]. The continuum hypothesis says that $2^{\aleph_0} = {\aleph_0}^+$, so we can prove the existence of lots of saturated models (of cardinality $\aleph_1$) in ZFC + CH.

To eliminate the need to assume CH for important results in model theory, the concept of a special structure was devised. For the record, here is the definition of this concept. However, I advise you not to get bogged down by the abstractness of this definition. Specialness should be thought of as a convenient, minor adaptation of the notion of saturation.

Definition. A structure $\mathfrak{A}$ is called **special** if it is the union of an elementary chain $\{\mathfrak{A}_\kappa \mid \kappa < Card(A)\}$, where the subscript κ is restricted to infinite cardinals, and each $\mathfrak{A}_\kappa$ is κ^+-saturated.

Every saturated structure is special, trivially—just let each $\mathfrak{A}_\kappa$ be $\mathfrak{A}$ itself. But it is easier for a structure to be special than saturated. More precisely, the following can be proved in ZFC:

Theorem 5.27.

(a) *Every theory with an infinite model has arbitrarily large special models.*

(b) *If a theory in a countable language has an infinite model, then it has a special model of cardinality $\beth_\omega$. ($\beth_\alpha$ is defined in Appendix C.)*

(c) *Theorem 5.26 holds with "saturated" replaced by "special."*

We now return to proving that specific theories are complete. We will present the two most famous results of this type.

Let RCOF denote the first-order theory of a real-closed ordered field, as described in the introduction to this chapter. By definition, a real-closed ordered field is a type of ordered ring. It is also possible to define and axiomatize the notion of a **real-closed field**, with no reference to an ordering. To do this, first define a field to be **formally real** if -1 cannot be written as a sum of squares. In a first-order theory, this requires an infinite axiom schema: for each n, the statement that -1 is not the sum of n squares. Essentially, this property replaces the ordering axioms. More precisely, a field is formally real if and only if it is **orderable**, meaning that an order can be defined on it that makes it into an ordered field. For instance, it's obvious that the fields $\mathbb{Q}$ and $\mathbb{R}$ are formally real, while $\mathbb{C}$ is not.

In addition to the axioms of a formally real field, include the schema that every polynomial of odd degree has a zero, as in RCOF. Finally, instead of saying that every nonnegative number has a square root, say that every number or its negative has a square root. Let RCF denote the resulting theory.

Exercise 25. Show that if x is any nonzero number in a real-closed field, exactly one of the numbers x and $-x$ has a square root.

For most purposes, the theories RCF and RCOF are interchangeable. In terms of structures, there is a simple correspondence between real-closed fields and real-closed ordered fields: every real-closed ordered field becomes a real-closed field simply by dropping the ordering relation, and every real-closed field can be turned into a real-closed ordered field in a unique way. To do this, define a number to be positive if it is nonzero and has a square root. So the unique, implicit ordering of any real-closed field is $\emptyset$-definable.

Exercise 26.

(a) Using the facts stated in the previous paragraph, show that RCOF is a **conservative extension** of RCF, meaning that it's an extension of RCF but proves no additional theorems in the (smaller) language of RCF.

(b) Show that RCF is complete if and only if RCOF is complete.

However, we will see in the next section that there are some notable differences between the theories RCF and RCOF.

Theorem 5.28 (Tarski). *The theories RCF and RCOF are complete.*

Proof. Unlike the theory of a dense unbounded ordering, these theories are not categorical in any infinite power. But the following weaker result holds: any two special real-closed fields of the same cardinality are isomorphic. This result, restricted to saturated real-closed fields, first appeared in [EGH, Theorem 2.1], using a classic back-and-forth argument. The proof for cardinality $\aleph_1$ is also given in Theorem 5.4.4 of [CK]. The generalization to special real-closed fields is not difficult. What makes this back-and-forth argument more complex than that of Theorem 5.24 is that every time we extend the isomorphism to one more member of the domain or range, we must also extend it to a real-closed field containing the new member.

Now, assuming that RCF is not complete, we can form two different complete extensions T_1 and T_2 of RCF. Then, since all models of T are infinite, T_i has a special model $\mathfrak{A}_i$ of cardinality $\beth_\omega$, for $i = 1, 2$. By the previous paragraph we would have $\mathfrak{A}_1 \cong \mathfrak{A}_2$, which would imply $T_1 = T_2$, a contradiction. ■

Exercise 27. Using this theorem and some results from Chapter 4, prove that the set of natural numbers is not $\emptyset$-definable in the ordered field of real numbers. (In the next section, we will see that $\mathbb{N}$ isn't even definable with parameters in this ordered field.)

The proof of Theorem 2.1 of [EGH] also yields the following:

Corollary 5.29. *In order for a real-closed (ordered) field to be κ-saturated, it suffices for it to be κ-saturated as a linear ordering.*

We now turn our attention to the other major category of "closed" fields:

Notation. Let ACF denote the first-order theory of an algebraically closed field. In addition to the field axioms, this theory requires an infinite axiom schema: for each positive integer n, an axiom that says every polynomial of degree n has a zero.

Also, let ACF_k denote the first-order theory of an algebraically closed field of characteristic k. Here, k may be 0 or any prime number. For example, ACF_3 is ACF plus the axiom $1 + 1 + 1 = 0$, which we may abbreviate as $3 = 0$ (although $\overline{3} = 0$ or even $\overline{3} = \overline{0}$ would be more correct). ACF_0 is ACF plus the axioms $p \neq 0$ for every prime p.

Theorem 5.30. *Each of the theories* ACF_k *is complete.*

Proof. This result is older and perhaps simpler than the previous theorem. The key lemma, due to Ernst Steinitz, is that an algebraically closed field is uniquely determined (up to isomorphism) by its characteristic and its **transcendence degree**—the cardinality of a maximal set of independent transcendental elements over its smallest subfield. But for uncountable fields, the transcendence degree is just the cardinality of the field itself. In other words, for each particular characteristic, the theory we are considering is categorical in every uncountable power. Therefore, by the Łoś–Vaught test, it is complete. ■

To demonstrate the power of this completeness result, here is an interesting fact of "ordinary mathematics" that was first proved by Ax, using Theorem 5.30. The proof requires somewhat more knowledge of algebraic concepts than we have been assuming.

Theorem 5.31. *Let $n \in \mathbb{Z}^+$ and let K be an algebraically closed field. Then every one-to-one polynomial function from K^n to K^n is onto.*

Proof. First we consider the special case where K is the algebraic closure of $\mathbb{Z}_p$ for some prime p. Assume f is such a polynomial, and let $\vec{y} = (y_1, y_2, \ldots, y_n) \in K^n$. We must show that there exists $\vec{x} \in K^n$ such that $f(\vec{x}) = \vec{y}$. Now, if L is a finite field and a is algebraic over L, then the field $L(a)$ is also finite. (Here, $L(a)$ denotes the smallest extension field of L that contains a.) Therefore, since every member of K is algebraic over every subfield of K, it follows by induction that every finite subset of K is contained in a finite subfield of K. So let K_0 be a finite subfield of K that contains $y_1, y_2, \ldots, y_n$ as well as all the coefficients of f. Then the restriction of f to ${K_0}^n$ is a one-to-one function from the finite set ${K_0}^n$ to itself. Therefore, there exists $\vec{x} \in {K_0}^n$ such that $f(\vec{x}) = \vec{y}$. So the theorem holds for fields of this type.

Now let a prime p and $n \in \mathbb{Z}^+$ be fixed. For each $m \in \mathbb{Z}^+$, there is a first-order sentence P_m that expresses "Every one-to-one polynomial with n input variables and n output variables, of degree $\leq m$, is onto." (Example: the polynomial $f(x, y) = (x^3 - 5xy, -2x + x^3y^2)$ has two input variables, two output variables, and degree 5.) We have just shown that each P_m is true in the algebraic closure of $\mathbb{Z}_p$, which is a model of ACF_p. So $\mathrm{ACF}_p \not\vdash \sim \mathrm{P}_m$. Therefore, since ACF_p is complete, $\mathrm{ACF}_p \vdash \mathrm{P}_m$. Since this holds for every m, the theorem is established for algebraically closed fields of prime characteristic.

It remains to prove the characteristic 0 case. Again, let n be fixed. It will suffice to show that $\mathrm{ACF}_0 \vdash \mathrm{P}_m$ for every m. Assume this is false. Then $\mathrm{ACF}_0 \vdash \sim \mathrm{P}_m$ for some m, since ACF_0 is complete. So consider some proof of $\sim \mathrm{P}_m$ from ACF_0. Only a finite number of axioms of the form $p \neq 0$ are used in this proof, so we can find a prime q that is larger than all these primes p. Then all of these sentences of the form $p \neq 0$ are theorems of ACF_q. But this implies that $\sim \mathrm{P}_m$ is a theorem of ACF_q, contradicting the previous paragraph. ∎

Corollary 5.32. *Let $n \in \mathbb{Z}^+$. Then every one-to-one polynomial function from $\mathbb{C}^n$ to $\mathbb{C}^n$ is onto.*

There are a few other interesting examples of complete theories of an algebraic nature. The theory of an abelian group in which every element has order p (for a fixed prime p) is categorical in every power in which it has a model. Therefore, the theory of an infinite group of this type is complete. Another example is the theory of a nontrivial divisible torsion-free abelian group, which is categorical in every uncountable power. Outside of algebra, there are few "substantial" complete theories. This is partly explained by Gödel's incompleteness theorem: an axiomatizable theory that includes Peano arithmetic cannot be complete.

We have now seen examples of various combinations of categoricity. The theory of a dense unbounded ordering is $\aleph_0$-categorical but not categorical in uncountable powers. For the theory of an algebraically closed field of fixed characteristic, it's the opposite. The theory of a real-closed field is not categorical in any infinite power, even though it is complete. And the previous paragraph mentioned a theory that is categorical in every power. In terms of infinite models of theories in countable languages, there are no other possibilities: a deep result known as **Morley's theorem** shows that if a theory in a countable language is categorical in one uncountable power, then it is categorical in every uncountable power.

5.6 Quantifier elimination

If a set (or relation or function) is definable in a structure, one important measure of the complexity of the set is based on the quantifier complexity of some formula that defines it. Thus, we can refer to sets that are **quantifier-free definable** (also called **Σ_0-definable** or **Π_0-definable**), **Σ_2-definable**, **Π_5-definable**, etc. For short, one simply refers to Σ_n sets and Π_n sets. Somewhat imprecisely, we will sometimes use these same symbols to denote the corresponding collections of definable sets. In other words, "$A \in \Sigma_n$" means the same thing as "A is Σ_n." A set that is both Σ_n and Π_n is called a Δ_n set. This "hierarchy" of complexity, devised by Kleene, provides one of the main approaches to understanding the structure of definable sets. There is also a similar but more limited hierarchy of $\emptyset$-definable sets. Here are a few simple facts about these hierarchies:

Proposition 5.33. *For any structure $\mathfrak{A}$:*

(a) *Σ_n (that is, the collection of Σ_n sets) is closed under finite unions and intersections. So is Π_n.*

(b) *The Δ_n sets form an algebra.*

(c) *Every Boolean combination of Σ_n sets is Δ_{n+1}.*

Proof. The proof of (a) is straightforward, using ideas mentioned in the proof of Theorem 1.5. (See the exercise below.) Part (b) follows immediately from (a). To prove (c), first note that the insertion of "dummy quantifiers" (also mentioned in Section 1.6) guarantees that every Σ_n set is Δ_{n+1}. But (b) says that the Δ_{n+1} sets are closed under Boolean combinations, so (c) follows. ■

Exercise 28. Prove this proposition in more detail, especially (a).

Thus the hierarchy of definable subsets of a structure looks like Figure 5.1, where each line indicates inclusion from left to right. Here, Γ_n denotes the Boolean combinations of Σ_n sets, but this notation is not standard. Clearly, every Π_n set is a Boolean combination of Σ_n sets (specifically, the complement of a Σ_n set), and vice-versa.

Example 20. The ordering on $\mathbb{R}$ is $\emptyset$-definable in the field of reals, since

$$x \leq y \leftrightarrow \exists z(x + z \cdot z = y)$$

is true in this structure, for any reals x and y. So, more specifically, the ordering relation is Σ_1, without parameters. We will soon see that $\leq$ and $<$ are not Σ_0, even with parameters.

Example 21. Sets (including relations and functions) that are definable in the standard model of arithmetic $\mathfrak{N}$ are called **arithmetical**. Since every natural number is defined by a term (numeral), definability and $\emptyset$-definability are the same in $\mathfrak{N}$. Given that $\mathfrak{N}$ is a model of PA, it is clear that every set that is representable in PA is arithmetical. Therefore, every recursive set is arithmetical.

But where do recursive sets fit in the **arithmetical hierarchy**? By the main lemma used to solve Hilbert's tenth problem (discussed in Section 3.4), every RE set is Σ_1. (This can also be obtained by first

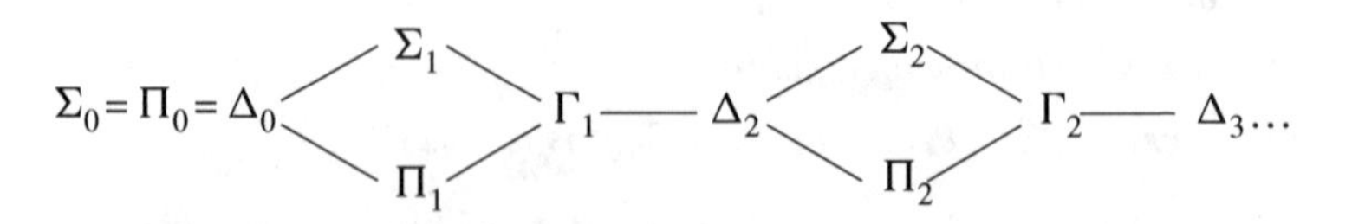

Figure 5.1. The hierarchy of definable subsets of a structure

showing that Kleene's T-predicate and the upshot function U are Σ_1, and then applying Theorems 3.5 and 3.10(e)). The converse of this is also true, since a Σ_0 set is obviously decidable. So a subset of $\mathbb{N}^k$ is RE if and only if it's Σ_1. Then, from Theorem 3.12, it follows that the recursive sets are precisely the Δ_1 ones. So recursive and RE sets form two of the lowest levels in the arithmetical hierarchy. (Some authors define the Σ_0 and Π_0 sets in $\mathfrak{N}$ to include all PR ones or even all recursive ones. This modification has no effect on the higher levels of the hierarchy, including Σ_1 and Δ_1.)

It is common to adjoin a superscript of 0 to the notation for the arithmetical hierarchy. Thus, recursive sets are Δ_1^0 (read "Delta-0-1"), and RE sets are Σ_1^0. This superscript indicates that the quantifiers allowed in the defining formulas are first-order, over *elements* of $\mathbb{N}$. (Logically, one could claim that this superscript is "one off," but that's just how it is.) In Section 6.5, we will define the analogous notation with a superscript of 1, as well as another approach to defining these hierarchies.

Definable sets of a structure are, for the most part, more "well behaved" than other sets. On the other hand, definable sets of high syntactic complexity are not usually easy to deal with. Therefore, model theorists are interested in theories whose models have particularly simple definable sets. The following definition describes the most desirable situation of this sort:

Definition. Let T be a theory in a language $\mathcal{L}$. We say that T has **quantifier elimination** if, for every $\mathcal{L}$-formula P, there is a quantifier-free $\mathcal{L}$-formula Q such that $T \vdash (\mathrm{P} \leftrightarrow \mathrm{Q})$.

Exercise 29. Prove that, in order for a theory to have quantifier elimination, it suffices for the definition to hold whenever P is a Σ_1 formula with just one quantifier.

Lemma 5.34. *Suppose T has quantifier elimination. Then, if* P *is not a sentence or the language of T has at least one constant symbol, the quantifier-free formula* Q *can be chosen to have only variables that are* *free in* P. *Otherwise,* Q *can be chosen to have* v_0 *(or any other specified* v_i*) as its only variable.*

Proof. Suppose T has quantifier elimination and P is a formula. So there's a quantifier-free formula Q such that $T \vdash (\mathrm{P} \leftrightarrow \mathrm{Q})$. Now, suppose some variable v_k is in Q but is not free in P. By the generalization theorem (Theorem 1.4), $T \vdash \forall v_k(\mathrm{P} \leftrightarrow \mathrm{Q})$, since T consists of sentences. Then, by the universal specification axiom of logic (as in Appendix A), $T \vdash (\mathrm{P} \leftrightarrow \mathrm{Q}')$, where Q′ results from Q by replacing every occurrence of v_k by some free variable of P or a constant symbol, (or by v_0, if P is a sentence and there are no constant symbols). Since we can do this for each such v_k, the lemma is proved. ■

There is often a relationship between quantifier elimination and completeness for a theory, although neither of these properties implies the other in general. We will now illustrate this relationship by revisiting some of the complete theories discussed in the previous section.

Theorem 5.35. *The theory of a dense unbounded ordering has quantifier elimination.*

Proof. Let T be this theory. In its language $\mathcal{L}$, all atomic formulas are of the form $v_j = v_k$ or $v_j < v_k$. By an **arrangement** of a finite nonempty set of variables, we mean a conjunction of atomic formulas that precisely specifies the order of those variables in a consistent way. For example, the formula

$$(v_2 < v_7) \wedge (v_7 = v_{12}) \wedge (v_{12} < v_3)$$

is an arrangement of $\{v_2, v_3, v_7, v_{12}\}$.

The main lemma for this theorem, whose proof can be found in [CK] or [Mar], is that every quantifier-free formula Q of $\mathcal{L}$ is equivalent, in T (or even in the theory of a total ordering), to either a contradiction or a disjunction of arrangements of the free variables of Q. Using this lemma, we proceed as follows:

By Exercise 29, we can restrict our attention to a formula P of the form $\exists v_k \mathrm{R}$, where R is quantifier-free. If R is equivalent to a contradiction, then so is P, and we have $T \vdash (\mathrm{P} \leftrightarrow v_0 < v_0)$. On the other hand, if R is equivalent to a disjunction of arrangements $\mathrm{A}_1 \vee \mathrm{A}_2 \vee \ldots \vee \mathrm{A}_n$, then we obtain

$$T \vdash \mathrm{P} \leftrightarrow (\exists v_k \mathrm{A}_1 \vee \exists v_k \mathrm{A}_2 \vee \ldots \vee \exists v_k \mathrm{A}_n),$$

because the existential quantifier "distributes" over disjunctions in first-order logic. But in any dense unbounded ordering (and hence in T), a formula of the form $\exists v_k \mathrm{A}_i$ is equivalent to the new arrangement B_i obtained by deleting all conjuncts that mention v_k from A_i. Therefore P is equivalent, in T, to the quantifier-free formula $\mathrm{B}_1 \vee \mathrm{B}_2 \vee \ldots \vee \mathrm{B}_n$. ■

Corollary 5.36. *The theory of a dense unbounded ordering is complete.*

Proof. By the theorem and Lemma 5.34, every $\mathcal{L}$-sentence P is equivalent, in T, to a quantifier-free formula with only one variable v_0, which in turn must be equivalent to a contradiction or a disjunction of arrangements of v_0. But the only arrangement of v_0 is $v_0 = v_0$, and a disjunction formed from this equation is clearly equivalent to the same equation. So either $T \vdash (\mathrm{P} \leftrightarrow v_0 < v_0)$, which implies $T \vdash \sim \mathrm{P}$; or $T \vdash (\mathrm{P} \leftrightarrow v_0 = v_0)$, in which case $T \vdash \mathrm{P}$. ■

Thus we have an alternative proof of Theorem 5.24. The proof using quantifier elimination is more concrete; specifically, the quantifier elimination process is effective and therefore provides an actual decision procedure for the theory. This is the main advantage of quantifier elimination over more abstract methods such as saturation. In the words of [CK], "the method is extremely valuable when we want to beat a particular theory into the ground." (But this cute remark is certainly not intended as a criticism of this powerful technique.)

Exercise 17 in the previous section referred to three other complete theories involving dense orderings. Interestingly, none of these theories has quantifier elimination:

Exercise 30. Prove that the theory of a dense ordering with endpoints does not have quantifier elimination. Specifically, show that the formula $\exists v_1(v_1 < v_0)$, which says that v_0 is not the least element, has no quantifier-free equivalent. (Hint: Note that the "main lemma" referred to in the proof of Theorem 5.35 also applies to this theory.)

By contrast, the proof of Corollary 5.36 shows that in the theory of a dense unbounded ordering, you can't say anything nontrivial about a single element, such as the assertion that it's the least one.

In spite of this exercise, the method of quantifier elimination can still be used to prove these three theories are complete. For example, let T be the theory of a dense ordering with endpoints. Now consider the same theory, augmented by two constant symbols a and b and axioms "a is the least element" and "b is the greatest element." This theory T' can be shown to have quantifier elimination and to be complete, using arguments very similar to the proofs of Theorem 5.35 and Corollary 5.36. It is also clear that every model of T can be turned into a model of T' in a unique way, by interpreting a and b in the obvious way. Therefore, by the reasoning of Exercise 26, T' is a conservative extension of T, and T is also complete.

The argument just given is part of a general phenomenon: every theory has a conservative extension with quantifier elimination. However, the construction of this conservative extension is often complicated and gives little or no useful information about the original theory.

We've just seen three examples of complete theories without quantifier elimination. The next exercise and theorem illustrate the opposite phenomenon:

Exercise 31. Let $\mathcal{L}$ be the language with no relation or function symbols, and two constant symbols a and b. Let T be the $\mathcal{L}$-theory whose only proper axiom says that every element equals a or b.

(a) Show that T is not complete, by finding two models of T that are not elementarily equivalent.

(b) Show that T has exactly two complete extensions (up to equivalence of theories).

(c) Show that T has quantifier elimination. (Hint: Use Exercise 29, Proposition 1.1, and the fact that, in T, the negation of any equation involving a variable is equivalent to a positive combination of equations.)

Theorem 5.37. *The theory ACF of algebraically closed fields has quantifier elimination.*

The proof of this theorem is not terribly difficult, but it requires some nontrivial concepts from algebra and we will not give it. It can be found in [Mar]. From the previous section, we know that ACF is not complete. It has an infinite number of non-equivalent completions, the theories ACF_k. But ACF is "close to" being complete, in the sense that all that's required to complete it is to add some *quantifier-free* axioms (to specify the characteristic). Clearly, if it needed the addition of axioms with quantifiers to become complete, it couldn't have quantifier elimination.

Exercise 32. Use the previous theorem to give an alternative proof that the theories ACF_k are complete. (Hint: Use the facts about prime fields that are given in Appendix D. From them, prove that for each k, all models of ACF_k are elementarily equivalent.)

Corollary 5.38. *In any algebraically closed field, the definable sets are precisely the finite or cofinite subsets of the field.*

Proof. Let S be a definable subset of an algebraically closed field $\mathfrak{A}$. By the theorem, that means $S = \{x \in A : \mathfrak{A} \models \mathrm{Q}[g_x^0]\}$, for some quantifier-free formula Q and some assignment g. But in the language of field theory, an atomic formula must be an equation between two polynomials. These polynomials may have variables other than v_0, but under g these variables are replaced by particular members of A, which can be viewed as part of the coefficients of the polynomials. In other words, if Q is atomic, then $S = \{x \in A : p(x) = q(x)\}$, where p and q are polynomials in one variable with coefficients in A. By the fundamental theorem of algebra, such an equation either has a finite number of solutions or is true for all members of the domain. So since any quantifier-free Q is a Boolean combination of atomic formulas, S must be the result of taking (finite) unions, intersections, and complements of finite sets. Therefore, it is finite or cofinite. ■

In particular, this corollary tells us that very few sets are (first-order) definable in the field of complex numbers. The reals, the pure imaginary numbers, and the unit circle are examples of undefinable sets. However, all of these sets become definable if the complex conjugation function is added to the structure.

A theory is called **strongly minimal** if the only definable sets in its models are finite or cofinite. By Exercise 19(c), this means that its models have no more definable sets than absolutely necessary. So this corollary says that ACF is strongly minimal. The fact that ACF has quantifier elimination also gives useful information about the definable *relations* in an algebraically closed field: they are precisely the so-called **constructible** relations, an important notion in algebraic geometry. This is the sort of link that leads to applications of model theory outside of foundations.

We now turn our attention to real-closed fields. It is here that we will see an interesting distinction between the theories RCF and RCOF.

Proposition 5.39. *The theory RCF does not have quantifier elimination.*

Proof. The proof of Corollary 5.38 actually shows that in any field whatsoever, the Σ_0 sets (with parameters) are just the finite and cofinite sets. So, if RCF had quantifier elimination, these would be the only definable sets in the real-closed field $\mathbb{R}$. But the first example in this section shows that the positive reals are definable in the field of reals. ■

Theorem 5.40. *The theory RCOF has quantifier elimination.*

As with ACF, we will not provide the proof of this result because it requires a significant amount of algebraic machinery. This theorem is due to Tarski and was his original path to proving completeness:

Corollary 5.41. *The theories RCF and RCOF are complete, in their respective languages.*

Proof. Every real-closed ordered field has characteristic 0 and therefore has a subfield isomorphic to $\mathbb{Q}$. From this it can be shown, as in Exercise 32, that all models of RCOF are elementary equivalent, so RCOF is complete. The completeness of RCF then follows from Exercise 26(b). ∎

So we have a situation here that fits the phenomenon described after Exercise 30. The theory RCF can't have quantifier elimination but, fortunately, it is not hard in this case to find a conservative extension that does. Then the completeness of both theories follows.

As with ACF, the fact that RCOF has quantifier elimination provides sharp information about the definable sets in $\mathbb{R}$ and other real-closed fields. Since the unique ordering of any real-closed field is definable in RCF, we will get the same definable sets whether or not the symbol $<$ is part of the language. But with this symbol included, the proof of Corollary 5.38 must be modified to allow atomic formulas of the form $p(x) < q(x)$, for polynomials p and q. And the solution set of such an inequality can be any finite union of intervals. Furthermore, when you take Boolean combinations of finite unions of intervals, what you get is finite unions of points and intervals. We have just more or less proved:

Corollary 5.42. *In any real-closed field, including $\mathbb{R}$, the definable sets are precisely the finite unions of points and intervals. In particular, the sets $\mathbb{N}$, $\mathbb{Z}$, and $\mathbb{Q}$ are not definable in the field of reals.*

So we see that RCF and RCOF are not strongly minimal, in contrast to ACF. In fact, this is obvious from the fact that their models are ordered (explicitly for RCOF, implicitly for RCF). If a theory T proves that some formula $\mathrm{P}(x, y)$ defines a total ordering, then every interval under that ordering must be definable in every model of T. Therefore, if T has any infinite models, it cannot be strongly minimal. Instead, we consider the following concept:

Definition. Let T be a theory in which it is provable that some binary relation symbol defines a total ordering. T is called **o-minimal** if the only definable sets in any model of T are finite unions of points and intervals.

So the previous corollary says that RCOF is o-minimal. A bit loosely, we can say the same for RCF. More generally, definable relations in a real-closed field are precisely the relations that are known as **semialgebraic**. As with algebraically closed fields, these model-theoretic results about real-closed fields are powerful tools for doing "ordinary mathematics." For instance, the o-minimality of RCOF can be used to prove that every semialgebraic (first-order definable) function from $\mathbb{R}$ to itself is piecewise continuous.

5.7 Additional topics in model theory

This chapter has barely scratched the surface of the deep and rich subject of model theory. In this section we briefly touch on a few other aspects of the field.

Axiomatizable and nonaxiomatizable classes

The relation $\models$ provides a fruitful way of passing back and forth between sets of first-order sentences and classes of first-order structures.

Notation. If T is a set of $\mathcal{L}$-sentences, $Mod(T)$ denotes the class of all models of T. We also write $Mod(\mathrm{P})$, where P is a single sentence.

In the other direction, if $\mathcal{C}$ is a class of $\mathcal{L}$-structures, $Th(\mathcal{C})$ denotes the set of all $\mathcal{L}$-sentences that are true in every structure in $\mathcal{C}$. We also write $Th(\mathfrak{A})$, where $\mathfrak{A}$ is a single structure.

Note that $Th(\mathfrak{A})$ is automatically a complete theory. The converse, that every complete theory is of the form $Th(\mathfrak{A})$, is essentially a restatement of the harder direction of the completeness theorem.

Definition. A class of $\mathcal{L}$-structures is called **axiomatizable** (respectively, **finitely axiomatizable**, **recursively axiomatizable**) if it is $Mod(T)$ for some T (respectively, finite T, recursive T).

Clearly, a finitely axiomatizable class must be of the form $Mod(\mathrm{P})$, where P is the conjunction of all the sentences in the finite set T.

Recall, from Chapter 4, that for *theories* we are using "axiomatizable" as an abbreviation for "recursively axiomatizable." We do that because there is no other reasonable meaning for the phrase "axiomatizable theory." But for classes of structures, it makes sense to give these terms different meanings.

To illustrate this new terminology, let's use the compactness theorem to prove a fact that was mentioned in Section 1.5 (Example 17):

Proposition 5.43.

(a) *The class of all fields of finite characteristic is not axiomatizable.*

(b) *The class of all fields of characteristic zero is recursively axiomatizable but not finitely axiomatizable.*

Proof.

(a) Assume $Mod(T)$ consists of all fields of finite characteristic. Now let

$$T' = T \cup \{1+1 \neq 0, 1+1+1 \neq 0, 1+1+1+1+1 \neq 0, \dots\},$$

where there is an inequality for each prime number. Since the characteristic of every field is either 0 or a prime number, every finite subset of T' has a model but T' itself does not. This violates the compactness theorem.

(b) The standard axiomatization of this class of fields, consisting of the field axioms plus the schema shown in braces in the proof of part (a), is certainly recursive. Now assume this class of fields is finitely axiomatizable. Then, for some sentence P, $Mod(\mathrm{P})$ is the class of all fields of characteristic zero. But then if T consists of $\sim$ P and all the field axioms, $Mod(T)$ is the class of all fields of finite characteristic, in violation of part (a). ■

Exercise 33.

(a) Prove that the class of all finite groups is not axiomatizable. (Hint: Recall, from Section 1.3, that statements of the form "There are at least n elements" and "There are exactly n elements" can be formalized in first-order logic, for each positive integer n.)

(b) Prove that the class of all infinite groups is recursively axiomatizable but not finitely axiomatizable.

Similarly, the class of all finite fields and the class of all finite rings are not axiomatizable, while the class of all infinite fields and the class of all infinite rings are recursively axiomatizable but not finitely axiomatizable. There is a great variety of similar but more sophisticated examples involving algebraic theories:

Proposition 5.44. *The class of all divisible groups is recursively axiomatizable but not finitely axiomatizable.*

Proof. As in Appendix D, the class of divisible groups is axiomatized by the recursive set of axioms $T \cup \{\mathrm{P}_n \mid n \in \mathbb{Z}^+\}$, where P_n is the statement $\forall x\, \exists y (y^n = x)$. To prove that this class is not finitely axiomatizable, it will suffice to prove, as in the previous proposition, that the class of groups that are not divisible isn't even axiomatizable.

So assume that the class of non-divisible groups has an axiomatization T'. Then the theory $T' \cup \{\mathrm{P}_n \mid n \in \mathbb{Z}^+\}$ is inconsistent, since only a divisible group can satisfy all the P_n's. Therefore, $T' \cup \{\mathrm{P}_1, \mathrm{P}_2, \dots, \mathrm{P}_k\}$ is inconsistent for some k. But then let G_k be the set of all rational numbers whose denominators, in lowest terms, have no prime factors greater than k. It is routine to verify that G_k, under addition (so that y^n becomes ny), is a model of this theory. By this contradiction, we're done. (Note the tacit use of completeness/compactness in this argument.) ■

Exercise 34.

(a) Complete this proof by by showing that $(G_k, +)$ is a group, is not divisible, and does satisfy the statements $\mathrm{P}_1, \mathrm{P}_2, \dots, \mathrm{P}_k$.

(b) Prove that the class of all torsion-free groups is recursively axiomatizable but not finitely axiomatizable. You will need to find a group in which every element except the identity has finite order greater than a given number k.

(c) Prove that the class of all torsion groups is not axiomatizable. (Hint: Consider the discussion of nonstandard analysis in Section 5.3, in which a new constant symbol is added to a language.)

Even when $\mathcal{C}$ is a nonaxiomatizable class, we call $Th(\mathcal{C})$ the first-order theory of $\mathcal{C}$. For instance, if we refer to the first-order theory of finite groups, we do not mean a particular list of axioms that define finite groups precisely, because there is no such list of axioms. Rather, we mean $Th(\mathcal{C})$, where $\mathcal{C}$ is the class of all finite groups. Exercise 37(c) below will highlight this point.

The study of theories of nonaxiomatizable classes has led to some significant research. For instance, the class of all finite groups and the class of all finite fields are both nonaxiomatizable, by identical reasoning. But while the first-order theory of finite groups is not decidable, a very deep result due to James Ax [Ax] shows that the first-order theory of finite fields is decidable. Ax's result is particularly striking because many simpler (axiomatizable or even finitely axiomatizable) theories, such as group theory, field theory, and Peano arithmetic, are not decidable.

Notation. Recall that $Thm(T)$ denotes the set of theorems of any theory T. We will also write $\overline{Thm}(T)$ for the set of *sentences* that are theorems of T.

The following exercises might look substantial, but for the most part they are quite trivial, once the new terminology is "unraveled."

Exercise 35. Let P be a sentence, and T a theory. Prove:

(a) $T \models \mathrm{P}$ if and only if $\mathrm{P} \in Th(Mod(T))$. Therefore, Gödel's completeness theorem (restricted to sentences) can be stated in the form $\overline{Thm}(T) = Th(Mod(T))$.

(b) $Mod(T) = Mod(\overline{Thm}(T))$.

(c) If $\mathcal{C}$ is any class of structures, $Th(\mathcal{C})$ must be closed under $\vdash$, that is, $\overline{Thm}(Th(\mathcal{C})) = Th(\mathcal{C})$.

Exercise 36. Prove that the following are equivalent, for any theories T_1 and T_2:

(a) T_1 and T_2 are equivalent, that is, $Thm(T_1) = Thm(T_2)$.

(b) $\overline{Thm}(T_1) = \overline{Thm}(T_2)$.

(c) $Mod(T_1) = Mod(T_2)$.

Exercise 37.

(a) Fill in the blank: if $\mathfrak{A}$ is any structure, then $Mod(Th(\mathfrak{A}))$ is the class of all structures that are ______ to $\mathfrak{A}$.

(b) Let $\mathcal{C}$ be an axiomatizable class of structures, such as the class of all groups or the class of all infinite rings. What is $Mod(Th(\mathcal{C}))$?

(c) Now let $\mathcal{C}$ be a nonaxiomatizable class of structures. What can be said about $Mod(Th(\mathcal{C}))$? For instance, suppose that $\mathcal{C}$ is the class of all finite fields. Does $Mod(Th(\mathcal{C}))$ include all finite fields? Does it include any infinite fields?

Thus we see that the operators *Mod* and *Th* are inverses to a limited extent: Exercise 35(a) tells us that $Th \circ Mod$ is the identity on sets of sentences that are closed under $\vdash$. Exercise 37(b) tells us that $Mod \circ Th$ is the identity on axiomatizable classes of structures.

By Exercise 37(c), there are **pseudofinite fields**, meaning infinite fields that are models of the first-order theory of finite fields. This seems odd, but actually it requires some thought to see why there are any infinite fields that are *not* pseudofinite. To establish this, we need to find a first-order sentence that is true in every finite field but not in every field. One key to this is the fact that any function from a finite set to itself is one-to-one if and only if it is onto. Of course, this is not true for infinite sets.

Example 22. Let's show that the field of real numbers is not pseudofinite. The polynomial $f(x) = x^3 - x$ (technically $x \cdot x \cdot x - x$, in the first-order language of a field), is a simple example of a function from $\mathbb{R}$ onto itself that is not one-to-one. We can express that f is onto in the usual way: $\forall y \exists x[y = f(x)]$. Similarly, the assertion that f is one-to-one can be formalized as $\forall x, y[f(x) = f(y) \to x = y]$.

Now let P be the formal statement that says that f is one-to-one if and only if it is onto. P is true in every finite field, but not in $\mathbb{R}$. So $\mathbb{R}$ is not a model of the first-order theory of finite fields.

Exercise 38. Show that the fields $\mathbb{Q}$ and $\mathbb{C}$ are not pseudofinite. You can use the same reasoning as in the previous exercise, but you might need to find different polynomials.

Unfortunately, it is not possible to give an example of a pseudofinite field without using more sophisticated methods, such as ultraproducts. By the way, readers of [Ax] must digest quasifinite fields, pseudofinite fields, and hyperfinite fields. But the effort is worthwhile: the main result about finite fields has some important algebraic consequences, such as a decision procedure for determing whether a system of Diophantine equations has, for all primes p, a solution modulo p (or a p-adic solution).

Stone spaces

Why is the compactness theorem called that? Its content certainly resembles a statement of topological compactness, and indeed we can make this precise: given a first-order language $\mathcal{L}$, let $\mathcal{S}$ be the set of all theories of the form $Th(\mathfrak{A})$, where $\mathfrak{A}$ is an $\mathcal{L}$-structure. Note that each member of $\mathcal{S}$ is therefore a complete $\mathcal{L}$-theory. Conversely, the completeness theorem says that each complete $\mathcal{L}$-theory is in $\mathcal{S}$.

Intuitively, $\mathcal{S}$ can also be thought of as the collection of all equivalence classes of $\mathcal{L}$-structures under $\equiv$. But these equivalence classes would be proper classes, and it is "iffy" to work with a collection of proper classes. This can be rectified by considering only $\mathcal{L}$-structures whose cardinality is no greater than that of $\mathcal{L}$.

For each sentence P of $\mathcal{L}$, let $U_{\mathrm{P}} = \{T \in \mathcal{S} \mid \mathrm{P} \in T\}$. The collection of all the U_{P}'s is closed under finite intersections, since $U_{\mathrm{P}} \cap U_{\mathrm{Q}} = U_{\mathrm{P} \wedge \mathrm{Q}}$. Therefore we can use the U_{P}'s as a basis for a topology on $\mathcal{S}$, called the **Stone space** of $\mathcal{L}$. So an open set in this space is any union of U_{P}'s.

Note that $U_{\mathrm{P}} = \mathcal{S} - U_{\sim \mathrm{P}}$, so each U_{P} is **clopen** (closed and open). Therefore the Stone space of $\mathcal{L}$ is **totally disconnected**, meaning that it has a basis of clopen sets. Recall that the only sets that must be clopen in a topological space are the whole space and $\emptyset$, and in **connected** spaces (such as $\mathbb{R}^n$), these are the only clopen sets. The Stone space is also Hausdorff (see the next exercise). Finally, it is not hard to show that the compactness theorem is equivalent to the compactness of the Stone space of first-order languages.

Exercise 39.

(a) Verify that the Stone space of $\mathcal{L}$ is Hausdorff. That is, any two distinct points are contained in a pair of disjoint open sets.

(b) Without assuming the completeness theorem, prove that the compactness theorem is equivalent to the compactness of the Stone space of all first-order languages.

In general topology, a Stone space is any compact, Hausdorff, totally disconnected space. The most well-known Stone space is the **Cantor set**, which has several homeomorphic (topologically equivalent) versions. It can be defined simply as the set of real numbers between 0 and $1/9$ (inclusive) that have a decimal expression using only 0's and 1's, under the relative topology induced by $\mathbb{R}$. A version that is easier to visualize is the set of real numbers between 0 and 1 that have a base 3 expression using only 0's and 2's. (See Figure 5.2. Here, the Cantor set is the complement in $[0, 1]$ of the union of all the sets A_n.) The Cantor set is also the product space $\{0, 1\}^{\mathbb{N}}$, where $\{0, 1\}$ is given the discrete topology. Finally, any set of the form $\mathcal{P}(A)$, where A is denumerable, has a natural Cantor set topology induced by the standard bijection between $\{0, 1\}^{\mathbb{N}}$ and $\mathcal{P}(\mathbb{N})$. We will say more about the Cantor set in Section 6.5.

Tarski's undefinability theorem

The main syntactic version of Tarski's truth theorem was presented in Chapter 4 (Theorem 4.14). Here is a stronger semantic result that is also sometimes referred to as Tarski's truth theorem:

Theorem 5.45 (Tarski's Undefinability Theorem). *Let $\mathfrak{N}$ be the standard model of arithmetic. Then $Th(\mathfrak{N})$ (as a set of Gödel numbers) is not arithmetical.*

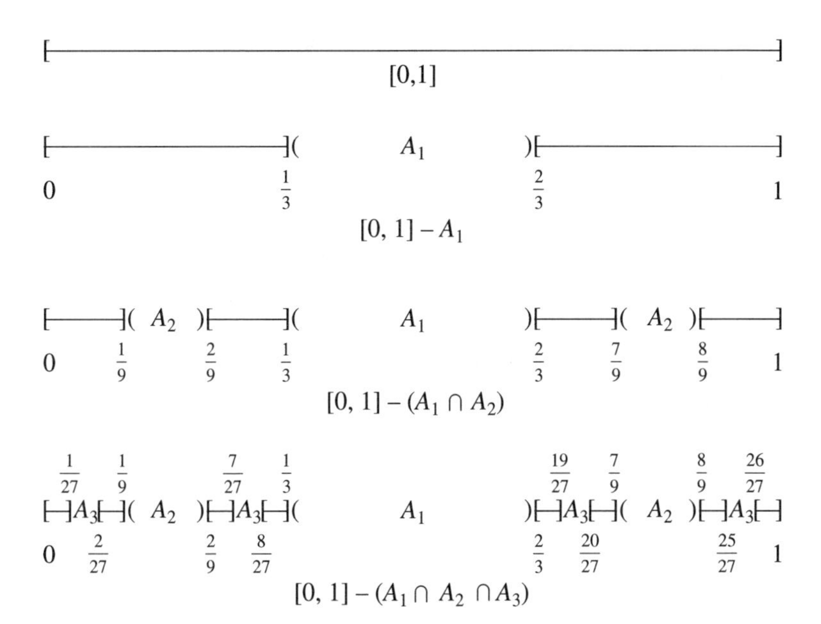

Figure 5.2. The first three stages in the construction of the Cantor set

Proof. Assume on the contrary that $Th(\mathfrak{N})$ is definable in $\mathfrak{N}$, so we have $Th(\mathfrak{N}) = \{n \in \mathbb{N} \mid \mathfrak{N} \models \mathrm{R}(n, b_1, \dots, b_k)\}$, for some formula R and some $b_1, \dots, b_k \in \mathbb{N}$. Replacing each b_i by its numeral changes R into a formula P(n) with one free variable. (In other words, definability in $\mathfrak{N}$ implies $\emptyset$-definability, as we mentioned earlier.) So, for any sentence Q, #Q $\in Th(\mathfrak{N})$ (that is, $\mathfrak{N} \models$ Q) if and only if $\mathfrak{N} \models \mathrm{P}(\overline{\#\mathrm{Q}})$.

Now apply the fixed point lemma (Lemma 4.7) to $\sim$ P, with T being PA, to obtain a sentence Q. So PA $\vdash [\mathrm{Q} \leftrightarrow \sim \mathrm{P}(\overline{\#\mathrm{Q}})]$. Since $\mathfrak{N}$ is a model of PA, we have $\mathfrak{N} \models [\mathrm{Q} \leftrightarrow \sim \mathrm{P}(\overline{\#\mathrm{Q}})]$. This clearly contradicts the previous paragraph. ■

Exercise 40. Assuming Tarski's undefinability theorem (but not the fixed point lemma), prove Theorem 4.14 for any theory T that is satisfied by $\mathfrak{N}$.

Since we know, from the previous section, that all recursive relations are definable in $\mathfrak{N}$, Tarski's undefinability theorem is also a strengthening of Corollary 4.12.

Second-order model theory

In Section 1.6 we described second-order logic. Now that we know something about model theory, we can enter into a more meaningful discussion of this subject.

Let $\mathcal{L}$ be a first-order language. We will denote the second-order language based on $\mathcal{L}$ by $\mathcal{L}^+$. Recall that the essential new ingredients of $\mathcal{L}^+$ are an infinite list of n-ary relation variables for each $n \in \mathbb{Z}^+$, terms based on these relation variables, and quantification over relation variables.

Now let $\mathfrak{A}$ be any $\mathcal{L}$-structure. Then $\mathfrak{A}$ can also be viewed as an $\mathcal{L}^+$-structure, and indeed there is no need to make a distinction between $\mathcal{L}$-structures and $\mathcal{L}^+$-structures. When $\mathfrak{A}$ is viewed as an $\mathcal{L}^+$-structure, a quantified n-ary relation variable is naturally interpreted as ranging over *all* possible subsets of A^n. This is an essential requirement of second-order semantics.

The next step is to define the second-order versions of the two relations denoted by $\models$. Of course, the definition of an assignment g becomes more complex in the second-order setting: an assignment must not only map each individual variable of $\mathcal{L}^+$ to an element of A, but must also map each n-ary relation variable of $\mathcal{L}^+$ to a subset of A^n. With this modification, the definition of the second-order version of $\mathfrak{A} \models \mathrm{P}[g]$ becomes straightforward. Then we can define $T \models \mathrm{P}$, where T and P consist of $\mathcal{L}^+$-formulas, just as in Section 5.2.

Section 5.3 presented the three main results of first-order model theory: the completeness theorem, the compactness theorem, and the LST theorem. Let us now consider these theorems in the context of second-order logic.

One standard, concise version of the completeness theorem is that $T \vdash \mathrm{P}$ if and only if $T \models \mathrm{P}$. But model theorists tend to view this theorem in a different light. To a model theorist, the basic ingredients of "a logic" are a formal language (with connectives, perhaps quantifiers, and grammatical rules for the formation of formulas), and some reasonable definition of structures and models for that language. In other words, the semantic notions $\mathfrak{A} \models \mathrm{P}$ and $T \models \mathrm{P}$ are basic, necessary parts of a logic; the syntactic notion $T \vdash \mathrm{P}$ is not.

In this context, the completeness theorem for a logic becomes the assertion that *there exists* a notion of deduction, based on some clear-cut, mechanical procedure for manipulating formulas of the language, such that $\vdash$ and $\models$ coincide. In this sense, the completeness theorem is false for second-order logic, and for most of the other important logics that extend first-order logic. In other words, there is no reasonable way to axiomatize second-order logic in such a way that the provable statements are precisely the valid ones. We will prove this shortly.

In contrast to the completeness theorem, the compactness theorem and the LST theorem make no mention of $\vdash$ in their statements.

Proposition 5.46. *The compactness theorem fails for second-order logic.*

Proof. Let T be the following second-order theory: for each n, T includes the (first-order) sentence that states that there are at least n distinct elements. Also, let R be a binary relation variable. T includes the statement that, for every R, if R defines a total ordering, then there is a least element and a greatest element in this ordering. This statement is true in every finite structure but in no infinite structures. Therefore, every finite subset of T is satisfiable, but T is not. ■

Corollary 5.47. *The completeness theorem fails for second-order logic.*

Proof. Recall that the compactness theorem for first-order logic is a simple corollary of the completeness theorem for first-order logic. If there were a notion of deduction for second-order logic (in which proofs are finite) making the completeness theorem hold, then the compactness theorem would hold as well. ■

Proposition 5.48. *The LST theorem fails for second-order logic.*

Proof. The defining properties of a complete ordered field can be stated (as a single sentence, if desired) in second-order logic. Every model of this sentence is isomorphic to the field $\mathbb{R}$ and so has cardinality $2^{\aleph_0}$. ■

An alternative proof of this proposition would be to consider second-order Peano arithmetic, as described in Section 1.5. Every model of this theory is isomorphic to the usual structure with domain $\mathbb{N}$, and so is denumerable.

So none of the major theorems of first-order logic holds for second-order logic. This situation highlights the specialness of first-order logic. In fact, a theorem due to Tom Lindström asserts that first-order logic is the only "reasonable" logic that satisfies both the compactness and LST theorems. Here, "reasonable" means that the relation $\models$ satisfies the obvious defining conditions with respect to the standard connectives and quantifiers. (For example, $\mathfrak{A} \models P \land Q$ should hold if and only if $\mathfrak{A} \models P$ and $\mathfrak{A} \models Q$.) Of course, this special status of first-order logic does not necessarily make first-order logic superior to other logics. On the contrary, the LST theorem demonstrates a serious limitation of first-order logic and is one of the primary reasons for considering stronger logics.

CHAPTER 6

Contemporary Set Theory

6.1 Introduction

Why should students of mathematics want to know something about contemporary set theory? Here is one answer: the independence results that sparked the modern era of set theory are surprising and fascinating, in some ways even more so than the incompleteness theorems discussed in Chapter 4. That's because the statements that were shown to be independent of set theory in the 1960s, notably the axiom of choice and the continuum hypothesis, are basic and natural mathematical statements that many people thought should be settled by a sensible set of axioms for set theory such as ZF. In contrast, none of the arithmetical statements that have been shown to be independent of Peano arithmetic, such as those discussed in Chapter 4, is in the realm of "mainstream" mathematics.

Another reason to know something about contemporary set theory is that the methods that were introduced to prove the independence results are extremely powerful and provide a great deal of insight into the cumulative hierarchy of sets. These methods fueled a huge increase in the quantity and sophistication of research in set theory. No other development since the discovery of Russell's paradox has provided nearly as much stimulation to the subject as the independence results and the techniques behind them.

Some more history of set theory

We now return to the historical narrative that was begun in Chapter 2. Recall that the second phase in the history of set theory, covering the first few decades of the twentieth century, began with the creation of more refined versions of set theory than those of Cantor and Frege, with the goal of avoiding contradictions. This period also brought the realization that ZFC is a very strong and versatile theory, apparently capable of interpreting and carrying out all of contemporary mathematics.

This period was also marked by continuing efforts to settle the two famous problems left open by Cantor. Recall that Cantor used the well-ordering principle to show that the ordering on sets by cardinality is total, but he failed to prove the well-ordering principle. The first proof of that was supplied by Zermelo in 1904, but he needed the axiom of choice. In the process, Zermelo was the first to identify and name this important postulate. Soon it was shown that AC is equivalent to the well-ordering principle, the totality of the ordering on cardinals, and several other useful postulates such as Zorn's lemma and the Hausdorff maximal principle. But all attempts to prove or disprove any of these statements from the more acceptable axioms of ZF failed.

Moreover, AC caused substantial controversy in the mathematical community. Since it says that one can define a set based on an infinite number of unspecified choices, it was unacceptable to most constructive-minded mathematicians. But most mathematicians of a more abstract bent viewed it as intuitively correct. AC also has pros and cons from a practical standpoint. It is needed to prove many useful and plausible results, for example, that every vector space has a basis (Theorem 2.20) and that every field has an algebraic closure. But it also has some inconvenient consequences such as the existence of nonmeasurable sets of reals, and some bizarre ones as well, notably the **Banach–Tarski paradox**: a solid sphere can be decomposed into a finite number of pieces that can be reassembled (using only "rigid motions," translation and rotation) into two spheres of the same radius as the original! (See [Jech73] or [Wag].)

There is an interesting weak version of the axiom of choice known as the **axiom of dependent choices** (DC). It states that if R is a binary relation on a set A and $Dom(R) = A$, then there exists an infinite sequence (a_n) such that $(a_n, a_{n+1}) \in R$ for every n in ω. In essence, it says that one can make an infinite sequence of choices when, at each stage, there is a nonempty set of possibilities that can depend on the choices made so far. This axiom provides most of the "benefits" of the full AC; in particular, it implies the usual AC (as defined in Section 2.3) restricted to countable sets x. At the same time, DC avoids most of the unpleasant consequences of AC, such as sets of reals that are not Lebesgue measurable. In spite of this, almost all mathematicians accept and use the full axiom of choice.

Naturally, early set-theorists who considered AC to be true would have liked to prove it, preferably in ZF or a weaker theory. As we will soon see, this problem was eventually settled, but in a different way. By the middle of the century, AC was accepted by the vast majority of mathematicians, but sets whose definition requires AC are always less concrete, harder to visualize, than sets defined without AC. Therefore, many mathematicians prefer to avoid AC when they can. For more about the axiom of choice, see [Moo] or [Jech73]. The former reference is an especially good source for the history of AC and of set theory in general.

In contrast to AC, which most mathematicians think is intuitively correct, few mathematicians have given compelling reasons for believing the continuum hypothesis (CH), let alone the generalized continuum hypothesis (GCH). One piece of evidence that Cantor presented for CH, a corollary of the Cantor–Bendixson theorem of 1884, is that CH does hold for closed subsets of $\mathbb{R}$: every uncountable closed set of reals has the same cardinality as $\mathbb{R}$. In 1916, Hausdorff extended this result to all Borel sets of reals. (We will define and discuss Borel sets in Section 6.5.) Since the majority of sets dealt with in real and complex analysis are Borel, this result is quite useful. However, as Hausdorff himself pointed out, "most" (in the sense of cardinality) sets of reals are not Borel: there are $2^{(2^{\aleph_0})}$ sets of reals, only $2^{\aleph_0}$ of which are Borel. (This is the same sense in which most real numbers are irrational.) So there is no obvious reason to expect this result to generalize to all subsets of $\mathbb{R}$. On the other hand, the fact that no one could find any counterexamples even to GCH gave these conjectures plenty of credence. CH and GCH were eventually settled in the same sense that AC was.

Some good general references for the material in this chapter are [Jech78], [JW], [Kan], [Kun], and [TZ].

6.2 The relative consistency of AC and GCH

The first part of the answer was provided in 1938 by Gödel, who proved that if ZF is consistent, then so is ZFC + GCH. In other words, if ZF is consistent, then neither AC nor GCH can be disproved in ZF. This was the first major **relative consistency** result obtained in set theory. Since the second incompleteness theorem tells us that ZF cannot prove its own consistency, *relative* consistency results are the best we can hope for without introducing controversial axioms. Gödel's result does not show that AC and GCH are provable or true, but it certainly provides evidence in favor of believing them. The method Gödel used to prove these results is very interesting and important in its own right, so we will describe it in some detail.

By the way, it turns out that AC is provable in ZF + GCH. (See [Coh] or [Dra].) So the references to AC in the previous paragraph are technically redundant.

We will frequently be talking about models of set theory. By definition, these must be structures of the form (A, R), where A is a nonempty set and R is a binary relation on A. In this chapter, R will always be the membership relation $\in$ on A, and so we will sometimes be a bit imprecise and refer to a set (or even a proper class) as if it were a structure for set theory. In particular, when we refer to a definable subset of A, we really mean a subset of A that is definable in $(A, \in)$.

Recall the definition of the cumulative hierarchy of sets in Section 2.5. Gödel decided to define a more "selective" hierarchy, the **constructible hierarchy**:

Notation. We define a set L_α for each ordinal α, by transfinite induction:

(a) $L_0 = \emptyset$.

(b) For every α, $L_{\alpha+1}$ consists of all definable subsets of L_α, that is, sets of the form $\{x \in L_\alpha \mid \mathrm{P}(x)\}$, where P is a formula of ZF with all quantified variables interpreted as ranging over L_α, and all free variables other than x interpreted as particular elements of L_α.

(c) If λ is a limit ordinal, then $L_\lambda = \bigcup_{\alpha<\lambda} L_\alpha$.

We say that x is **constructible** if $x \in L_\alpha$ for some α. The proper class of all constructible sets is denoted L. Even though L is not a set, there is no problem talking about L in ZF, just as there is no problem talking about the class of all ordinals. The statement that every set is constructible (formally, $\forall x\, \exists\alpha(x \in L_\alpha)$) is called the **axiom of constructibility**, written simply as $V = L$. We will write ZFL as an abbreviation for ZF $+$ $(V = L)$.

Proposition 6.1.

(a) *L_α is transitive, for every α. So is L.*

(b) *If $\alpha < \beta$, then $L_\alpha \subset L_\beta$.*

Exercise 1. Prove this proposition. The proof is nearly identical to that of Proposition 2.21.

It is worthwhile to give some thought to the meaning of transitivity. To say that a set A is transitive means that sets in the structure $(A, \in)$ are what they "ought to be." To understand this, suppose that A is not transitive. Then it could occur that $\mathbb{R} \in A$, but no real numbers are in A. This would mean that in $(A, \in)$, $\mathbb{R}$ "looks like" the empty set. So the statement that "the set of real numbers is in A" would be technically true, but quite misleading.

The constructible hierarchy differs from the cumulative hierarchy in just one way: instead of throwing in all subsets of the previous stage at successor stages, now we only include definable subsets of the previous stage. Since every subset of a finite set is definable, it follows that

$L_\alpha = V_\alpha$ for $\alpha \leq \omega$. Then $L_{\omega+1}$ must contain all arithmetical subsets of ω such as the set of all primes, the set of all counterexamples to Goldbach's conjecture, etc. But while $V_{\omega+1}$ is uncountable, $L_{\omega+1}$ must be denumerable because L_ω is denumerable. It then follows by transfinite induction that L_α is denumerable whenever α is denumerable.

Exercise 2. Verify all the claims made in the last four sentences.

Proposition 6.2. *Each ordinal α is in $L_{\alpha+1} - L_\alpha$.*

Exercise 3. Prove this proposition, by induction on α. The key point is that α is always a definable subset of L_α.

So we can think of L as a thinner V-shape than V, again with the ordinals as a "spine" (Figure 6.1). Gödel reasoned that L is similar enough to V that all the axioms of ZF should still be true "in L," and that because the construction of L is "stingy" about adding subsets of a given set, GCH should be true "in L." More formally:

Definition. For any formula P of set theory, the **relativization** of P to L, denoted P^L, means the formula P with every quantified variable restricted to L. In other words, every subformula of P of the form $\forall x$ Q

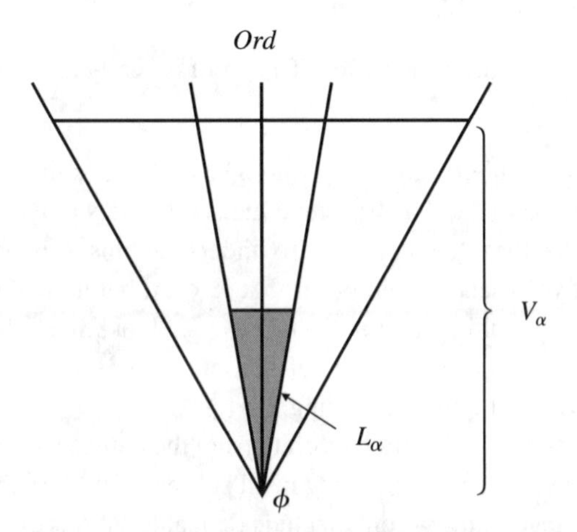

Figure 6.1. The constructible hierarchy

(respectively, $\exists x$ Q) is replaced by $\forall x(x \in L \to \mathrm{Q})$ (respectively, $\exists x(x \in L \wedge \mathrm{Q})$).

Theorem 6.3. *If* P *is any proper axiom of* ZF, *then* $\mathrm{ZF} \vdash \mathrm{P}^L$.

Proof. We give the proof for two of the axioms, starting with one of the simplest ones. Except for replacement, none of the axioms is particularly difficult to deal with.

Let P be the pairing axiom. Then P^L is literally

$$\forall x, y \in L\, \exists z \in L\, \forall u \in L(u \in z \leftrightarrow u = x \vee u = y).$$

Because L is transitive, this just comes to $\forall x, y \in L(\{x, y\} \in L)$. To prove this, assume $x, y \in L$. Then x is in some L_β and y is in some L_γ. Letting α be the larger of β and γ, we have $x, y \in L_\alpha$. Then $\{x, y\}$ is clearly a definable subset of L_α, so $\{x, y\} \in L_{\alpha+1}$.

Now let P be the power set axiom. Then P^L is

$$\forall x \in L\, \exists y \in L\, \forall u \in L[u \in y \leftrightarrow (u \subseteq z)^L].$$

If we again apply the transitivity of L a couple of times, this simplifies to $\forall x \in L[(\mathcal{P}(x) \cap L) \in L]$. Now, for each u in $\mathcal{P}(x) \cap L$, there's a least α such that $u \in L_\alpha$. By replacement, we can form the set of all these α's and then obtain their least upper bound β. Then $\mathcal{P}(x) \cap L = \{u \in L_\beta \mid u \subseteq x\} \in L_{\beta+1}$. ■

The proof of Theorem 6.3 illustrates an important concept "from both sides." Suppose that $x, y, z \in L$. Then this proof shows that the formula $(z = \{x, y\})^L$ is equivalent (in ZF) to $z = \{x, y\}$. We describe this situation by saying that the formula $z = \{x, y\}$, as well as the term $\{x, y\}$, is **absolute for** L. That is, a formula $\mathrm{P}(x_1, x_2, \ldots, x_n)$ is absolute for L if

$$\mathrm{ZF} \vdash \forall x_1, x_2, \ldots, x_n \in L(\mathrm{P} \leftrightarrow \mathrm{P}^L).$$

One can refer to absoluteness of a formula or term for any transitive class.

Example 1. It turns out that many basic terms and formulas of set theory, including most of the axioms of ZFC, are absolute for every transitive class. Examples include the union axiom and the term $\bigcup x$; extensionality; regularity; the terms $\emptyset$ and (x, y); the formula $x \subseteq y$; and the properties of being an ordinal, being a limit ordinal, and being a finite ordinal. Many of these examples are explained by part (a) of the following exercise.

Exercise 4.

(a) Let P be a formula of set theory in which every quantifier is restricted to a set, that is, of the form $\forall v_i \in v_j$ or $\exists v_i \in v_j$. Prove that P is absolute for every transitive class.

(b) Prove that the formulas "x is transitive" and "x is an ordinal" are absolute for every transitive class.

In contrast, the proof of Theorem 6.3 shows that the term $\mathcal{P}(x)$ is not absolute for L. Specifically, the formula $(y = \mathcal{P}(x))^L$ is equivalent to $y = \mathcal{P}(x) \cap L$ rather than $y = \mathcal{P}(x)$, whenever $x, y \in L$. Indeed, we would not be able to prove that the power set axiom holds in L if the power set operation were absolute. For example, we can prove that $(\mathcal{P}(\omega) \cap L) \in L$, but not that $\mathcal{P}(\omega) \in L$. Other important formulas that are not absolute for L are the property of being a von Neumann cardinal, AC, CH, and GCH. By the way, perhaps you can see a link between the nonabsoluteness of $\mathcal{P}$ and Skolem's paradox.

Proposition 6.4. $\mathrm{ZF} \vdash (V = L)^L$.

Proof. $(V = L)^L$ is literally

$$(\forall x \in L)(\exists \alpha \in L)[(\alpha \in Ord)^L \wedge (x \in L_\alpha)^L].$$

We know that every ordinal is in L and the formula $\alpha \in Ord$ is absolute for every transitive class. The formula $x \in L_\alpha$ is not absolute for every transitive class, but it can be shown to be absolute for L. Therefore, $(V = L)^L$ is equivalent to $(\forall x \in L)(\exists \alpha)[\alpha \in Ord \wedge (x \in L_\alpha)]$, which is trivially true. ∎

Theorem 6.5.

(a) *If* P *is any theorem of* ZFL, *then* ZF $\vdash$ P^L.

(b) *If* ZF *is consistent, then so is* ZFL.

Proof.

(a) The previous theorem and proposition take care of all the proper axioms of ZFL. The relativization of any logical axiom to L (or to any class or set, for that matter) is still an axiom or at worst an easy theorem of logic. This is obvious for tautologies but also straightforward for the axioms in Appendix A that involve quantifiers.

It remains to show that our rule of inference modus ponens preserves relativizations to L. But this is trivial: $(\mathrm{P} \to \mathrm{Q})^L$ is the same as $\mathrm{P}^L \to \mathrm{Q}^L$, so from that and P^L we can conclude Q^L by modus ponens.

(b) Assume ZFL is inconsistent. Then some contradiction of the form $\mathrm{P} \vee \sim \mathrm{P}$ is a theorem of this theory. By (a), $(\mathrm{P} \vee \sim \mathrm{P})^L$ is provable in ZF. This is still a contradiction, so ZF is inconsistent. ∎

This theorem should be thought of as a result *about* ZF rather than a result proved *in* ZF. That puts it, and all relative consistency results, in the category of metamathematics. From Chapter 4, we know that such results can generally be formalized and proved in Peano arithmetic, and that is the case for all results of this type that are stated in this chapter.

Theorem 6.6. GCH *(and hence* AC*) is provable in* ZFL.

Proof. Working in ZFL, it is helpful to first prove AC. The idea is simple: a straightforward transfinite induction on α shows that L_α can be well ordered. The key idea is that a well-ordering of L_α induces a well-ordering of all formulas of ZF with constants from L_α, and hence a well-ordering of $L_{\alpha+1}$. Since every set in L is a subset of some L_α, every set in L can be well ordered. (Actually, a stronger result can be proved (in ZF) in this way: there is a single formula of ZF that well-orders all of L.)

The proof of GCH is quite technical. Here is an outline of the proof that CH holds in L; similar reasoning establishes GCH. In the cumulative hierarchy, all subsets of ω are included at the $(\omega + 1)$-st stage, $V_{\omega+1}$. In contrast, only denumerably many subsets of ω are in $L_{\omega+1}$. How far along in the constructible hierarchy might we still be getting new subsets of ω? The answer to this question is the main (nontrivial!) lemma needed for this proof: if $x \in L$ and $x \subseteq \omega$, then $x \in L_\alpha$ for some *countable* α. The number of countable ordinals is $\aleph_1$, and there can be only a countable number of subsets of ω added at each such stage, by the last part of Exercise 2. Therefore, the cardinality of $\mathcal{P}(\omega)$ in L is at most $\aleph_0 \cdot \aleph_1$, which equals $\aleph_1$ (using AC). Since the cardinality of $\mathcal{P}(\omega)$ in L must be greater than $\aleph_0$, it must be $\aleph_1$, as desired. ■

Corollary 6.7. *If* ZF *is consistent, then so is* ZF + GCH.

If Gödel had wanted to be bold, he could have proposed $V = L$ as a new axiom for set theory. This would have settled AC and GCH, by making them provable, and would have created a much "tidier" set theory than ZF or ZFC. Gödel may have briefly considered this, but eventually came to believe that $V = L$ and even CH are probably false. Many set-theorists since Gödel have felt that $V = L$ is not a plausible axiom of set theory, because L seems too limited or "meager" to contain the full cumulative hierarchy. So ZFL is usually considered to be an interesting and important version of set theory, but not a serious candidate for *the* standard version of set theory.

6.3 Forcing and the independence results

After Gödel's achievement, the status of AC, CH, and GCH remained open for another 25 years. Then, in 1963, Paul Cohen proved that if ZF is consistent, then so are ZFC + ($\sim$ CH) and ZF + ($\sim$ AC) + ($\sim$ CH). Not long thereafter, a construction due to Robert Solovay established the relative consistency of ZF + CH + ($\sim$ AC). These results were in a sense more surprising than Gödel's relative consistency results, because most mathematicians believe that AC is true. Cohen's breakthrough becomes even more impressive when one considers that Cohen was not even a specialist in foundations before tackling these problems.

When all these relative consistency results are combined, they become independence results, in the sense of Section 1.4: if ZF is consistent, then CH is independent of both ZFC and ZF + ($\sim$ AC), and AC is independent of both ZF + CH and ZF + ($\sim$ CH).

We now present an outline of the powerful method that Cohen devised to prove these results. For Cohen's own account of his work, see [Coh]. For a more thorough and modern account of the independence results of set theory, see [Jech78] or [Kun]. Before tackling AC and CH, Cohen first proved the weaker result that $V = L$ is not provable in ZF (assuming ZF consistent). We have seen that L satisfies all the axioms of ZFL. Cohen's idea was to make $V = L$ false by taking L and throwing in a very "undefinable" set (of natural numbers, say) that is not constructible. Cohen called the type of set he wanted to adjoin **generic**, and reasoned that if the set had this property, including it would not make any of the axioms of ZF become false.

At this point it will be helpful to clarify some of the results presented in the previous section. We showed that ZF proves the relativization of every theorem of ZFL to L. Intuitively, this says that ZF proves that the structure $(L, \in)$ is a model of ZFL. But this cannot be literally correct, because the completeness theorem says that a first-order theory has a model if and only if it's consistent. Thus, if ZF could prove that ZFL has a model, then ZF could also prove that ZFL is consistent, which would mean (by Gödel's incompleteness theorem) that ZF is inconsistent.

The resolution of this problem is that $(L, \in)$ is not a true model of ZFL, because L is a proper class rather than a set. Therefore, the completeness theorem does not apply to this "class-structure." Instead, we describe the situation of Theorem 6.5(a) by saying that L is an **inner model** of ZFL, meaning that L can be proved to be a transitive proper class in which all the axioms of ZFL hold. ZF has a trivial inner model, namely V. Theorem 6.5(a) provided the first example of a possibly nontrivial inner model of set theory. (We say "possibly" because we cannot prove that $V \neq L$.)

If A is a transitive *set* and $(A, \in)$ is a model of ZF, we call A a **transitive model** of ZF. So the only difference between a transitive model and an inner model is that the domain of a transitive model is a set rather than a proper class. The existence of a transitive model of ZF is a stronger assumption than the consistency of ZF.

Cohen's idea was to construct an inner model of ZF that is just a little bit bigger than L. But, conceptually, it is easier to explain this construction if one starts with an actual (set) model of ZF, and more specifically a transitive one. It turns out that if ZF has a transitive model, then it has a particularly nice one:

Theorem 6.8. *If* ZF *has a transitive model, then the intersection of all transitive models of* ZF *is a transitive model of* ZFL. *This model is of the form* L_λ, *where* λ *is a denumerable limit ordinal, and is therefore denumerable.*

The model described in this theorem is called the **minimal model**; we will denote it $\mathfrak{M}$. Cohen showed that if a generic set G of natural numbers were adjoined to $\mathfrak{M}$, the resulting **generic extension** $\mathfrak{M}[G]$ would have to be a model of $\text{ZF} + \text{GCH} + (V \neq L)$. The definition of the universe of $\mathfrak{M}[G]$ is identical to the inductive definition of the universe L_λ of $\mathfrak{M}$, except that the set G is included when we get to stage $\omega + 1$.

So let us now consider how to construct a generic subset G of ω. Cohen's intention was to construct a set satisfying no special properties. More precisely, G (and therefore $\mathfrak{M}[G]$) should satisfy a property if and only if it is "forced" to satisfy it by a *finite* amount of information. Cohen used this idea to define the method of **forcing**:

Definitions. A **condition** (in the context of one generic subset of ω) is a finite, consistent set of statements about particular numbers being or not being in G. For instance, $\{3 \in G, 57 \notin G, 873 \notin G\}$ is a condition.

We will use the letters c and d to denote conditions. We say that d is an **extension** of c if $d \supseteq c$ (that is, $c \subseteq d$). This defines an important partial ordering on the set of conditions.

Suppose we want to decide how many primes are in our generic set G. Intuitively, a condition could force G to have at least a hundred primes. A condition could force G to have at least a million primes. On the other hand, no condition could force G to have fewer than a million primes. This seems to imply that G must have an infinite number of primes. Similarly, G should have an infinite number of nonprimes, an infinite number of even numbers, an infinite number of odd numbers, etc. Thus, a generic set differs from every definable set.

You might still wonder how a finite amount of information about G could force G to have an infinite number of primes. This works because of the very clever definition of forcing, which is by induction on the structure of the formula being forced. The usual definition uses $\wedge$, $\sim$, and $\forall$ as basic logical operators; the other connectives and the quantifier $\exists$ are then viewed as abbreviations. Here are the inductive clauses. The symbol $\Vdash$ is read "forces":

$c \Vdash (\mathrm{P} \wedge \mathrm{Q})$ if and only if $c \Vdash \mathrm{P}$ and $c \Vdash \mathrm{Q}$.

$c \Vdash \sim \mathrm{P}$ if and only if no extension of c forces P.

$c \Vdash \forall x \mathrm{P}(x)$ if and only if $c \Vdash \mathrm{P}(a)$ for each element a of the appropriate domain.

These clauses do not constitute a full definition of forcing. For one thing, the clause for $\forall$ requires clarification to be rigorous. More seriously, we have not defined $c \Vdash \mathrm{P}$ when P is atomic. In fact, this is the most complicated part of the definition of forcing, and we will not give it. But here is one simple part of it: for any condition c and natural number n, $c \Vdash (n \in G)$ if and only if $(n \in G) \in c$. (We are speaking somewhat imprecisely here, since "$n \in G$" is not a formula of ZF. As we will see shortly, a constant for each element of $\mathfrak{M}[G]$ will be added to the language. So we really mean $c \Vdash (\overline{n} \in \overline{G})$ if and only if $(n \in G) \in c$, where $\overline{n}$ and $\overline{G}$ are the constants for n and G. We will repeat this minor sin several times in this section.)

Exercise 5.

(a) Prove that $c \Vdash (n \notin G)$ if and only if $(n \notin G) \in c$.

(b) Using $\sim(\sim P \wedge \sim Q)$ as the unabbreviated form of $P \vee Q$, prove that $c \Vdash (P \vee Q)$ if and only if $\forall d \supseteq c \exists e \supseteq d(e \Vdash P \text{ or } e \Vdash Q)$.

(c) Using $\sim \forall x \sim P$ as the unabbreviated form of $\exists x P$, prove that $c \Vdash \exists x P(x)$ if and only if $\forall d \supseteq c \exists e \supseteq d \exists a(e \Vdash P(a))$, where a is an element of the appropriate domain.

(d) Using $\sim(P \wedge \sim Q)$ as the unabbreviated form of $P \to Q$, prove that $c \Vdash (P \to Q)$ if and only if $\forall d \supseteq c[d \Vdash P \text{ implies } \exists e \supseteq d(e \Vdash Q)]$.

Here is another fact to bear in mind: adding a generic subset of ω will not change the truth of any formula that is just about natural numbers. In other words, if P is an arithmetical formula (a formula of Peano arithmetic, interpreted in ZF in the standard way), then $c \Vdash P$ is always equivalent to P, no matter what c is. More generally, the truth of arithmetical statements cannot change in any of the models that can be obtained by forcing. This is considered by many to support a Platonist view of the standard model of Peano arithmetic $\mathfrak{N}$: that there is only one such structure (even though there are, of course, nonstandard models of PA), and questions about it have absolute answers.

Here are some simple but important properties of forcing that follow directly from the definition:

Proposition 6.9.

(a) **(Consistency of Forcing).** *$c \Vdash P$ and $c \Vdash \sim P$ can never hold simultaneously.*

(b) *Given any P and c, some extension of c either forces P or forces $\sim P$.*

(c) *If $c \Vdash P$ and $d \supseteq c$, then $d \Vdash P$.*

Proof. Parts (a) and (b) follow immediately from the clause for negation in the definition of forcing. Part (c) is left as an exercise. ■

Exercise 6. Prove part (c) of this proposition, by induction on the structure of P. You may assume that it holds when P is atomic.

Exercise 7.

(a) Prove that $c \Vdash \sim\sim \mathrm{P}$ if and only if $c \Vdash \mathrm{P}$. (The reverse direction of this is trivial. Prove the forward direction by induction on the structure of P. Again, you may assume the result holds when P is atomic.)

(b) Prove that $c \Vdash (\mathrm{P} \rightarrow \mathrm{Q})$ if and only if $\forall d \supseteq c(d \Vdash \mathrm{P}$ implies $d \Vdash \mathrm{Q})$. You may use Exercise 5(d) as a starting point.

Now we are in a better position to see why a generic set G must have an infinite number of primes. In fact, let's show that this statement is forced by the empty condition:

Proposition 6.10. $\emptyset \Vdash [\forall m\, \exists n(n > m \wedge n \textit{ is prime } \wedge n \in G)]$, *where* m *and* n *have* ω *as their domain.*

Proof. By applying the defining clauses of forcing, we obtain:

$\emptyset \Vdash \forall m\, \exists n(n > m \wedge n \text{ is prime } \wedge n \in G)$

iff $\forall m[\emptyset \Vdash \exists n(n > m \wedge n \text{ is prime } \wedge n \in G)]$

iff $\forall m, c\, \exists d \supseteq c,\ \exists n[d \Vdash (n > m \wedge n \text{ is prime } \wedge n \in G)])$, by Exercise 5(c).

To see why this last statement is true, let m and c be given. Choose a prime number n greater than m such that $(n \notin G)$ is not in c. Such an n must exist because c contains only a finite amount of information. Then let $d = c \cup \{n \in G\}$. Since $(n > m \wedge n$ is prime) is then a true arithmetical statement, d must force it. And since $(n \in G)$ is in d, d also forces this conjunct. ■

We still haven't explained how to decide which numbers are actually in G. Is 12 in G, or not? Here is where it is more convenient to work with the minimal model $\mathfrak{M}$ than with the entire class L. Since $\mathfrak{M}$ is denumerable, $\mathfrak{M}[G]$ will be too. Therefore, if we augment the language of ZF by adding a constant for every element of $\mathfrak{M}[G]$, then this augmented language is still denumerable, and we can enumerate the sentences of this language in an infinite sequence $(\mathrm{P}_1, \mathrm{P}_2, \ldots.)$.

It is crucial that we set up this enumeration *before* we actually construct G; this works because of the very orderly way that the constructible hierarchy is formed. Strictly speaking, the constants we add are for all *potential* members of $\mathfrak{M}[G]$; Cohen called this set of constants the **label space** for $\mathfrak{M}[G]$. When G, and thereby $\mathfrak{M}[G]$, is finally defined, many of these constants will turn out to denote the same set, but this creates no problems. For example, there must be constants denoting the sets $G \cup \{1\}$ and $G - \{1\}$ as well as G, but G must equal one of these other sets.

Since ω is denumerable, so is the set of all conditions, and we can also enumerate these in a sequence $(c_1, c_2, \ldots)$. Now we proceed as follows. Let $d_0 = \emptyset$. Inductively, having defined the condition d_n, define d_{n+1} thus: if d_n forces $\sim \mathrm{P}_n$, let $d_{n+1} = d_n$. Otherwise, there must be an extension of d_n that forces P_n. So let d_{n+1} be the first condition in the sequence (c_k) that forces P_n and is an extension of d_n.

A sequence (d_n) formed in this way is called a **generic sequence** of conditions. Every sentence in the sequence (P_n) is eventually decided (that is, forced true or false) by such a sequence. In particular, every sentence of the form $n \in G$, where n is a particular natural number, is decided. Thus a generic sequence determines a generic set G, and so the structure $\mathfrak{M}[G]$ is defined. One remarkable feature of $\mathfrak{M}[G]$ is that the statements that are true in it are precisely those that are forced by some d_n.

Loosely speaking, GCH holds in $\mathfrak{M}[G]$ because it holds in $\mathfrak{M}$ and we haven't added enough new sets to change this; and $V = L$ fails in $\mathfrak{M}[G]$ because the set G is "very undefinable" and therefore not constructible in this model. We have thus given a bare bones outline of a proof of the following result:

Theorem 6.11. *If* ZF *has a transitive model, then* $\mathfrak{M}[G]$ *is a model of* ZF + GCH + $(V \neq L)$, *and so* $V = L$ *is independent of* ZF + GCH.

Since the assumption that ZF has a transitive model is stronger than the assumption that ZF is consistent, this theorem is not quite what we want. That is why it is more fruitful to adjoin the generic set G to the inner model L, rather than to $\mathfrak{M}$. This may be viewed as a purely *syntactic* approach to forcing in which models are not even mentioned. Using this approach, here is what Cohen proved.

Theorem 6.12. *The following facts are provable in* ZF *(for each individual* P *and* Q*, not with* P *and* Q *being quantified variables):*

(a) *For each axiom* P *of* ZF*, including logical axioms,* $\emptyset \Vdash \mathrm{P}$. *Also, if* $\emptyset \Vdash \mathrm{P}$ *and* $\emptyset \Vdash (\mathrm{P} \rightarrow \mathrm{Q})$*, then* $\emptyset \Vdash \mathrm{Q}$. *In other words, modus ponens preserves the property that* $\emptyset \Vdash \mathrm{P}$. *Therefore,* $\emptyset \Vdash \mathrm{P}$ *for each theorem* P *of* ZF.

(b) $\emptyset$ *forces* GCH *and* $V \neq L$.

Part (b) of this theorem is specific to the set of conditions we have been using, conditions designed to produce a single generic subset of ω. But part (a) holds no matter what generic set the forcing conditions are designed to produce, and is therefore an important lemma for all results based on forcing.

Now we may reason as follows: suppose that ZF + GCH + $(V \neq L)$ is inconsistent. Then there is a proof of a contradiction $\mathrm{P} \wedge \sim \mathrm{P}$ in this theory. For each step of this proof, it is provable in ZF that $\emptyset$ forces it, by Theorem 6.12. In particular, it is provable in ZF that $\emptyset$ forces $\mathrm{P} \wedge \sim \mathrm{P}$. But by Proposition 6.9(a), $\emptyset \Vdash (\mathrm{P} \wedge \sim \mathrm{P})$ is a contradiction. Thus ZF is inconsistent. This reasoning, which is similar to the proof of Theorem 6.5(b), can be formalized in PA.

Cohen went on to prove analogous results for AC and CH, in each case using a much more complex set of conditions and label space than the one described above, with labels for many generic sets rather than just one, and a corresponding set of conditions. Starting from $\mathfrak{M}$, he obtained models of ZFC+($\sim$ CH) and ZF+($\sim$ AC). As above, one can eliminate the use of models in these proofs and turn them into relative consistency proofs that are formalizable in PA. Here is a summary of all the independence results we have discussed:

Theorem 6.13 (Main Independence Result, Provable in PA). *If* ZF *is consistent, then:*

(a) $V = L$ *is independent of* ZF + GCH.

(b) CH *and* GCH *are independent of both* ZFC *and* ZF + ($\sim$ AC).

(c) AC *is independent of both* ZF + CH *and* ZF + ($\sim$ CH).

The reader who consults [Coh] should be aware that Cohen used a different inductive definition of forcing from what we have presented. Let's write $\Vdash^{\mathrm{C}}$ to denote Cohen's forcing. Then the relationship between the two versions is that our $d \Vdash \mathrm{P}$ is equivalent to $d \Vdash^{\mathrm{C}} \sim\sim \mathrm{P}$, but weaker than $d \Vdash^{\mathrm{C}} \mathrm{P}$. So the forward direction of Exercise 7(a) fails for Cohen's forcing. (The reverse direction still holds, because Proposition 6.9 is still true for Cohen's forcing.) Because of this, Cohen's forcing was called **strong forcing** and what we have presented was called **weak forcing**. Since Exercise 7(a) is not correct for strong forcing, it is clear that Theorem 6.12 cannot hold for strong forcing either. Cohen used strong forcing as his primary notion, but then essentially used weak forcing (by inserting "$\sim\sim$" before every formula) to state his main results. Eventually, weak forcing became the standard way to define forcing, so I have conformed to this usage.

6.4 Modern set theory and large cardinals

The third main phase in the development of set theory began with Cohen's monumental achievement. In 1963, the two problems that had dominated set theory for most of the century were settled—or were they? The independence of AC and CH from ZF obviously ended certain efforts, but it did not end interest in those postulates. Since almost all mathematicians had long believed that AC is a correct principle of reasoning, it became nearly universal to include it as an axiom and not worry about justifying it from more basic principles. But Cohen's work actually stimulated, rather than diminishing, efforts to provide compelling evidence for or against CH and GCH.

Indeed, the invention of forcing has led to decades of very fruitful research in set theory and the foundations of mathematics. Within a couple of years, numerous people refined and simplified the method. In Cohen's original work, a condition consisted of partial (but not necessarily finite) information about the makeup of one or more sets. Soon it was realized that just about any partially ordered set of objects can be used as conditions. Then the notion of a generic set G and a generic extension $\mathfrak{M}[G]$ of a given transitive model $\mathfrak{M}$ can be defined directly from the ordering on the conditions, without even defining forcing explicitly. This generalization made the method much more versatile.

A more dramatic reinterpretation of forcing was the method of **Boolean-valued models**, created by Scott, Solovay, and Petr Vopěnka. Boolean-valued models were a brilliant application of the older notion of many-valued logic, and became the preferred method of proving independence results for many set theorists.

Forcing, in its several forms, became an extremely powerful tool for constructing models and proving independence results. For one thing, the independence of CH and GCH turned out to be only the beginning of the story. There are surprisingly few limitations on the cardinality of $\mathbb{R}$, or the cardinalities of power sets in general. If ZFC is consistent, it stays consistent with the addition of all sorts of combinations of statements about the cardinalities of power sets.

To make this more precise, we need a few definitions. For the rest of this section and the next one, we always work in ZFC (rather than ZF), unless noted otherwise. We also use the von Neumann (initial ordinal) definition of cardinals, so ω_α and $\aleph_\alpha$ are synonymous.

Definitions. A cardinal ω_α with $\alpha \neq 0$ is called a **successor cardinal** or a **limit cardinal** according to whether α is a successor or a limit ordinal. A **strong limit cardinal** is an uncountable cardinal κ such that $2^\mu < \kappa$ whenever $\mu < \kappa$.

It is clear that a strong limit cardinal must be a limit cardinal, and that GCH implies the converse.

Definitions. Let λ be a limit ordinal. The **cofinality** of λ, denoted $cf(\lambda)$, is the smallest cardinality of any cofinal (unbounded above) subset of λ. An infinite cardinal κ is called **regular** if $cf(\kappa) = \kappa$, and **singular** if $cf(\kappa) < \kappa$.

Example 2. The set $\{\omega_n : n \in \omega\}$ is denumerable but is cofinal in ω_ω. So ω_ω has cofinality ω, making it a singular cardinal.

An alternative definition of singularity that does not require thinking of cardinals as ordinals is to say that an infinite cardinal κ is singular if there is a collection $\{A_i \mid i \in I\}$ such that $I \prec \kappa$ and each $A_i \prec \kappa$, but $\bigcup_{i \in I} A_i \sim \kappa$.

Exercise 8. Show that $cf(\lambda)$ is always a regular cardinal.

Example 3. Since (by AC), the union of a countable collection of countable sets is countable, ω_1 is regular. This reasoning generalizes to show that ω_α is regular whenever α is a successor ordinal. AC is required in this argument. In fact, if ZF is consistent, it cannot prove the existence of uncountable regular cardinals.

Exercise 9. Show that ω is a regular cardinal.

We are now ready to clarify our earlier remark that ZFC imposes few restrictions on the cardinalities of power sets. First of all, it is obvious that $\kappa < \nu$ implies $2^\kappa \leq 2^\nu$. That is, the mapping from κ to 2^κ must be nondecreasing. So this is one simple restriction. It turns out that the only other restriction on this mapping for *regular* κ is provided by the following theorem and corollary:

Theorem 6.14 (König's Theorem). *If $A_i \prec B_i$ for each $i \in I$, then $\bigcup_{i \in I} A_i \prec \prod_{i \in I} B_i$.*

Proof. Assuming the hypothesis, we first want $\bigcup_{i \in I} A_i \preceq \prod_{i \in I} B_i$. That is, we want to show there's a one-to-one function from $\bigcup_{i \in I} A_i$ to $\prod_{i \in I} B_i$. See the next exercise.

Now let $f : \bigcup_{i \in I} A_i \to \prod_{i \in I} B_i$. We must show that f cannot be surjective. For each $i \in I$, let $f_i = (\pi_i \circ f)|_{A_i}$, where $|$ denotes restriction of the domain, and π_i is the projection map to the ith coordinate of $\prod_{i \in I} B_i$. Then $f_i : A_i \to B_i$, and since $A_i \prec B_i$, we can choose $b_i \in B_i - Rng(f_i)$. It is then clear that the element of $\prod_{i \in I} B_i$ whose ith coordinate is b_i for every i is not in the range of f. ■

Exercise 10. Using AC, prove the claim made in the first paragraph of this proof.

The first part of the proof of König's theorem requires AC, to define the required element of $\prod_{i \in I} B_i$. In fact, König's theorem with each $A_i = \emptyset$ says precisely that the Cartesian product of nonempty sets is nonempty, which is one of the standard forms of AC. In other words, König's theorem is another equivalent of AC.

Exercise 11. Prove that König's theorem is also a generalization of Cantor's theorem. You may use the fact that $\mathcal{P}(x) \sim 2^x$ for every set x (Proposition C.1(d), Appendix C).

Corollary 6.15. *For every infinite κ, $cf(2^\kappa) > \kappa$.*

Proof. Assume the contrary. Then there's a function $f : \kappa \to 2^\kappa$ such that $\bigcup(Rng(f)) = 2^\kappa$. In other words, $\sum_{\alpha<\kappa} f(\alpha) = 2^\kappa$. But by König's theorem, $\sum_{\alpha<\kappa} f(\alpha) < (2^\kappa)^\kappa = 2^{\kappa \cdot \kappa} = 2^\kappa$. ∎

Here is the main result on powers of regular cardinals, proved by William Easton [Eas] in 1964. The proof of this theorem was the first significant use of a proper class of conditions, instead of a set of conditions.

Theorem 6.16. *Assume that $\mathfrak{A}$ is a transitive model of* ZF + GCH, *and some formula* P(x, y) *defines a (proper class) function in $\mathfrak{A}$ from the class of all regular cardinals to the class of all cardinals. We write $\nu = F(\kappa)$ for* P(κ, ν). *Suppose further that:*

(a) *F is nondecreasing, that is, $\kappa < \mu \to F(\kappa) \leq F(\mu)$, for all regular κ and μ in $\mathfrak{A}$, and*

(b) *$cf(F(\kappa)) > \kappa$ for every regular κ in $\mathfrak{A}$.*

Then there is a transitive extension $\mathfrak{B}$ of $\mathfrak{A}$ that is also a model of ZFC, *with the same cardinals and cofinalities as $\mathfrak{A}$, in which $2^\kappa = F(\kappa)$ for every regular κ.*

It is also possible to give a version of this result that does not refer to models. Easton's theorem tells us everything we could want to know about powers of regular cardinals: there are two specific restrictions, and that's all. For instance, the cardinality of $\mathbb{R}$ could be enormous, but then GCH could hold for all sufficiently large sets. Or we could have $2^{\aleph_0} = 2^{\aleph_1} = 2^{\aleph_2} = \aleph_5$, and then $2^{\aleph_3} = \aleph_{17}$, etc.

The status of powers of singular cardinals is not as clear. It is tempting to think that complete knowledge of the powers of regular cardinals would determine the powers of singular cardinals, by a sort of continuity argument, but this line of reasoning does not succeed. Powers of singular cardinals are determined by powers of regular ones under the assumption that there is no nontrivial elementary embedding of L into itself. Without this assumption, many questions involving the possibilities for powers of singular cardinals are open, although some important and surprising limitations have been proved by Jack Silver and others.

Exercise 12. Does the assumption that every limit cardinal is a strong limit cardinal imply GCH (in ZFC)?

Large cardinals and the consistency of ZF

While the method of forcing is extremely powerful and has produced an enormous number of important results, it is worth noting that most of these are *negative* results, in the sense that they tell us that various statements cannot be proved (assuming that ZF or some other theory is consistent). What about positive results? For the remainder of this chapter we will discuss attempts to increase our knowledge about sets, including efforts to show that ZF is consistent, and efforts to establish the status of CH and GCH. From the work of Gödel and Cohen, we know that these efforts must go beyond the axioms of ZFC to be productive.

If one wants to make a case for the consistency of ZF, it is natural to search for a model of ZF—not an inner model like L, but a model that is a set. Now, the definition of the cumulative hierarchy makes it natural to ask whether some piece of that hierarchy, some V_α, might already be a transitive model of set theory.

Proposition 6.17 (Provable in ZF). *For any limit ordinal $\lambda > \omega$, every proper axiom of* ZF*, except possibly replacement, is true in V_λ. This result also holds with "*ZF*" replaced by "*ZFC*" in both places.*

Exercise 13. Prove this proposition. The proof is similar to that of Theorem 6.3: all of these axioms and all of the terms involved in them, including $\mathcal{P}(x)$, are absolute for V_λ.

It is much harder for replacement to be true in V_α. For example, let's see why replacement fails in $V_{\omega+\omega}$. The formula $\forall n \in \omega\, \exists! y(y = \omega + n)$ is true in this structure. However, the conclusion that would follow by replacement, namely $\exists b\, \forall y(y \in b \leftrightarrow \exists n \in \omega(y = \omega+n))$, fails because the ordinal $\omega + \omega$ is not in $V_{\omega+\omega}$. Similarly, replacement must fail in V_α whenever α is not a cardinal. Nor is the situation remedied simply by using V_κ, where κ is any old uncountable cardinal:

Exercise 14.

(a) Show that replacement fails in V_{ω_ω}. Generalize your proof from ω_ω to any singular cardinal.

(b) Show that replacement also fails in V_{ω_1}. (Hint: Consider an instance of replacement in which a is the set of all well-orderings on ω.) Generalize your proof from ω_1 to any successor cardinal.

Definition. A regular strong limit cardinal is called **strongly inaccessible**, or simply (an) **inaccessible**.

Lemma 6.18. *If κ is inaccessible and $\alpha < \kappa$, then $\beth_\alpha < \kappa$.*

Exercise 15. Prove this lemma, by transfinite induction on α.

Theorem 6.19. *If κ is inaccessible, then V_κ is a model of* ZF. *Therefore, the existence of an inaccessible implies the consistency of* ZF.

Proof. Let κ be inaccessible. By the previous proposition we only need to verify replacement in V_κ.

Assume $a \in V_\kappa$ and $\mathrm{P}(x, y)$ is a formula such that

$$\forall x \in a \exists!\, y \mathrm{P}(x, y)$$

is true in V_κ. Let $b = \{y \in V_\kappa \exists x \in a[\mathrm{P}(x, y) \text{ is true in } V_\kappa]\}$. Clearly, $b \subseteq V_\kappa$. We want to show that $b \in V_\kappa$.

For each $x \in a$, let $f(x)$ be the rank of the unique y such that $P(x, y)$ is true in V_κ. Since $a \in V_\kappa$ and κ is a limit ordinal, $a \in V_\alpha$ for some $\alpha < \kappa$. Thus $a \subseteq V_\alpha$ because V_α is transitive. Now, a simple induction shows that $Card(V_\alpha) \leq \beth_\alpha$, so $Card(a) \leq \beth_\alpha$. Therefore, $Card(a) < \kappa$ by Lemma 6.18. Since, $f : a \to \kappa$, $Card(a) < \kappa$, and κ is regular, the range of f must be bounded in κ. Let $\beta \in \kappa$ be greater than the LUB of the range of f. Then $b \in V_\beta$, so $b \in V_\kappa$, as desired.

■

The converse of this theorem holds for cardinals but not for ordinals. By considering Skolem functions, it can be shown that if any V_α is a model of ZF, the smallest such α has cofinality ω and so is not an inaccessible cardinal. The reasons that inaccessible cardinals have been studied so extensively as a source of models of set theory are that they are defined in a simple way, and that many people (including Gödel) have proposed plausible arguments for their existence.

Since the existence of inaccessible cardinals implies the consistency of ZF, the search for them must be elusive at best. The existence of even a **weakly inaccessible** cardinal (a regular limit cardinal) cannot be proved in ZFC, or in ZF + GCH for that matter, if ZF is consistent. Certainly, every "obvious" limit cardinal, such as ω_ω, is singular.

Here is an enlightening characterization of inaccessibility:

Proposition 6.20.

(a) *A cardinal κ is weakly inaccessible if and only if it is regular and it is a **fixed point** of the $\aleph$ function, meaning that $\aleph_\kappa = \kappa$.*

(b) *A cardinal κ is inaccessible if and only if it is a regular fixed point of the $\beth$ function.*

Exercise 16. Prove this proposition.

Think about the equations $\aleph_\kappa = \kappa$ and $\beth_\kappa = \kappa$. It might seem that $\aleph_\alpha$ and $\beth_\alpha$ must be huge compared to α. But in fact it is easy to satisfy these equations. Just start with any ordinal, say 0, iterate the $\aleph$ function or the $\beth$ function on it ω times, and then take the union (LUB) of this countable set of cardinals. But even though it is easy to prove the existence of fixed points of these operations, the size of such a fixed point, even of the $\aleph$ function, is difficult to imagine. Of course, a fixed point constructed in this way has cofinality ω and so is not weakly or strongly inaccessible.

And yet, inaccessibles are among the smallest of the so-called **large cardinals**. Here is one natural way to postulate significantly larger ones: by Proposition 6.20(a), a weak inaccessible is a regular cardinal κ such that there are κ smaller cardinals. By analogy, a **hyper-inaccessible** cardinal is a regular κ such that there are κ smaller inaccessibles. Such a cardinal must be inaccessible but is much bigger than an "ordinary" inaccessible. The existence of a hyperinaccessible cannot be proved even from the assumption that there exists a proper class of inaccessibles.

We can continue in this manner: a **hyper-hyperinaccessible** cardinal is a regular κ such that there are κ smaller hyperinaccessibles, and so on. This process can be iterated transfinitely:

Definition. The notion of an α-**hyperinaccessible** cardinal is defined by induction on α:

(a) κ is 0-hyperinaccessible if it is inaccessible.

(b) κ is $(\alpha + 1)$-hyperinaccessible if it is regular and there are κ smaller α-hyperinaccessibles.

(c) For $\text{Lim}(\lambda)$, κ is λ-hyperinaccessible if it is α-hyperinaccessible for every $\alpha < \lambda$.

So, for instance, a 2-hyperinaccessible is the same as a hyper-hyperinaccessible. As early as 1911, Paul Mahlo used this idea of defining large cardinals as fixed points. We will not give the rigorous definition of a **Mahlo** cardinal, but we remark that if κ is Mahlo, then it is α-hyperinaccessible for every $\alpha \leq \kappa$. Once Mahlo cardinals are defined, one can define a whole hierarchy of **hyper-Mahlo** cardinals, using an inductive definition similar to the previous one.

Mahlo cardinals, inaccessible cardinals, and even weakly inaccessible cardinals are much too large to visualize, at least for most of us. And yet, these are all in the category known, a bit playfully, as **small large cardinals**. Roughly speaking, a **large large cardinal** is one for which there is no indication of how to build it up "from below." For example, an inaccessible can be described from below, at least vaguely, by the idea of starting at $\aleph_0$ and going beyond anything that can be reached using power sets and replacement. The most well-known type of large large cardinal was defined by Stanislaw Ulam in 1930. (One could surmise that Ulam liked large, powerful things, since he was also one of the main inventors of the hydrogen bomb.)

Definition. An uncountable cardinal κ is called **measurable** if there is a function $M : \mathcal{P}(\kappa) \to \{0, 1\}$ such that:

(a) $M(\kappa) = 1$;

(b) $M(\{a\}) = 0$ for every $a \in \kappa$;

(c) Whenever $Card(I) < \kappa$ and $\{A_i : i \in I\}$ is a collection of disjoint subsets of κ, $M(\bigcup_{i \in I} A_i) = \sum_{i \in I} M(A_i)$.

The postulate that a measurable cardinal exists is called the **axiom of measurable cardinals**, abbreviated MC.

The function M is called a **nontrivial, 2-valued, κ-additive measure** on κ. It is very different from the standard Lebesgue measure on (some, not all) subsets of $\mathbb{R}$, which we will discuss in the next section. So the word "measurable" applied to cardinals should not be confused with the same word applied to sets of reals. Think of a measure on a cardinal as a partition of *all* subsets of κ into two categories "small" and "large" (measure 0 and 1, respectively). Part (c) of the definition, κ-additivity, is easily interpreted as follows: given such a collection $\{A_i : i \in I\}$, at most one of the A_i's can be large, and their union is large if and only if one of them is large.

By the way, if (c) of this definition is relaxed to include only finite collections of disjoint sets, then the set of large subsets under M is called a **nonprincipal ultrafilter** on κ. Using AC, it can be proved that there is a nonprincipal ultrafilter on every infinite set. With the original stronger (c), the set of large subsets becomes a **κ-complete** nonprincipal ultrafilter on κ. A principal ultrafilter is one in which some one-point set is large, so (b) of the definition fails. Principal ultrafilters are not particularly interesting or useful.

Exercise 17. Prove that if κ is measurable, then:

(a) Every subset of κ of cardinality less than κ is small.

(b) κ is regular.

A measurable cardinal must be inaccessible and in fact hyper-Mahlo, but neither that nor their definition indicates why measurable cardinals are larger than "small large." One way to get a feel for this is to consider the mind-boggling variety (an "exotic zoo," according to [Roi]) of large cardinals that have been postulated. A *partial* list (in increasing order, for the most part) of types of cardinals greater than inaccessible and Mahlo cardinals, but still within the "small large" category, could include weakly compact cardinals, indescribable cardinals, ineffable cardinals, Erdös cardinals, Jónsson cardinals, Rowbottom cardinals, and Ramsey cardinals! Several of these types of cardinals are **partition cardinals**, meaning that they are defined to satisfy some infinitary analog of Ramsey's theorem. For example, a **weakly compact cardinal** is an uncountable κ such that $\kappa \to (\kappa)^2_2$. Ramsey cardinals are at the top of this list, but if κ is measurable, then there are κ Ramsey cardinals less than κ. This tells us that measurable cardinals dwarf these other types of large cardinals. See [Kan] for a definitive treatment of large cardinals. [Dra] is also very good, although somewhat outdated. Both books have concise and informative charts at the back.

There is not much intuitive or intrinsic evidence for (or against) the existence of measurable cardinals. Some well-respected set theorists have expressed strong convictions that they do exist, while a smaller number have voiced the opposite opinion. But most set theorists who work with large cardinals take a more cautious, formalist position and do not claim to have strong arguments for their actual existence. A very thoughtful and readable account of the arguments for and against the existence of large cardinals and other postulates such as CH, as well as the axioms of ZFC themselves, is given in [Mad]. (By the way, Scott proved that $V = L$ implies that there are no measurable cardinals. But few people view this as strong evidence against MC.)

Measurable cardinals are the most well-known and important of the large large cardinals, but they are not the largest ones that have been proposed and studied. Categories of even larger large cardinals that have been considered, again in increasing order, include strong cardinals, Woodin cardinals, weakly hyperWoodin cardinals, hyperWoodin cardinals, superstrong cardinals, supercompact cardinals, extendible cardinals, almost huge cardinals, huge cardinals, and superhuge cardinals. (It may seem puzzling to use the word "supercompact" for something very large. These cardinals derive their name from the fact that certain infinitary languages based on them satisfy the compactness theorem.) Even those who have defined and postulated the existence of these very large sets have, in many cases, made it clear that they can provide no intuitive motivation whatsoever for their existence. The definitions of all of these large large cardinals are beyond the scope of this book.

We have now provided two lists of large cardinals in "increasing order." Exactly what does this mean? One curious and appealing feature of the many types of large cardinals that have been postulated is that they have turned out, at least so far, to be totally ordered "by consistency strength." To make this more precise: if P and Q are statements of set theory, let's write $P \geq Q$ to mean that $\mathrm{Con}(\mathrm{ZFC}+\mathrm{P}) \rightarrow \mathrm{Con}(\mathrm{ZFC}+\mathrm{Q})$ is provable (in ZFC, say, although PA will normally suffice). Trivially, $\geq$ is a preordering. Much less obviously, it also appears to be total when restricted to large cardinal axioms. In many cases, a direct implication between the axioms actually holds, which immediately yields the corresponding relative consistency statement and perhaps more. For instance, let WCC be the postulate that a weakly compact cardinal exists. There must be many weakly compact cardinals below the first measurable cardinal. From this it follows (in ZFC) that MC implies both WCC *and* $\mathrm{Con}(\mathrm{ZFC}+\mathrm{WCC})$. Therefore, $\mathrm{MC} \geq \mathrm{WCC}$, and, under the assumption that (ZFC + WCC) is consistent, $\mathrm{WCC} \not\geq \mathrm{MC}$. (Why?)

Exercise 18. Given that every measurable cardinal is inaccessible and there must be a weakly compact cardinal less than the first measurable cardinal, prove that MC implies the consistency of (ZFC+WCC). Hint: The concept of absoluteness could be helpful.

In a sense, the totality of this ordering is not surprising. After all, the class of all ordinals is totally ordered, in fact well ordered. Therefore (in ZFC), so is the class of all cardinals. But the point is that if large cardinal assumptions were mere speculations with no connection to "reality," there is no reason to think that they would form a hierarchy in the same way that actual cardinals do. In the words of W. Hugh Woodin [Woo, pages 689–690], "To the uninitiated the plethora of large cardinal axioms seems largely a chaotic collection founded on a wide variety of unrelated intuitions. An enduring mystery of large cardinals is that empirically they really do seem to form a well-ordered hierarchy." In other words, the totality of this ordering on large cardinal axioms suggests that the study of large cardinals can be a fruitful path to advances in set theory. Furthermore, the neat hierarchy of large cardinals creates a useful "scale" for measuring the strength of all sorts of set-theoretic postulates, not just large cardinal axioms.

It was mentioned earlier that one of the main reasons for considering new axioms for set theory, including large cardinal assumptions, was to settle CH and GCH. Interestingly, this has not occurred. If ZFC + MC is consistent, then CH is independent of this theory. In fact, it was realized fairly soon after the discovery of forcing that, subject to very modest restrictions, no large cardinal axioms can settle CH. This was noted independently by Cohen and by Azriel Lévy and Solovay. (See [Bro, page 86] or [LS].) What does this tell us? Both Gödel and Cohen expressed the opinion that CH is false, although Cohen later adopted a more formalist point of view—that CH is not meaningful enough to be true or false. More recently, Feferman [Fef99] expressed a similar opinion: "I am convinced that CH is an inherently vague problem that *no* new axiom will settle in a convincingly definite way." It remains to be seen whether this will become the prevalent belief regarding CH.

At the end of this chapter we will discuss some ongoing, deep work by Woodin, aimed at settling CH. This work has produced some interesting circumstantial evidence against CH, but nothing so far that is likely to be convincing to many mathematicians.

6.5 Descriptive set theory

If axioms postulating the existence of very large cardinals are difficult to motivate, aren't needed to prove the consistency of set theory (since an inaccessible is good enough), and can't settle the continuum hypothesis, then why have they been studied so thoroughly? The simplest reason is that they lead to substantial and beautiful mathematics. (In [Mad], this is called extrinsic justification, as opposed to intrinsic or intuitive justification. The clearest reasons for considering CH as an axiom are also extrinsic.) They have all sorts of interesting consequences, which are not only about abstract objects but include some very specific results about sets of real numbers and even sets of integers. For example, Harvey Friedman found a rather natural mathematical statement that is equivalent (in ZFC) to the existence of n-hyper-Mahlo cardinals for every n in ω. He also found a similar statement that cannot be proved in ZFC plus the existence of Ramsey cardinals, assuming this theory is consistent, but can be proved in ZFC + MC [Har, pages 3–4].

The purpose of this section is to illustrate more of these consequences of large cardinal postulates, while also outlining a very interesting and fruitful branch of set theory. We will begin by defining some important generalizations of the arithmetical hierachy on sets of natural numbers. Let $\mathcal{L}$ be the language of second-order arithmetic, which has second-order variables, written in upper case, as well as lower case variables for numbers. For simplicity, let's assume that the only second-order variables are unary relation variables—in essence, variables for sets of numbers. This restriction does not diminish the power of second-order arithmetic, because the bijections B_n of Appendix C allow us to encode n-ary relations as unary relations.

Let P be a formula of $\mathcal{L}$, with j free first-order variables and k free second-order variables. Then, in the standard model of arithmetic $\mathfrak{N}$, P defines (without parameters) a subset of $\mathbb{N}^j \times [\mathcal{P}(\mathbb{N})]^k$. If P has no second-order quantifiers we call the defined subset **arithmetical**, and we can define a hierarchy of these sets just as before. In particular, if $j = 0$ we get a hierarchy of Σ_n^0, Π_n^0, and Δ_n^0 relations on $\mathcal{P}(\mathbb{N})$. This may be viewed as a "hybrid" hierarchy: it categorizes relations on the second-order universe, but allows only first-order quantification.

If P is allowed to have second-order quantifiers (but still no parameters), then the subset of $\mathbb{N}^j \times [\mathcal{P}(\mathbb{N})]^k$ defined by P is called **analytical**. In other words, the analytical sets are precisely the $\emptyset$-definable sets in the second-order structure $\mathfrak{N}$. To define the hierarchy of analytical sets, we count only second-order quantifiers when measuring the complexity of a prenex formula, and we use a superscript of 1 to distinguish analytical sets from arithmetic ones. For instance, a formula of the form $\forall u, v\, \exists w, A\, \forall x\, \exists y\, \forall B\, \exists z(\dots)$, where $(\dots)$ is quantifier-free, would be Σ^1_2, as would the subset defined by such a formula. (If desired, a prenex formula about $\mathfrak{N}$ can always be transformed into an equivalent formula in which all the second-order quantifiers precede all the first-order ones and the sequence of second-order quantifiers alone is unchanged.) Thus, we can refer to Σ^1_n, Π^1_n, and Δ^1_n relations on $\mathbb{N}$ (another "hybrid" hierarchy) or on $\mathcal{P}(\mathbb{N})$. Note that, by definition, the Σ^1_0 and Π^1_0 sets are precisely the arithmetical ones.

The analytical hierarchy on $\mathbb{N}$ is an important extension of the arithmetical hierarchy. In particular, the Δ^1_1 relations on $\mathbb{N}$ are called **hyperarithmetical** and are studied in recursion theory and foundations. The set of true sentences of Peano arithmetic, which we know is not arithmetical by Theorem 5.45, can be shown to be hyperarithmetical. This class of relations can also be obtained by extending the levels of the arithmetical hierarchy transfinitely, up to a special countable ordinal known as **Church–Kleene** ω_1.

Exercise 19. Using ideas from Theorem 1.6, prove that every arithmetical relation on $\mathbb{N}$ is Δ^1_1.

From this point on, we will concentrate on hierarchies of relations on an uncountable domain. So far in this section, we've referred to $\mathcal{P}(\mathbb{N})$ as that domain. Other common choices are $2^{\mathbb{N}}$, $\mathbb{N}^{\mathbb{N}}$, and $\mathbb{R}$. All four of these sets have the same cardinality and can be interpreted as the domain of second-order variables in $\mathfrak{N}$. (This interpretation is less natural for $\mathbb{R}$ than for the other sets, but it can be done.) Also, the first three have natural topologies that make them "look like" subsets of $\mathbb{R}$. The first two are versions of the Cantor set, while $\mathbb{N}^{\mathbb{N}}$, with the discrete topology on $\mathbb{N}$ and the product topology on the whole space, is called **Baire space** and is homeomorphic to ("looks like") the space of irrational numbers in $\mathbb{R}$. Henceforth we will usually take $\mathbb{R}$ to be our uncountable domain, partly because it is more concise to refer to "sets of reals" than "sets of subsets of $\mathbb{N}$" or "sets of functions from $\mathbb{N}$ to $\mathbb{N}$."

The arithmetical and analytical hierarchies on an uncountable domain such as $\mathbb{R}$ are too "meager" for many purposes; for starters, they contain only denumerably many sets. The simplest way to fix this is to allow second-order parameters in the defining formulas. (Recall that natural number parameters don't help to define new subsets.) Thus we are led to:

Definition. The **finite Borel hierarchy** of relations on $\mathbb{R}$ is defined in the same way as the arithmetical hierarchy, except that second order parameters are allowed. We use the boldface notation $\mathbf{\Sigma}_n^0$, $\mathbf{\Pi}_n^0$, and $\mathbf{\Delta}_n^0$ for the levels of this hierarchy.

Definition. The **projective hierarchy** of relations on $\mathbb{R}$ is defined in the same way as the analytical hierarchy, except that second order parameters are allowed. We use the boldface notation $\mathbf{\Sigma}_n^1$, $\mathbf{\Pi}_n^1$, and $\mathbf{\Delta}_n^1$ for the levels of this hierarchy.

Thus, a relation on $\mathbb{R}$ is projective if and only if it's definable, with parameters, in the second-order structure $\mathfrak{N}$. At this point, we have discussed five hierarchies of relations on $\mathbb{R}$, each based on the type of formula used to define its members. At times, it is also helpful to have more geometric ways of describing some of these hierarchies:

Definition. Let A be a set, and m and n natural numbers with $m < n$. Then a function from A^n to A^m that simply "drops" $n - m$ specific coordinates of each n-tuple is called a **projection (mapping)**.

For instance, the function $f(u, v, w, x, y, z) = (v, x)$ is a projection from A^6 to A^2, assuming A is the domain of all six variables. We used projections from $\mathbb{N}^k$ to $\mathbb{N}$ to define the primitive recursive functions.

Lemma 6.21.

(a) *A relation on* $\mathbb{R}$ *is* $\mathbf{\Pi}_n^1$ *if and only if its complement is* $\mathbf{\Sigma}_n^1$ *(and vice-versa).*

(b) *A relation on* $\mathbb{R}$ *is* $\mathbf{\Sigma}_{n+1}^1$ *if and only if it is a projection of a* $\mathbf{\Pi}_n^1$ *relation (that is, the image of a* $\mathbf{\Pi}_n^1$ *relation under a projection).*

Proposition 6.22. *A relation on* $\mathbb{R}$ *is projective if and only if it can be obtained from a* $\mathbf{\Sigma}_0^1$ *relation by a finite sequence of complementation operations and projections. Therefore, the projective relations are the smallest collection of relations that contains all the* $\mathbf{\Sigma}_0^1$ *ones and is closed under complementation and projection.*

Exercise 20. Prove this lemma and proposition.

This explains why projective sets are called projective. The projective hierarchy is often defined as in this lemma and proposition, in which case "$\mathbf{\Sigma}_0^1$" is usually defined to mean "open." (As we've already seen in Example 21 of Section 5.6, there is often more than one reasonable way to define the bottom level of a hierarchy, with the differences disappearing at higher levels.) By the way, these two results remain true with "projection" replaced by "image under a continuous function."

The previous lemma and proposition also hold for arithmetical relations on $\mathbb{N}$, with lightface instead of boldface notation, naturally.

The last hierarchy we will define is the "boldface" analog of the hyperarithmetical hierarchy.

Definition. A collection of subsets of some set A is called a **σ-algebra** on A if it is closed under complementation and *countable* unions.

Note that a σ-algebra must also be closed under countable intersections, and we could just as well have used those instead of countable unions in the definition. Every σ-algebra is an algebra, but not conversely—consider the collection of all finite or cofinite subsets of $\mathbb{N}$.

Definition. For any topological space, the collection of its **Borel sets** is the σ-algebra generated by its open sets, that is, the smallest σ-algebra containing all the open sets.

This the simplest way to define the Borel sets, but we also want to put them in a hierarchy:

Definition. Let $k \in \mathbb{N}$ be fixed. The **Borel hierarchy** on $\mathbb{R}^k$ (and other topological spaces) is defined by transfinite induction for $0 < \alpha < \omega_1$. For any $A \subseteq \mathbb{R}^k$:

(i) A is $\mathbf{\Sigma}^0_1$ iff it is open. ($\mathbf{\Sigma}^0_0$ and $\mathbf{\Pi}^0_0$ may be left undefined.)

(ii) A is $\mathbf{\Pi}^0_\alpha$ iff its complement is $\mathbf{\Sigma}^0_\alpha$.

(iii) For $\alpha > 1$, A is $\mathbf{\Sigma}^0_\alpha$ iff it is a countable union of sets in $\bigcup_{\beta<\alpha} \mathbf{\Pi}^0_\beta$.

A set that is both $\mathbf{\Sigma}^0_\alpha$ and $\mathbf{\Pi}^0_\alpha$ is said to be $\mathbf{\Delta}^0_\alpha$.

Analogously to Lemma 6.21, it can be shown that this definition of the Borel hierarchy is equivalent, for $0 < n < \omega$, to our earlier definition of the finite Borel hierarchy. The main differences between the Borel hierarchy and the projective hierarchy are that the Borel hierarchy uses countable unions instead of projections to build the next level, and the Borel hierarchy is transfinite. The slightly complex clause (iii) in this definition is needed only at limit ordinals; for successor ordinals, we could simply say that a set is $\mathbf{\Sigma}^0_{\alpha+1}$ if and only if it's a countable union of $\mathbf{\Pi}^0_\alpha$ sets.

Proposition 5.33 holds for all of the hierarchies we have defined. (See the next exercise for part of this, regarding the Borel sets.) Therefore, Figure 5.1 in Section 5.6 applies to all of them, except that the Borel and hyperarithmetical hierarchies continue transfinitely to the right. Furthermore, all of these are *proper* hierarchies, meaning that each inclusion shown is proper. So every level adds new sets.

Exercise 21.

(a) Show that every open ball in $\mathbb{R}^k$ can be written as a countable union of closed balls. (Hint: the points with rational coordinates are dense in $\mathbb{R}^k$.)

(b) From (a) and the known fact that every open set in $\mathbb{R}^k$ is a countable union of open balls, prove that every $\mathbf{\Sigma}^0_1$ set is $\mathbf{\Sigma}^0_2$.

(c) Assuming that Proposition 5.33(a) holds for the Borel hierarchy, prove this slightly stronger version of Proposition 5.33(c) for $\mathbb{R}^k$: if $0 < \beta < \alpha < \omega_1$, then every Boolean combination of $\mathbf{\Sigma}^0_\beta$ sets is $\mathbf{\Delta}^0_\alpha$.

Theorem 6.23 (Lebesgue). *For each* $\mathbb{R}^k$,

$$\bigcup_{0<\alpha<\omega_1} \mathbf{\Sigma}^0_\alpha = \bigcup_{0<\alpha<\omega_1} \mathbf{\Pi}^0_\alpha = \textit{the collection of all Borel sets.}$$

Proof. By definition, all open sets are in the first union. That both of these unions are algebras and contain all open sets follows by part (c) of the above exercise. The regularity of ω_1 guarantees that these two unions are closed under countable unions, so they are σ-algebras. Therefore, each of these unions contains all Borel sets. Finally, the (nontrivial) fact that the Borel hierarchy is a proper hierarchy implies that no proper subcollection of either of these unions can contain all Borel sets. ∎

We need to have the subscript α run through all countable ordinals to obtain a σ-algebra, and therefore to get all the Borel sets. The hierarchies we have defined in which the subscript only goes up to ω yield algebras but not σ-algebras. It is possible to extend the analytical and projective hierarchies transfinitely to obtain σ-algebras, but this is not usually done.

If you're a bit bewildered by having a half dozen different hierarchies under discussion you can now relax a bit, since we will be restricting our attention primarily to the Borel and projective ones. Here is a table showing the relationships among these hierarchies on sets of reals:

2nd-order Parameters	Quantifiers & Levels		
	Over $\mathbb{N}$, ω levels	Over $\mathbb{N}$, transfinite	Over $\mathbb{N}$ and $\mathbb{R}$, ω levels
No	Arithmetical	Hyperarithmetical	Analytical
Yes	Finite Borel	Borel	Projective

Classical descriptive set theory

The study of the Borel and projective hierarchies (referred to as the "classical" hierarchies) is known as (classical) **descriptive set theory**. It was begun by the French mathematicians Emile Borel, René-Louis Baire, and Henri Lebesgue in around 1900, and developed much more fully by the Russian Nikolai Luzin and his students a decade or two later. The "effective" (lightface) hierarchies were not considered until much later; they were developed by Kleene in the 1950s. As we will see, there are many difficult problems involving low levels of the projective hierarchy. For an in-depth treatment of descriptive set theory, see [Kec].

Here is Mikhail Suslin's classic result that places the Borel sets in the projective hierarchy:

Theorem 6.24 (Suslin). *The Borel subsets of $\mathbb{R}^k$ are precisely the $\mathbf{\Delta}^1_1$ ones.*

This theorem is analogous to the result that the hyperarithmetical relations on $\mathbb{N}$, defined via a transfinite hierarchy, are precisely the Δ^1_1 ones. As a corollary, the $\mathbf{\Delta}^1_1$ sets form a σ-algebra. More generally, the $\mathbf{\Delta}^1_n$ subsets of $\mathbb{R}^k$ form a σ-algebra, for each n and k.

A great deal of effort has gone into the study of $\mathbf{\Sigma}^1_1$ sets, also known as **analytic** sets. (It is important not to confuse the terms "analytic" and "analytical.") Let us now list some of the results about these sets that can be obtained without special postulates. These results focus on three areas: cardinality, measure, and topology.

We mentioned the Cantor–Bendixson theorem in Section 6.1. More precisely, it states that every uncountable closed set of reals is the union of a countable set and a **perfect** set, meaning a nonempty closed set with no isolated points. Closed intervals and Cantor sets are simple examples of perfect sets. A set that is either countable or has a perfect subset is said to have the **perfect subset property**. Since it is not hard to show that every perfect subset of $\mathbb{R}$ has cardinality $2^{\aleph_0}$, CH holds for sets with the perfect subset property. After Hausdorff extended the perfect subset property to all Borel sets in 1916, Suslin extended it further:

Nikolai Luzin (1883–1950) did not show special talent for mathematics as a child. In fact, he was a rather poor mathematics student, probably because the instruction he received (in Tomsk, Russia) was based on rote learning and thus was quite uninspiring. But even at Moscow University, Luzin was only an average student. He was strongly influenced by the difficult political and economic conditions of the time, and had some personal crises in which he even considered suicide. For some time, he gave up mathematics in favor of medicine and/or theology, but in about 1909 he finally made a firm, permanent commitment to mathematics.

Luzin obtained important results in many branches of mathematics, including topology, measure theory, functional analysis, differential equations, and especially descriptive set theory. In that field, he greatly advanced and refined the ideas of the earlier French school. Whereas Borel and his contemporaries had freely used the axiom of choice, Luzin chose to concentrate on "effective" sets, ones that could be defined without AC. Luzin and his students tackled the major problems in descriptive set theory for two decades, and thus Luzin is viewed as the primary founder of this branch of foundations. He was also a dynamic lecturer with an outgoing personality, making him extremely popular with students.

In 1936, Stalin's government launched a vicious smear campaign against Luzin, accusing him of "anti-Soviet propaganda." His greatest offense was that he published most of his research in foreign publications. Fortunately, Luzin's international reputation was sufficiently prestigious to protect him from serious harm. The most significant consequence of this campaign was that, for many decades thereafter, Russian mathematicians published almost all their work in Russian journals.

Theorem 6.25.

(a) **(Bernstein).** *Not all sets of reals have the perfect subset property.*

(b) **(Suslin).** *All analytic sets have the perfect subset property.*

Surprisingly, part (b) does not extend to the "dual" class $\mathbf{\Pi}_1^1$, as we will soon see. The main other early cardinality result of projective set theory, due to Wacław Sierpiński, is that every $\mathbf{\Sigma}_2^1$ set is the union of $\aleph_1$ Borel sets. Therefore, the cardinality of a $\mathbf{\Sigma}_2^1$ set can only be countable, $\aleph_1$, or $2^{\aleph_0}$.

Exercise 22. Prove part (a) of the previous theorem, in this stronger form (as proved by Bernstein): there is a set of reals that has cardinality $2^{\aleph_0}$, contains no perfect subset, and intersects every perfect subset. (Hint: $\mathbb{R}$ and the collection of perfect subsets of $\mathbb{R}$ both have cardinality $2^{\aleph_0}$. Therefore, you can use an argument like the proof of Theorem 2.20, based on AC, to construct a set that intersects every perfect set and its complement.)

The early French pioneers of descriptive set theory were also the founders of measure theory. Let's review the definitions of the two most basic types of measure. For simplicity, we will define these measures on the unit interval [0, 1]. It is not much more difficult to define them on $\mathbb{R}$ or $\mathbb{R}^k$. Using a bounded interval avoids the minor inconvenience of having sets whose measure is $+\infty$.

Definition. A **Borel measure** on [0, 1] is a function $m_B : \mathcal{B} \to [0, 1]$, where $\mathcal{B}$ is the collection of all Borel subsets of [0, 1], satisfying:

(i) If A is an interval, $m_B(A)$ is the length of A.

(ii) For any A in $\mathcal{B}$, $m_B(A') = 1 - m_B(A)$, where A' is the complement of A.

(iii) m_B is $\aleph_1$-additive (or **countably additive** or **σ-additive**): the measure of a countable union of disjoint members of $\mathcal{B}$ is the infinite sum of their individual measures.

Borel proved that there is a unique Borel measure on [0, 1]. A few years later, Lebesgue noted that every subset of a set of measure 0 should logically also have measure 0, but the Borel sets do not provide for this. For example, the usual Cantor set (in either of the first two forms defined in Section 5.7) has Borel measure 0, but its cardinality is $2^{\aleph_0}$, the same as $\mathbb{R}$'s. This may seem paradoxical, but it is not hard to prove. Therefore, there are $2^{2^{\aleph_0}}$ subsets of the Cantor set. Not all of them are Borel because, by a straightforward argument, there are only $2^{\aleph_0}$ Borel sets. This led Lebesgue to a way of extending Borel measure to more sets:

Definitions. For any $A \subseteq [0, 1]$:

(a) A **null set** is a subset of a Borel set of measure 0.

(b) A is **(Lebesgue) measurable** if $A \Delta B$ is a null set, for some Borel set B.

Exercise 23.

(a) Show that the measurable sets form a σ-algebra. Therefore, they form the smallest σ-algebra containing all open sets and all null sets.

(b) Show that if A is measurable in "two different ways" (that is, $A \Delta B$ and $A \Delta B'$ are both null), then B and B' must have the same Borel measure.

Definition. If A is measurable, with $A \Delta B$ being null, then the **Lebesgue measure** of A is defined to be $m_B(B)$. Part (b) of the above exercise guarantees that this is a well-defined function.

Here are two classic results about Lebesgue measure:

Theorem 6.26.

(a) *Not every subset of* $[0, 1]$ *is measurable.*

(b) **(Luzin).** *Every Boolean combination of analytic subsets of* $[0, 1]$ *is measurable.*

It can be proved that Lebesgue measure is **translation invariant**, meaning that the measure of a set does not change if the set is merely moved to the left or right. This is certainly a sensible thing to expect of a measure. Using the axiom of choice, Giuseppe Vitali showed how to define a set that cannot be measurable in any translation-invariant extension of Lebesgue measure. So there is no reasonable way to extend Lebesgue measure to all subsets of $[0, 1]$. This may be viewed as a stronger version of (a) of the above theorem. Bernstein's set of Exercise 22 must also be nonmeasurable, and the Banach–Tarski paradox is also based on nonmeasurable sets.

To state the main topological results about analytic sets, we need some definitions. We will give these without much elaboration; if you have little or no experience with topology, you might find it difficult to digest this material without outside reading.

Definitions. Let A be a subset of a topological space. The **interior** of A, $\mathrm{Int}(A)$, is the union of all open sets contained in A. $\mathrm{Int}(A)$ must be open and is therefore the largest open set contained in A.

Similarly, the **closure** of A, $\mathrm{Cl}(A)$, is the intersection of all closed sets containing A. $\mathrm{Cl}(A)$ must be closed and is therefore the smallest closed set containing A.

A is called **nowhere dense** if $\mathrm{Int}(\mathrm{Cl}(A)) = \emptyset$, and **meager** or **first category** if it is a countable union of nowhere dense sets.

Finally, A has the **Baire property** if $A \Delta U$ is meager, for some open set U.

The fundamental result about these concepts is the **Baire category theorem**, which states that a nonempty open set in $\mathbb{R}^k$ cannot be meager.

Lebesgue measurability and the Baire property are very analogous concepts. Null sets and meager sets are two versions of "small" sets—the former measure-theoretic, the latter topological. The Cantor set is meager (in fact, nowhere dense) as well as being null, so a cardinality argument again implies that there are subsets of the Cantor set that are meager but not Borel. The sets with the Baire property also form a σ-algebra, which means that "open" can be replaced by "Borel" in the definition of the Baire property. In other words, both measurable sets and sets with the Baire property are types of "almost Borel" sets. Furthermore, we have these classic results:

Theorem 6.27. *For any positive integer k:*

(a) *Not every subset of* $\mathbb{R}^k$ *has the Baire property.*

(b) **(Luzin and Sierpiński).** *Every Boolean combination of analytic subsets of* $\mathbb{R}^k$ *has the Baire property.*

We already know that the perfect subset property and measurability fail for the set described in Exercise 22. Predictably, this set doesn't have the Baire property either.

The last three theorems have a clear, common theme: analytic sets are well behaved, in at least three important ways, called the **regularity properties**—measurability and the perfect subset and Baire properties. Naturally, Luzin and his contemporaries would have liked to extend the regularity properties to more levels of the projective hierarchy, and ideally to all projective sets. But they were not able to extend any of them, or disprove them, even for the next level. By 1925, Luzin came to what he called the "totally unexpected" conclusion that the question of whether these properties hold for all projective sets was "unsolvable." This bold prediction turned out to be correct, even when restricted to the next level of projective sets:

Theorem 6.28 (Gödel). *Assuming* $V = L$*:*

(a) *There is a* Δ^1_2 *well-ordering of* $\mathbb{R}$.

(b) *There is a* Π^1_1 *set of reals without the perfect subset property.*

(c) *There is a* Δ^1_2 *set that is not measurable and does not have the Baire property.*

Proof. Here is a brief outline of the proofs of (a) and (c). For details, see [Kan, 13.10 and 13.12]. As we mentioned in the proof of Theorem 6.6, there is a single formula of ZF that well-orders all of L. When restricted to $\mathbb{R}$, the binary relation A defined by this formula can be shown to be Σ^1_2. With the assumption $V = L$, A becomes Δ^1_2, since then $(x, y) \notin A$ is equivalent to $[(y, x) \in A \wedge x \neq y]$, a Σ^1_2 formula.

It is not hard to show that this same set A does not have the Baire property and is not measurable. As defined, $A \subseteq \mathbb{R} \times \mathbb{R}$, but the argument can be modified to get a similar subset of [0, 1]. ■

This theorem appeared in Gödel's original 1938 relative consistency paper [Go38], except for the part about the Baire property, which was proved later by Jan Mycielski. However, Gödel's notes give primary credit for the first part of (c) to Ulam.

Note that the sets described in this theorem are not boldface. So the sets in (b) and (c) are very "sharp" counterexamples—not just projective but analytical—to the goal of extending the regularity properties. Of course, since $V = L$ implies GCH, we can't expect the set defined in (b) to provide a counterexample to CH. However, it is known to be (relatively) consistent with ZFC that CH fails and there exist $\mathbf{\Pi}_1^1$ sets of power $\aleph_1$.

Almost three decades later, a powerful result in the opposite direction was obtained, using the new technique of forcing:

Theorem 6.29 (Solovay). *Assuming* MC *(the existence of a measurable cardinal), every* $\mathbf{\Sigma}_2^1$ *set has the regularity properties.*

This theorem is the primary example of a large cardinal hypothesis implying very specific and basic consequences about sets of real numbers. It was an unexpected and exciting development. Another consequence of MC, due to Donald A. Martin, is that every $\mathbf{\Sigma}_3^1$ set is a union of $\aleph_2$ Borel sets. Therefore, the cardinality of every $\mathbf{\Sigma}_3^1$ set must be $2^{\aleph_0}$, or no greater than $\aleph_2$.

In light of Solovay's theorem, one might expect that MC or some other large cardinal hypothesis could be used to extend the regularity properties to more levels of the projective hierarchy. But this is not the case. Here are some of the relative consistency results that show that Solovay's theorem is "best possible":

Theorem 6.30 (Silver). *If* (ZFC + MC) *is consistent, then it remains consistent with the addition of the axiom that there is a nonmeasurable* Δ_3^1 *set without the Baire property, and a* Π_2^1 *set without the perfect subset property.*

Theorem 6.31 (Solovay). *If* (ZFC + *"An inaccessible cardinal exists"*) *is consistent, then so is* (ZFC + *"Every projective set has the regularity properties"*).

Theorem 6.32 (Solovay). *The theory* (ZFC + *"An inaccessible cardinal exists"*) *is consistent if and only if* (ZF + DC + *"Every set of reals has the perfect subset property"*) *is consistent.*

Theorem 6.33 (Solovay). *The theory* (ZFC+MC) *is consistent if and only if* (ZFC + *"There is a countably additive extension of Lebesgue measure to every subset of* $[0, 1]$*"*) *is consistent.*

The last two theorems are in the category of **equiconsistency results**, biconditional relative consistency results. Of course, the extension of Lebesgue measure mentioned in the last theorem cannot be translation invariant, by Vitali's theorem.

We mentioned in the last section that Scott proved that MC implies $V \neq L$. Frederick Rowbottom then obtained the stronger result that the existence of just a Ramsey cardinal implies that $\mathcal{P}(\omega) \cap L$ is countable, so "most" subsets of ω are not constructible. Scott and Silver later sharpened this result by obtaining a nonconstructible Δ^1_3 subset of ω, from the same assumption. (The hypothesis for all of these results can be weakened further to the existence of a Jónsson cardinal.)

This nonconstructible Δ^1_3 subset of ω is a very special set called $0^\#$ ("zero sharp"), whose properties were first investigated by Silver. Even though $0^\#$ is "merely" a set of natural numbers, the existence of $0^\#$ is usually treated as a sort of large cardinal axiom. By the above, this postulate is implied by the existence of a Jónsson cardinal. In turn, if $0^\#$ exists, then so do indescribable cardinals. The existence of $0^\#$ was also proved by Kenneth Kunen to be equivalent to the existence of a nontrivial elementary embedding from L into itself.

6.6 The axiom of determinacy

Along with large cardinal assumptions, the most important category of proposed new axioms for set theory has to do with the existence of winning strategies for infinite games. Let us first examine the simpler situation regarding finite games. Consider a competitive game played between two players ("I" and "II"), with these characteristics:

(1) The game proceeds from a "starting position," with the players taking turns making "moves." Player I moves first.
(2) The number of legal moves at each point is finite.
(3) The game must terminate after a finite number of moves.
(4) There is no element of chance, as in dice games, or secrecy, as in most card games.
(5) There are a finite number of possible outcomes, which are linearly ordered in a consistent way for both players. For instance, these could be "I wins," "tie," and "II wins." The sequence of moves leading to the final position determines the outcome.

Such a game is called a **finite two-person game of perfect information**. Examples are tic-tac-toe, checkers, and chess. (A game of checkers or chess can technically continue forever, but must end after a finite number of moves if sensible rules, such as the fifty move rule in chess, are invoked.) A **strategy** for one of the players is a function that specifies his or her move in every position that could possibly arise for that player when playing according to that strategy.

The fundamental theorem of finite game theory, due to von Neumann, states that in every such game, there is a particular outcome and an "optimal" strategy for each player, which guarantees that outcome or better. For instance, most children learn that tic-tac-toe is always a tie unless somebody makes a mistake. It is not hard to learn a strategy for either player that will never lose. In contrast, no one knows the ideal outcome or optimal strategies for chess. But the theorem guarantees that a chess game between perfect players would always be a win for White, or always a draw, or (theoretically possible but extremely unlikely) always a win for Black.

In the special case of a game whose only outcomes are "I wins" and "II wins" (a **win-lose game**), the theorem simply says that one of the players must have a **winning strategy**—a strategy that always wins for that player. One of the few actual board games of this type is known as Hex. It is easy to show that going first cannot hurt in this game, which implies that Player I has a winning strategy, but no one knows such a strategy. Since a strategy can be coded as a natural number, this is an interesting existence result: we can prove that there's a natural number with a certain (decidable) property, but nobody actually knows one. More information about Hex is available at http://mathworld.wolfram.com/GameofHex.html.

Infinite games

Since the 1930s, mathematicians have been interested in infinite analogs of these win-lose games, in which the game continues for an infinite sequence of moves (a **play**). Let's assume that the set of legal moves is some fixed set M with two or more elements. So the set of plays is M^{ω}, which we'll also call P. The theory of these games does not change significantly if the number of legal moves can vary during the course of the game, as in chess and many other games.

The set of plays is topologized by giving M the discrete topology and then putting the product topology on M^{ω}. The set of plays extending any particular finite sequence of moves is clopen in this topology, and these clopen sets form a basis. Therefore, P is a totally disconnected space. If M is finite, P is the Cantor set. If M is denumerable, P looks like Baire space. In either of these cases, P can be viewed as a set of reals. For simplicity, we will assume unless stated otherwise that $M = \{0, 1\}$, in which case each play can be mapped to a real number simply by putting a decimal point in front of the sequence of bits.

For each $A \subseteq P$, let G_A be the infinite game in which I wins if the play is in A, and II wins otherwise. Typically, we apply adjectives to games that should technically modify A; for example, a Borel game means a game in which A is a Borel set. The notion of a strategy is defined as for finite games, except that now the domain of a strategy is a denumerable set: for I (respectively, II), it is a set of sequences of bits of even (respectively, odd) length. A strategy for I (respectively, II) is a winning strategy for that player in G_A if every play following that strategy is in A (respectively, in $P - A$).

We say that G_A (or A itself) is **determined** if one of the players has a winning strategy. The postulate that every infinite game with $M = \{0, 1\}$ is determined is called the **axiom of determinacy** (AD). Since the finite game version of AD is provable, one could hope that AD would be provable as well. But the construction of a nondetermined game, using the axiom of choice, is straightforward and was already well known when Mycielski and Hugo Steinhaus first proposed AD as a postulate in 1962. In other words, ZFC + AD is inconsistent. However, ZF + AD and even ZF + DC + AD are consistent, as far as anyone knows, and are interesting alternatives to ZFC. We will discuss some of the consequences of the full AD below. By the way, if we consider games in which there are ω_1 moves in every position, then we can prove the existence of a nondetermined game without using AC.

Exercise 24. Using AC (specifically, just a well-ordering of $\mathbb{R}$), prove that there is a nondetermined game with $M = \{0, 1\}$. (Hint: The set of all plays has cardinality $2^{\aleph_0}$. So do the sets of strategies for each player, and the set of plays that follow any particular strategy. Therefore, an argument similar to the proof of Theorem 2.20 creates a set A that "avoids" every strategy for either player.)

Exercise 25. Prove that AD implies that every game in which M is countable is determined. The idea is to show that every such game has an "equivalent" game with $M = \{0, 1\}$.

Because AD contradicts AC, we will take ZF rather than ZFC as our base theory for the rest of this chapter, unless noted otherwise.

Theorem 6.34. *Assuming* AD, *every subset of* $\mathbb{R}^k$ *has the regularity properties.*

Of course, determinacy itself is a sort of regularity property, albeit not one of the three classic ones. This theorem, which was proved by Mycielski and several others, provides interesting extrinsic evidence for considering AD as a new axiom. In ZF + AD, the usual "undesirable" sets of reals cannot exist. By the way, the Banach–Tarski paradox is based on nonmeasurable sets, so this paradox also disappears in ZF + AD. Since DC is almost as useful as AC for most types of mathematics, it would appear that ZF + AD + DC is a very appealing alternative to ZFC as a formal theory for general mathematics. In fact, a fairly useful version of choice is already provable in ZF + AD: every countable Cartesian product of sets of reals is nonempty.

In spite of the appeal of ZF + AD + DC, there have been no serious proposals to give it the kind of status that ZFC has. There are three obvious reasons for this. One is that it contradicts the full axiom of choice, which almost all mathematicians consider to be true. Another is the lack of intrinsic evidence for AD—almost nobody has provided plausible intuitive motivation for the postulate that all games are determined. Mycielski and Steinhaus did try: "If both players are infinitely clever and know what P is, then owing to the complete information during every play, the result of the play cannot depend on chance." From this, they argue that one of the players should be able to force a win no matter how the opponent plays. But this point of view has had little or no support since it was espoused in 1962. The final reason is that full AD is an extremely strong postulate, as evidenced by the next two results:

Theorem 6.35 (Solovay). *Assuming* AD, $\aleph_1$ *and* $\aleph_2$ *are measurable cardinals.*

Theorem 6.36 (Woodin). (ZF + AD) *is consistent if and only if* (ZFC + *"There are infinitely many Woodin cardinals"*) *is consistent.*

These results make it clear that AD is a significantly stronger hypothesis than MC, since a Woodin cardinal is much larger than a measurable one. Therefore, more attention has been given to restricted determinacy postulates than to full AD. Before discussing weaker determinacy postulates, let us consider which infinite games can be proved in ZFC to be determined. The first important result, due to David Gale and Frank Stewart [GS], was that every open or closed game (with M unrestricted) is determined. This result requires AC; in fact, it is equivalent (in ZF) to AC. In the same paper, Gale and Stewart provided the construction of a nondetermined game, and asked whether all Borel games are determined. This question soon became the major open problem in the field. Progress came slowly, one level at a time, with Σ^0_4 determinacy proved by Paris in 1972. Three years later, Martin [Mart] settled the issue, using a brilliant induction through the Borel hierarchy to "reduce" every Borel game to an open one:

Theorem 6.37 (In ZFC). *All Borel games (with M unrestricted) are determined.*

Incidentally, Friedman had previously shown that Borel determinacy, even restricted to $M = \{0, 1\}$, cannot be proved in Zermelo set theory (with choice). In other words, Martin's theorem requires replacement, specifically to prove the existence of V_{ω_1}, even though Borel determinacy for $M = \{0, 1\}$ is a statement about relatively simple sets of reals. This was one of the first examples given of a "natural" mathematical statement about low level objects, whose proof requires replacement.

Exercise 26. Without using any of the results mentioned in this section, prove that G_A must be determined if A or $P - A$ is countable.

Martin and Alexander Kechris also proved, in ZFC, that Σ^1_n determinacy implies that every Σ^1_{n+1} set has the regularity properties. By Theorem 6.28, it follows that Borel determinacy is the "best possible" result of this type in ZFC. Even the strong postulate MC does not prove much more determinacy. Martin showed that MC implies Σ^1_1 determinacy, but Theorem 6.30 and the Martin–Kechris result tell us that ZFC + MC, if consistent, can't prove Σ^1_2 determinacy.

Exercise 27. Prove, in ZF, that Σ^1_n determinacy and Π^1_n determinacy are equivalent. Show that the same equivalences hold within the Borel hierarchy and the other hierarchies we have discussed. What is the general principle underlying these equivalences?

Note that this exercise would follow immediately if determinacy were preserved under complementation (meaning that whenever G_A is determined, so is G_{P-A}), as are measurability and the Baire property.

But this preservation property is false. In fact, it's easy to show (with AC) that the collection $\{A \subseteq P \mid G_A \text{ is determined}\}$ is not closed under complements, finite unions, or finite intersections. So the set of determined games has no nice closure properties, in contrast to the various algebras and σ-algebras we've been working with.

Since Borel determinacy (and nothing more within the projective hierarchy) is provable in ZFC, and the full AD is too strong to be considered as an axiom, it makes sense to focus on **projective determinacy** (PD) as a postulate. Consequently, much of the determinacy research since 1975 has involved the strength and consequences of PD. From the Martin–Kechris result stated above, we see that PD implies that every projective set has the regularity properties. So PD precludes "undesirable" sets, within the projective hierarchy at least. Another appealing feature of PD is that, in contrast to the full AD, it does not appear to contradict the axiom of choice.

However, PD has some of the same drawbacks that AD does as a potential axiom. It has not received significantly more intuitive justification than the full AD, and it is also an extremely strong postulate. In the words of [Woo, page 571], "PD is not only *not* obviously true, it is not even obviously consistent. It is, however, a fruitful axiom, but (a logician's joke) so is the axiom $0 = 1$." Here are a couple of results that give some indication of the strength of PD:

Theorem 6.38 (Martin and John Steel). *The existence of infinitely many Woodin cardinals implies* PD.

The proof of this theorem is similar to Martin's proofs of Borel determinacy, and $\mathbf{\Sigma}^1_1$ determinacy from MC. It uses induction to "reduce" every projective game to an open game. In all three cases, the argument requires the existence of a large set: V_{ω_1} for Borel, a measurable cardinal for $\mathbf{\Sigma}^1_1$, and lots of Woodin cardinals for PD.

Theorem 6.39 (Woodin). PD *is equivalent to the existence, for every natural number n, of a countable, transitive, "countably iterable" model of* (ZFC + *"There exist n Woodin cardinals"*). *(Countable iterability is a technical term from the so-called inner model program.)*

Compare these last two results with Theorem 6.36. The above theorem of Woodin, showing that PD implies the *consistency* of the existence of many large cardinals, is not as strong as showing that PD implies the existence of these same cardinals. But it still puts PD in the realm of very strong postulates involving extremely large, abstract sets. MC, which at one time was the most daring large cardinal axiom under serious investigation, doesn't even come close to implying PD (assuming that ZFC + MC is consistent). On the other hand, Woodin cardinals are nowhere near the largest that are now considered, and the hypothesis of the above theorem of Martin and Steel is quite a bit weaker than earlier versions of the same result.

By the way, the last two theorems demonstrate an interesting, delicate balance. We could not expect PD to imply the existence of a model of ZFC with infinitely many Woodin cardinals. (Why not?)

Woodin's program

The last paragraph of Section 6.5 mentioned Woodin's efforts to shed new light on the continuum hypothesis. To close this chapter, here is a very brief outline of his profound, complex program. For more details, see [Woo].

Notation. For any infinite cardinal κ, $H(\kappa)$ denotes the collection of all sets whose transitive closure has cardinality less than κ.

For instance, $H(\omega)$ is just the collection of all hereditarily finite sets. The sets in $H(\omega_1)$ are called **hereditarily countable**. In Section 2.5 we proved that $H(\omega) = V_\omega$, but if κ is uncountable, $H(\kappa)$ is normally much smaller than (and a proper subset of) V_κ.

Woodin's first key point is that CH (but not GCH, of course) can be viewed as a statement about the structure $H(\omega_2)$—that is, about $(H(\omega_2), \in)$. This is not difficult to see. If we define a real number to be a subset of $\mathbb{N}$, which is fine for purposes involving cardinality, then every real is in $H(\omega_1)$. But the whole set of reals is in $H(\omega_2)$ if and only if CH holds. Therefore, a complete understanding of $H(\omega_2)$ would settle CH.

From this, Woodin continues [Woo, page 569]: "This suggests an incremental approach. One attempts to understand in turn the structures $H(\omega)$, $H(\omega_1)$, and then $H(\omega_2)$. A little more precisely, one attempts to find the relevant axioms for these structures."

The structure $H(\omega)$ is denumerable and is equivalent to the standard model of arithmetic $\mathfrak{N}$, in the sense that each structure can be "nicely" embedded in the other. From this, Woodin argues that $H(\omega)$ is well understood, and that Peano arithmetic forms the "relevant" or "canonical" axioms for this structure: they are clearly true, and they are "empirically complete." That is, PA seems to suffice for settling almost all important *mathematical* (as opposed to metamathematical) questions of number theory. The known exceptions, such as those discussed in Section 4.4, are outside of mainstream mathematics.

Woodin also points out that neither inner models such as L nor forcing extensions can alter the truth of arithmetical statements. As we mentioned in Section 6.3, this is strong evidence for the view that $H(\omega)$ and $\mathfrak{N}$ are uniquely defined structures, and that questions about them have absolute answers.

The structure $H(\omega_1)$ is equivalent to the second-order structure over $\mathfrak{N}$, something like $(\mathcal{P}(\mathbb{N}), \mathbb{N}, +, \cdot, \in)$. Of course, this is much harder to analyze than $H(\omega)$. Note that projective relations on $\mathbb{R}$ are not members of this second-order structure; rather, they are its definable subsets.

Having acknowledged that PD is "*not* obviously true," Woodin goes on to present a variety of evidence that PD, together with Peano arithmetic and some simple set-theoretic axioms, provides the "canonical," "correct" axiomatization for $H(\omega_1)$ and the projective sets [Woo, page 575]. The strongest argument for this position is probably extrinsic: PD yields the most complete and useful known theory for $H(\omega_1)$. In particular, we've mentioned that PD implies the regularity properties for all projective sets. Also, in the presence of PD, the axiom of choice is not needed for the mathematical study of the projective sets. So Woodin claims that PD is empirically complete for $H(\omega_1)$, in the same sense that Peano arithmetic is for $\mathfrak{N}$. He also supports the plausibility of PD by describing several disparate set-theoretic postulates that imply PD. (Recall Theorem 6.38, for example.)

The second part of [Woo] is devoted to analyzing $H(\omega_2)$. In analogy to the equivalences between $H(\omega)$ and $\mathfrak{N}$ and between $H(\omega_1)$ and second-order $\mathfrak{N}$, one might expect $H(\omega_2)$ to be equivalent to third-order $\mathfrak{N}$. That would be the structure $(\mathcal{P}(\mathcal{P}(\mathbb{N})), \mathcal{P}(\mathbb{N}), \mathbb{N}, +, \cdot, \in)$ or, equivalently, $(\mathcal{P}(\mathbb{R}), \mathbb{R}, \mathbb{N}, +, \cdot, \in)$. However, $H(\omega_2)$ is equivalent to this third-order structure if and only if CH holds. Otherwise, $H(\omega_2)$ can be very different from, and "fundamentally simpler" than, third-order $\mathfrak{N}$. Part of Woodin's innovation was to realize that there can be no "strongly canonical" theory for third-order $\mathfrak{N}$, and to focus on $H(\omega_2)$ instead.

Not surprisingly, $H(\omega_2)$ is vastly harder to analyze than $H(\omega)$ and $H(\omega_1)$. A substantial roadblock is the fact, mentioned near the end of Section 6.4, that large cardinal axioms cannot settle CH. Therefore, if we are looking for "strongly canonical" or "empirically complete" axioms for $H(\omega_2)$, those axioms "cannot be implied by any (consistent) large cardinal hypothesis remotely related to those currently accepted as large cardinal hypotheses" [Woo, page 682].

Therefore, Woodin reasons that first-order logic is too weak to provide a satisfactory axiomatization of $H(\omega_2)$, and instead he considers **strong logics**. The idea behind strong logics is simple: recall that Gödel's completeness theorem says that $T \vdash$ P if and only if $T \models$ P. Here, $\vdash$ refers to provability in first-order logic, while $\models$ refers to semantic entailment in *all* first-order structures. If we instead base the definition of $\models$ on some proper subclass of all first-order structures, and then redefine $\vdash$ accordingly (keeping the biconditional), we get a strong logic. Fewer structures means a stronger logic, in that more things are provable from a given T. The main strong logic that Woodin considers is called **Ω-logic**. Briefly, ZFC $\vdash_\Omega$ P if and only if P is true in every countable transitive model of ZFC that satisfies a technical condition called A_Ω-closure.

In order to obtain desirable properties of Ω-logic, one needs to assume that there is a proper class of Woodin cardinals. This is a significantly stronger assumption than the one used by Martin and Steel to prove PD. In particular, Woodin needs this assumption to prove that Ω-logic possesses two properties called **generic soundness** and **generic invariance**. These properties guarantee that "any axioms we find [in Ω-logic] will yield theories for $H(\omega_2)$ whose 'completeness' is immune to attack by Cohen's method of forcing, just as is the case for number theory" [Woo, page 682].

With these considerations in mind, here is the main result of [Woo]:

Theorem 6.40. *Assume that there is a proper class of Woodin cardinals, and that* Q *is a sentence such that* ZFC + Q *is true in some structure of the form* $(V_\kappa, \in)$*, and* Q *also proves or disproves, in* Ω*-logic, every statement of the form* $(H(\omega_2), \in) \models$ P*, where* P *is any sentence of set theory. Then the continuum hypothesis is false.*

In other words, if there is a proper class of Woodin cardinals, and if there is a reasonable finite and complete axiomatization for $H(\omega_2)$ in Ω-logic, then CH is false. The crucial axiom Q need not be a statement about $H(\omega_2)$; it can refer to arbitrary sets.

Much of the second part of [Woo] is devoted to discussing possibilities for this axiom Q. The most obvious candidates are **forcing axioms**, which are in essence postulates asserting generalizations of the Baire category theorem. By their very nature, forcing axioms imply that CH is false. Of particular interest is a forcing axiom known as **Martin's maximum**, the strongest consistent version of a well-known postulate called **Martin's axiom**. Martin's maximum implies that $2^{\aleph_0}$ is precisely $\aleph_2$. It also implies PD, indicating that it's a very strong axiom, and perhaps even a natural one. Woodin has proposed a variant of Martin's maximum known as **Woodin's Martin maximum**, called axiom (*) in [Woo], as a viable candidate for the role of the axiom Q in the above theorem. As of 2003, he has come very close to proving this. All that's missing is to show that this postulate is consistent with the necessary large cardinal assumptions.

Clearly, this complex program, with its strong and specialized assumptions, is not a conclusive piece of evidence against CH. The continuum hypothesis remains a very elusive problem that is nowhere near being settled. What Woodin (and others) have developed is a strategy and methods that might provide more insight into CH than anything previously devised. However, [Woo] ends on a cautious note: "The universe of sets is a large place. We have just barely begun to understand it."

CHAPTER 7

Nonstandard Analysis

7.1 Introduction

Why should students of mathematics want to know something about nonstandard analysis? Here is one answer: nonstandard analysis provides a powerful tool for understanding and using the main ideas of calculus, analysis, and other branches of mathematics. This tool can be very helpful to students and teachers of mathematics as well as those doing mathematical research.

Contrary to what many mathematicians may believe, nonstandard analysis is much more than a clever, unorthodox approach to calculus and real analysis. Nonstandard methods have been useful in almost every branch of mathematics and even subjects outside of mathematics. For example, the anthology [Cut] includes articles about applications of nonstandard methods to probability theory, functional analysis, mathematical physics, computational group theory, and ordinary differential equations. The more recent collection of papers [Ark] includes many of the same subjects as well as martingales and stochastic processes, asymptotic analysis, hydromechanics, and "hyperfinite mathematical finance." The even more recent work [LW] provides, among many other topics, nonstandard treatments of Brownian motion and mathematical economics (including Nash equilibria).

Another answer to this question is that nonstandard analysis may be viewed as the culmination of the longest debate in the history of mathematics, a debate that is well over two thousand years old. Every mathematics student can appreciate knowing something about the ebbs and flows of this ancient controversy. Part of the beauty of nonstandard analysis is that it demonstrates that there are no losers in this debate, and that the use of infinitesimals, banished from rigorous mathematics in the latter part of the nineteenth century, is perfectly legitimate after all. The rest of this section is devoted to the history of this debate.

"Limits vs. infinitesimals" through the ages

As early as 400 B.C., the mathematicians of ancient Greece were aware of a way of reasoning that could provide short, ingenious solutions to a variety of problems in geometry. Here is a typical example:

Proposition 7.1. *The area of a circle equals one-half the product of the radius and the circumference.*

Proof. Consider a circle of radius r, with a regular n-sided polygon inscribed in it. (See Figure 7.1.) Then the polygon consists of n congruent isosceles triangles, each with the same height h and base b. Thus we have

$$\text{Area of polygon} = n(bh/2) = h(nb)/2 = hP/2,$$

where P is the perimeter of the polygon.

Now imagine that we make the number of sides of the polygon *infinite*, so that each triangle has height r and an *infinitesimal* base. Then the polygon fills the circle, so the area of the circle equals the area of the polygon. Also, the perimeter of the polygon now equals the circumference of the circle. Thus the formula above involving the polygon becomes what we are trying to prove for the circle. ■

This particular "proof" is usually attributed to Johannes Kepler, but it is similar to arguments used by Democritus and other early Greek mathematicians.

We have previously discussed the ancient Greeks' "horror of the infinite." So it should come as no surprise that no important mathematicians of that epoch, even those who devised such proofs, considered

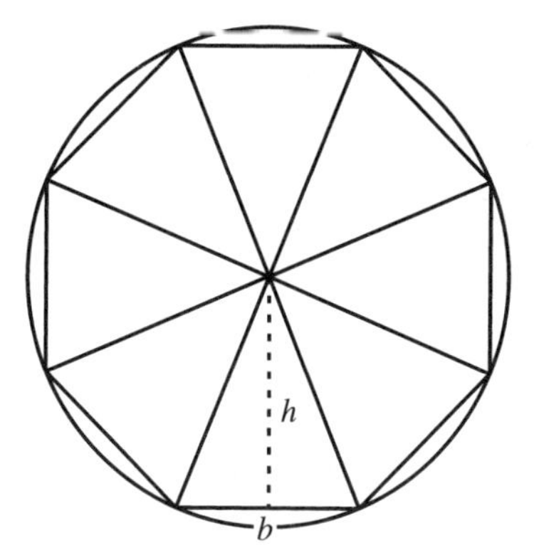

Figure 7.1. Approximating the area of a circle with triangles

them to be rigorous. The same is true of almost all mathematicians since then, including contemporary ones. Currently, if we wanted to make this proof acceptable, we would instead talk about the number of sides *approaching* infinity, and use the idea of limits. The Greeks did not use these terms, but they had a method, called the **method of exhaustion**, that was quite similar to the modern rigorous use of limits. First clearly articulated by Eudoxus around 370 B.C., the method of exhaustion was the standard tool used by the Greeks to provide formal justification for their results involving areas and volumes of curved regions and solids—results that some consider the origin of integral calculus.

The Greeks' ambivalence toward infinitesimals is epitomized by the work of Archimedes. Archimedes made great strides in computing areas and volumes, and his published results in this subject include meticulous, rigorous, beautifully written proofs, usually using the method of exhaustion. Indeed, the phrase "Archimedean rigor" has been used for over two thousand years to describe the pinnacle of elegance and precision in mathematical writing.

However, a proof of a formula by the method of exhaustion gives no indication of how the formula was discovered in the first place. Mathematicians after Archimedes were mystified as to how

Archimedes (287?–212 B.C.) was the most productive mathematician of ancient Greece and one of the greatest of all time. He was also a superb engineer and scientist who invented various devices (catapults, hooks to lift ships out of the water, etc.) that helped the island colony of Syracuse fend off the Romans' siege for several years. It was reported that he used a system of mirrors reflecting sunlight to set Roman ships on fire, but most historians doubt this. Another anecdote has it that Archimedes, while at a public bathhouse, suddenly saw how he could use principles of water displacement to determine whether King Hieron's crown was pure gold. He then ran home, naked, shouting, "Eureka" ("I have found it").

Even Archimedes' death was extraordinary. When the Romans finally overtook Syracuse, he was apparently drawing geometric figures in the sand, oblivious to the war going on around him. According to one version of the story, Archimedes saw a Roman soldier walking toward him and asked him not to interrupt his work, whereupon the enraged soldier killed him. The Roman general Marcellus, who greatly admired Archimedes and had given his soldiers orders not to harm the old man, built him a tomb displaying one of Archimedes' favorite figures: a sphere inscribed inside a cylinder.

Archimedes derived some of his area and volume results. This mystery was not solved until 1906, when a document by Archimedes called "The Method" was discovered on a palimpsest—a parchment that had been partially erased around 1300 A.D. so that it could be reused for religious texts. This document describes an ingenious procedure in which one or more solids could be mentally sliced into infinitesimally thin pieces, and then a balance could be achieved, slice by slice, according to the basic physical theory of moments. Archimedes would then use the equation expressing this balance to compute the volume of one of the solids. This **method of equilibrium**, also called his **method of discovery**, is one of the cleverest "integration tricks" ever devised. In particular, this is the method he used to derive the formula for the volume of a sphere. (See [Edw] or [Eves].) And yet, Archimedes did not even mention (let alone use) this method in any of his works that were intended for the public, because he did not consider it to be rigorous.

After the death of Archimedes there were, incredibly, no significant advances in the computation of areas and volumes for about 1800 years. By 100 B.C. the Roman civilization dominated Europe and parts of Asia and Africa. The Romans were skilled engineers but had little interest in theory, including most branches of mathematics. The fall of the Roman Empire in 476 A.D. led to the Dark Ages, several centuries of no enduring intellectual achievement whatsoever. It is possible that the great works of antiquity would have become lost forever during this period, were it not for the resurgence of academic activity that began in the Arab world in about 800. Slowly, the spirit of inquiry returned to Europe, and by the early 1400s the Renaissance had begun. The invention of the printing press in about 1450 led to the first widespread distribution of Euclid's *Elements* and other classic works of Greek mathematics, including those of Archimedes. At around the same time, Constantinople fell to the Turkish Empire, causing many Arab scholars to flee to Europe, and the modern age of mathematics blossomed.

The 1600s were quite possibly the most important century in the history of mathematics. In Italy, Germany, France, Holland, England, and Scotland, mathematicians made all sorts of discoveries. Area and volume problems were especially popular during this period. Although the rigorous work of Euclid and Archimedes was highly respected, mathematicians of the 1600s were more interested in efficiency and results. Therefore, they tended to avoid using the method of exhaustion in their writings, although they often explained that their results could be proved rigorously in that way. Kepler summed up the general attitude of the period: "We could obtain absolute and in all respects perfect demonstrations from these books of Archimedes themselves, were we not repelled by the thorny reading thereof." Instead, almost all of the mathematicians of this period made free use of infinitesimals in their work. We have already given an example of Kepler's style of reasoning in Proposition 7.1. Kepler was a brilliant and productive mathematician, but his use of infinitesimals was sometimes freewheeling to the point of sloppiness. For example, his beautiful second law of planetary motion, that any one planet's orbit must sweep out equal areas in equal time periods, was derived in spite of two errors that happened to cancel each other out ([Edw], pp. 100–101). At the same time, Bonaventura Cavalieri was solving area and volume problems using his **method of indivisibles**, essentially a simplified version of Archimedes's method of equilibrium (which, you may recall, was unknown at that time).

The invention of analytic geometry in the 1630s by Descartes and Fermat provided an enormous impetus to the development of modern calculus. The importance of this direct link between symbolic entities (equations and inequalities) and pictorial entities (lines, circles, parabolas, and so on) would be difficult to overestimate. In modern parlance, one could say that analytic geometry is the most important mathematical tool for communication between the "left brain" and the "right brain." Armed with this powerful new method, mathematicians of the middle part of the seventeenth century discovered many new techniques of integration, and also discovered the basics of differential calculus.

Finally, Isaac Newton (in the 1660s) and Leibniz (a decade or so later, but apparently independently) organized the advances of their predecessors into a coherent body of material and thereby are viewed as the co-inventors of modern calculus. In particular, they were the first to appreciate the value of the fundamental theorem of calculus, although Isaac Barrow more or less discovered and proved the theorem several years before they did, and Evangelista Torricelli saw the basic idea even earlier. By demonstrating the close inverse relationship between differentiation and integration, this powerful theorem provided almost trivial solutions to integration problems that previously took years to solve. To see that this is no exaggeration, first evaluate an integral like

$\int_1^4 x^{-1/2}\,dx$ using the standard "antidifferentiation" method, and then try to evaluate the same integral as the limit of Riemann sums. It is only through the fundamental theorem that calculus becomes a viable subject.

Like Archimedes, Newton and Leibniz appreciated the value of both sides of the "limits vs. infinitesimals" debate. Newton's calculus was based on the concept of the **moments** of variable quantities, meaning "their indefinitely small parts, by addition of which they increase during each indefinitely small period of time." This is a parametric approach, in the sense that all variables are viewed as functions of time. He computed a variable's **fluxion** (its derivative with respect to time) by dividing its moment by a moment of time. Then other derivatives were found as quotients of fluxions. (That is, if we know dy/dt and dx/dt, we can divide these to find dy/dx.) It seems clear that Newton's moments are infinitesimals, and therefore that his approach to calculus is completely based on the use of infinitesimals. But in later writings Newton wrote that "there's no need of introducing infinitely small (numbers)" to carry out his development of calculus, because the use of moments could be replaced by the use of limits of ordinary quantities. He did not provide a rigorous definition of what he meant by a limit, but he did state (presumably after verifying that it was so) that his work could be made "agreeable to the ancients."

Leibniz's version of calculus was even more completely based on infinitesimals, in the form of **differentials**. He introduced the notation dy/dx for derivatives, and to him dx and dy had separate meaning, as infinitesimal changes in the variables x and y. Leibniz was not as brilliant a scientist or problem-solver as Newton, but he had a better feel for concepts and notation. Many of his discoveries stemmed from clever geometric reasoning using differentials, often based on the **characteristic triangle** formed by the horizontal segment dx, the vertical segment dy, and the hypotenuse ds, which represents an infinitesimal part of a curve or graph. Recall also that several important formulas of calculus, such as the chain rule and the formulas for integration by substitution and integration by parts, are much easier to understand and work with in Leibniz's notation than in any other notation. But Leibniz also pointed out that the use of infinitesimals was not essential for his results, because "the whole matter can always be referred back to assignable quantities" (that is, ordinary quantities), and verified with something like the method of exhaustion. He made it clear that differentials could be viewed as merely a convenient fiction, and that he did not necessarily believe in the actual existence of infinitesimal quantities.

The eighteenth century was the heyday of infinitesimals, referred to facetiously as the "golden age of nothing." While most of the mathematicians of the seventeenth century had tempered their use of infinitesimals with a measure of "Archimedean caution," the great mathematicians of the 1700s used infinitesimal and infinitely large quantities freely, with little or no concern about whether these quantities exist or whether their results could be established rigorously. The champion of this approach was certainly Leonhard Euler. Euler's use of infinitesimal and infinite quantities was more freewheeling than that of any of his predecessors, including Kepler; but Euler's intuition was so superb that he seems to have made no serious errors with his "iffy" procedures. For example, Euler freely applied the binomial expansion with an infinite exponent N, and often simplified the resulting infinite series by claiming that, since N is infinite,

$$\frac{N-1}{N} = \frac{N-2}{N} = \frac{N-3}{N} = \cdots = 1.$$

The most serious critic of the lack of rigor in eighteenth century mathematics was the English Bishop George Berkeley. Although he was not a mathematician, Berkeley had a keen understanding of mathematical principles, as demonstrated by his clever essay "The Analyst" ([BJ], [Str]). The "infidel mathematician" in question was Edmund Halley, the astronomer after whom Halley's Comet was named and a staunch supporter of Newton. Primarily, Berkeley was criticizing the standard procedure for computing a derivative in which one first computes the difference quotient $[f(x+h) - f(x)]/h$, which requires that h be nonzero for the fraction to be defined, but then sets h to zero to arrive at the correct answer. Clearly, the reasoning involved in this pro-

Leonhard Euler (1707–1783) was born and raised in Basel, Switzerland. By the age of twenty he had displayed enough mathematical talent and ability to receive an appointment at the new St. Petersburg Academy in Russia. During the course of his career, Euler also had academic positions in Basel and Berlin, but he spent the bulk of his career—over thirty years, including the last seventeen years of his life—in St. Petersburg. Euler's personality was not to everyone's liking, and he felt much more accepted in St. Petersburg than elsewhere.

Euler was a true genius, the most important mathematician of the eighteenth century, probably the most versatile mathematician of all time, and definitely the most prolific. He published over 500 books and papers in his lifetime, and another 360 or so were published posthumously. He worked in every branch of mathematics and had extensive knowledge of many sciences, history, theology, languages, and more. Euler had an amazing memory, on which he relied greatly during the last fifteen years of his life. His productivity was barely diminished during that period, even though he was totally blind!

It is difficult to synopsize Euler's mathematical achievements. He made enormous advances in calculus, the theory of sequences and series, and many areas of applied mathematics. He made important contributions in algebra, number theory, and geometry. He crystallized the modern concept of a function, and his solution to the Königsberg Bridge problem helped create the subject of topology. Euler also invented a great deal of useful mathematical notation, including e for the main constant of exponential functions, i for the imaginary unit, $f(x)$ for functions, and Σ for summation.

cedure is suspect, even if one does believe in infinitesimals! Berkeley wrote, "And what are these same evanescent increments [moments]? They are neither finite quantities, nor quantities infinitely small, nor yet nothing. May we not call them the ghosts of departed quantities?" And: "He who can digest a second or third fluxion ... need not, methinks, be squeamish about any point in divinity."

Most of the result-oriented mathematicians of the 1700s ignored Berkeley's critique. Others responded with inadequate defenses of the foundations of calculus. But, gradually, serious advances were made. In 1754, Jean d'Alembert published an article elaborating on Newton's later view that derivatives should be viewed as the limits of ratios, rather than simply ratios. At the end of the century Joseph Louis Lagrange published a comprehensive treatment of calculus that used Taylor series to define derivatives, in an attempt to eliminate the use of both infinitesimals and limits from the foundations of calculus. Recall that if a function has a Taylor series at a given point, then it is straightforward to compute its derivatives of all orders (at that point) from the Taylor series. The main problem with this approach, which was discovered fairly soon, is that not every infinitely differentiable function (let alone every differentiable function) has a valid Taylor series.

In the early part of the nineteenth century, Bernard Bolzano gave the first acceptable definition of continuity, avoiding infinitesimals in favor of limit language that would today be considered correct but not quite rigorous. A few years later, in the 1820s, Augustin-Louis Cauchy used a similar concept of limits to define derivatives as well as continuity. Cauchy also defined an infinitesimal to be a quantity that becomes arbitrarily small (that is, approaches 0), rather than one that is *statically* very small. But it took until about 1875 for Weierstrass to give the modern ϵ-δ definitions of limits and continuity (Example 7, Section 1.3) that completely avoid vague words such as "approaches," and thereby achieve full rigor. In fact, Weierstrass's definitions could be called the first major advance in rigor involving calculus concepts since the method of exhaustion was devised some 2200 years earlier!

Weierstrass's definitions caused infinitesimals to be banished from serious mathematics. For nearly a century thereafter, the ϵ-δ definition of limits was *the* correct tool for developing calculus and analysis. And yet, in spite of their prohibited status, infinitesimals continued to be used freely by scientists and engineers, and in informal mathematical reasoning. Weierstrass's definitions, like their "predecessor in spirit" the method of exhaustion, did not have sufficient intuitive appeal to replace the use of infinitesimals. With this in mind, several twentieth-century mathematicians attempted to provide a rigorous basis for the use of infinitesimals, with the major breakthrough occurring in 1960. It was then that Abraham Robinson created nonstandard analysis, meaning the rigorous development of calculus, analysis, and other branches of mathematics with infinitesimals available as actual objects.

There are many types of references on nonstandard analysis. For Robinson's original treatment of the subject, see [Robi]. The collections [Ark], [Cut], and [LW] have good general introductions as well as many applications. [HK] and [Kei86] are beginning calculus texts based on the nonstandard point of view. This approach to the teaching of calculus deserves more attention and is the subject of [Tod]. [Kei76] is the instructor's manual for [Kei86], but it can also stand alone as an interesting introduction to nonstandard methods. Finally, we mention [Robe] as a good treatment of **internal set theory**, an important axiomatic approach to nonstandard analysis developed by Edward Nelson.

7.2 Nonarchimedean fields

We assume familiarity with some of the basic properties of fields and ordered fields. Throughout this section, K always denotes an ordered field. Technically, what we mean is that K is the universe of a structure $(K, +, \cdot, <)$ that satisfies the usual list of first-order axioms for an ordered field. Recall that every ordered field has characteristic 0, and therefore its prime field, its smallest subfield, is the unique subfield that is isomorphic to $\mathbb{Q}$. So we will talk as if $\mathbb{Q}$ (and therefore $\mathbb{Z}$ and $\mathbb{N}$) are subsets of K.

Definitions. For $x \in K$, we call x **infinitesimal** or **small** if $|x| < 1/n$ for every $n \in \mathbb{N}$. We say x is **unlimited** or **infinite** or **large** if $1/x$ is small. And we say that x is **medium** if it is neither small nor large. Also, a **limited** number means one that is small or medium.

Suppose $x, y \in K$. We say x is **close to** y, written $x \approx y$, if $x - y$ is small. The **monad** of x is defined to be

$$\mu(x) = \{y \in K \mid y \approx x\}.$$

The terms "small," "medium," and "large" introduced here are not common, but they are appealing because they are so natural and simple.

Exercise 1. Prove that $x \approx y$ is an equivalence relation. Therefore the monads, being the equivalence classes under $\approx$, partition K.

Definition. An ordered field is called **Archimedean** if 0 is the only small number in it.

Theorem 7.2. *For any ordered field K, the following are equivalent:*

(a) *$\mathbb{Z}$ is unbounded (above and below) in K.*

(b) *There are no large numbers in K.*

(c) *K is Archimedean.*

(d) *$\mathbb{Q}$ is dense in K.*

(e) *K is isomorphic (as an ordered field) to a subfield of $\mathbb{R}$.*

Proof. (a) implies (b): A large number would automatically be either an upper bound of $\mathbb{Z}$ or a lower bound of $\mathbb{Z}$.

(b) implies (c): If x were a nonzero small number, then $1/x$ would be large.

(c) implies (d): Assume that K is Archimedean and that the conclusion is false, so we have $x, y \in K$ with $x < y$ but no rational between x and y. Since K has no nonzero small numbers, it has no large numbers either. Thus there are rationals greater than x, so $\{u \in \mathbb{Q} \mid u \leq x\}$ is a bounded set of real numbers, which must have a supremum r (in $\mathbb{R}$). Similarly, $\{u \in \mathbb{Q} \mid u \geq y\}$ has an infimum s.

It is clear that $r \leq s$. If $r < s$, then there is a rational between r and s, which must also be between x and y. On the other hand, if $r = s$, then $y - x$ is small but nonzero, contradicting the assumption that K is Archimedean.

(d) implies (e): Assume $\mathbb{Q}$ is dense in K. For each $x \in K$, there must be rationals greater than x, since there are rationals between x and $x + 1$. So we can define a function $f : K \to \mathbb{R}$ by $f(x) = \text{SUP}(\{u \in \mathbb{Q} \mid u \leq x\})$. Then f is a an isomorphic embedding of K in $\mathbb{R}$. That is, it is an order-preserving field homomorphism. The proof of this is left as an exercise.

(e) implies (a): We will show that $\mathbb{Z}$ is unbounded in $\mathbb{R}$. The extension to any subfield of $\mathbb{R}$ and to any ordered field that is isomorphically embedded in $\mathbb{R}$ is automatic. Assume, on the contrary, that $\mathbb{Z}$ has an upper bound in $\mathbb{R}$. Then, by completeness, it has a least upper bound x. If $n \in \mathbb{Z}$, then $n + 1 \in \mathbb{Z}$, so $n + 1 \leq x$. Therefore, $n \leq x - 1$. This shows that $x - 1$, which is less than x, is an upper bound of $\mathbb{Z}$, a contradiction. Similar reasoning shows that $\mathbb{Z}$ has no lower bound in $\mathbb{R}$. ■

Exercise 2. Complete the proof that (d) implies (e) in this theorem.

Corollary 7.3. *The ordered field $\mathbb{R}$ is Archimedean.*

The use of the term "Archimedean" for such fields is based on the fact that Archimedes used a condition equivalent to (a) of this theorem as a postulate about the real numbers. Some authors call these fields **Eudoxian** instead, since the correctness of the method of exhaustion is equivalent to the Archimedean condition.

In nonstandard analysis, the focus is on nonarchimedean fields. We now list some basic properties of these fields and the types of numbers we have defined in them:

Exercise 3. Fill in each blank with the most restrictive of the following phrases that is correct: "small"; "medium"; "large"; "limited"; "small or large"; "medium or large"; or "a member of K." All of these statements refer to an arbitrary ordered field:

(a) The sum of two small numbers must be ________.

(b) The product of two small numbers must be ________.

(c) The sum of two medium numbers must be ________.

(d) The product of two medium numbers must be ________.

(e) The quotient of two medium numbers must be ________.

(f) The sum of two large numbers must be ________.

(g) The difference of two large numbers must be ________.

(h) The quotient of two large numbers must be ________.

Exercise 4. Prove any three of the parts of the previous exercise. You might want to use the following example as a guide.

Example 1. Let us show that the product of a small number and a limited number must be small. Assume x is small and y is limited. This condition on y means that $|y| < m$ for some $m \in \mathbb{N}$. Fix such an m. By definition of smallness, $|x| < 1/n$ for every $n \in \mathbb{N}$. Therefore, $|x| < 1/mn$ for every $n \in \mathbb{N}$. So $|xy| = |x||y| < m(1/mn) = 1/n$ for every $n \in \mathbb{N}$. So xy is small.

Exercise 5.

(a) State at least six other correct results of the form, "The ________ of a ________ number and a ________ number must be a ________ number." Here, the first blank should be filled by one of the words "sum," "difference," "product," or "quotient." The remaining blanks should be filled with one of the words "small," "medium," or "large," with *different* words filling the middle two blanks. For example, one permissible statement, if it is correct, would be "The product of a small number and a large number must be a medium number."

(b) Prove two of your statements from part (a).

If we want to provide a setting for doing calculus with infinitesimals, it makes sense to work in a nonarchimedean field that is an extension of $\mathbb{R}$. Here is perhaps the most straightforward construction of such a field. Let $\mathbb{R}[t]$ be, as usual, the ring of polynomials in one variable, with real coefficients. Since $\mathbb{R}[t]$ is an integral domain, it has a field of fractions $\mathbb{R}(t)$. This field is known as the field of rational functions in one variable, with real coefficients. But technically, the elements of $\mathbb{R}(t)$ are not rational functions but equivalence classes of rational expressions. For example, $(t^2 + 1)/5$ and $(t^3 + t)/5t$ define different functions (because 0 is not in the domain of the second one), but they are in the same element of $\mathbb{R}(t)$. $\mathbb{R}$ may be viewed as a subfield of $\mathbb{R}(t)$ by identifying any number r with the equivalence class of $r/1$.

$\mathbb{R}(t)$ is not automatically an ordered field, but it can be turned into one. In fact, there are an infinite number of ways to do this, corresponding to each possibility of where to "put" t (more precisely, the equivalence class of $t/1$) relative to the ordering on $\mathbb{R}$. For example, there is a unique way to make $\mathbb{R}(t)$ into an ordered field in which t is negative and large. There is a unique way to make $\mathbb{R}(t)$ into an ordered field in which t is "right before" the number 4.73; more precisely, that means that $4.73 - t$ is a positive small number. No matter where we put t, $\mathbb{R}(t)$ becomes a nonarchimedean field extension of $\mathbb{R}$. For simplicity, suppose that $\mathbb{R}(t)$ is an ordered field in which t is positive and large.

Note that this condition on t completely fixes the ordering on $\mathbb{R}(t)$, as follows: first of all, a nonzero polynomial is positive if and only if its leading coefficient is positive. Then, a fraction is positive if and only if its numerator and denominator have the same sign. This condition is easily shown to be well defined on equivalence classes of rational expressions. Thus, we have a well-defined specification of the positive elements of $\mathbb{R}(t)$, and this defines the ordering on $\mathbb{R}(t)$.

Exercise 6.

(a) Prove the well-definedness condition asserted above.

(b) Define the ordering on $\mathbb{R}(t)$ in which t is a negative small number.

7.3 Standard and nonstandard models

Unfortunately, the nonarchimedean field $\mathbb{R}(t)$ is not "rich" enough to use as a framework for calculus with infinitesimals. It is adequate as long as we only want to work with polynomials and rational functions. But $\mathbb{R}(t)$ does not allow functions as simple as radicals to be defined; for instance, t does not have a square root in $\mathbb{R}(t)$. Furthermore, there is no apparent way to define the important transcendental functions of calculus—trigonometric, exponential, and logarithmic functions—in $\mathbb{R}(t)$. Therefore, a more sophisticated construction is needed if we really want to carry out all of basic calculus in a setting with infinitesimals. So we want to define not just a nonarchimedean ordered field, but a structure including such a field in which all of calculus and more can be carried out.

Definition. Let S be a set. We then define $V_n(S)$, where $n \in \mathbb{N}$, inductively: $V_0(S) = S$, and $V_{n+1}(S) = V_n(S) \cup \mathcal{P}(V_n(S))$. The **superstructure** over S, denoted $V(S)$, is the set $\bigcup_{n\in\mathbb{N}} V_n(S)$. For any $a \in V(S)$, the **rank** of a is the least n such that $a \in V_n(S)$.

Compare this to the definition of the cumulative hierarchy in Section 2.5. The superstructure over $\emptyset$ is the same as V_ω, and any superstructure $V(S)$ may be viewed as the ωth stage of the cumulative hierarchy in a universe with a set of urelements S, assuming that the elements of S are not themselves sets. This assumption is usually made when a superstructure is defined. So, when working with superstructures, we don't take the "pure" approach to set theory in which everything is a set (although we could take that approach and make things work just as well). Note that the rank function for a superstructure is "one off" from the rank function defined in Section 2.5. For example, in the cumulative hierarchy, the rank of $\emptyset$ is 0, but in a superstructure its rank is 1, even if $S = \emptyset$. This is because, in a superstructure, we want the elements of S (and nothing else) to have rank 0. However, this should not be viewed as an issue of much importance.

It is easy to see that $V(S)$ must be transitive (in the sense that $x \in A \in V(S)$ implies $x \in V(S)$), provided that S consists of urelements or is itself a transitive set. As mentioned in Section 6.2, this guarantees that sets in superstructures are what they "ought to be."

Essentially, we want to use the structure $(V(\mathbb{R}), \in)$ as our "standard" model. But since $V(\mathbb{R})$ has two types of members, real numbers and sets, we will view it as a structure for a two-sorted language $\mathcal{L}_0$ (with variables $x, y, z, \ldots$ for real numbers and $A, B, C, \ldots$ for sets). Also, in order to make it more convenient to work with $\mathbb{R}$ as an ordered field, let us also include the operations $+$ and $\cdot$ and the relation $<$ (on $\mathbb{R}$) as part of the structure, and therefore as symbols in $\mathcal{L}_0$.

Definition. The **standard model of analysis** is the structure

$$\mathfrak{M} = (\mathbb{R}, V(\mathbb{R}) - \mathbb{R}, +, \cdot, <, \in).$$

From facts that were mentioned in Chapter 2, it follows that all the objects encountered in ordinary real and complex analysis are in $\mathfrak{M}$. For example, every set of real numbers is of rank 1. The power set of $\mathbb{R}$ is of rank 2. The Cartesian product $\mathbb{R} \times \mathbb{R}$ is of rank 3 (by virtue of the Kuratowski definition of ordered pairs). The collection of all functions from $\mathbb{R}$ to $\mathbb{R}$ is of rank 4. Every function from $\mathbb{R} \times \mathbb{R}$ to $\mathbb{R}$, including addition and multiplication, is of rank 5. The *set* of all complex numbers is of rank 3, since by definition it is precisely $\mathbb{R} \times \mathbb{R}$. But the *field* of complex numbers, which includes the addition and multiplication operations on that set, could have rank as high as 10 or so, depending on exactly how we define it.

Not every object we might ever want to discuss in mathematicians is in the standard model. For example, we cannot expect an arbitrary group or topological space to be in $\mathfrak{M}$. If we want to use nonstandard reasoning about such a structure, we must augment S so that the superstructure $V(S)$ contains the structure under consideration as well as $\mathbb{R}$. From now on, we will assume that $\mathfrak{M}$ is large enough to include all standard objects that are being discussed. For more sophisticated purposes, one might want to assume that $\mathfrak{M}$ is a transitive model of ZFC rather than just a superstructure.

Our next goal is to define a **nonstandard extension** of $\mathfrak{M}$, which we will also refer to as a **nonstandard model of analysis**. There are several methods of constructing such an extension, all of which require some form of the axiom of choice. The most powerful and versatile method is the method of **ultrapowers**, a very important tool in model theory. We will not describe the ultrapower construction here; it can be found in any model theory book such as [CK]. Instead, we will explain a simpler way to define nonstandard models via the compactness theorem, which was Robinson's original approach to the subject.

We have already outlined this method in Section 5.3. To define a true nonstandard model, we need to apply compactness to a very rich theory based on the structure $\mathfrak{M}$. Let $\mathcal{L}$ be the language obtained from $\mathcal{L}_0$ by adding to it a constant symbol for each member of $\mathfrak{M}$. Note that $\mathcal{L}$ is a very large language—its cardinality is $\beth_\omega$.

Now, let T_1 be the $\mathcal{L}$-theory $Th(\mathfrak{M})$. That is, T_1 consists of all true sentences about $\mathfrak{M}$ in the language $\mathcal{L}$. Like $\mathcal{L}$, T_1 has large cardinality. It includes not only most true statements of "ordinary mathematics" but also all true statements involving particular real numbers, particular sets of real numbers, etc. Some authors call this T_1 the **complete diagram** of $\mathfrak{M}$ or the **elementary diagram** of $\mathfrak{M}$.

Next, let T consist of all sentences in T_1 in which every quantified variable is restricted to a set. That is, for a sentence of T_1 to be in T, every quantifier in it must be of the form $\forall v_i \in v_j$, $\exists v_i \in v_j$, $\forall v_i \in c$, or $\exists v_i \in c$, where c is some constant of $\mathcal{L}$ that denotes a set. We will refer to an $\mathcal{L}$ formula of this type as a **restricted** formula.

Finally, add to $\mathcal{L}$ one more new constant symbol h. Then form a new theory T' by adding to T the new axioms $h \in \mathbb{R}$ and $h > 0$ plus the axiom schema $\{h < 1/n \mid n \in \mathbb{Z}^+\}$.

Proposition 7.4. *T' has a model.*

Proof. By the compactness theorem, it suffices to show that every finite subset F of T' has a model. Given such an F, we already know that $\mathfrak{M}$ satisfies all the statements in F that are in T. Then, if we interpret the constant h as some positive real number that is less than all the fractions $1/n$ appearing in inequalities of the form $h < 1/n$ in F, it is clear that $\mathfrak{M}$ becomes a model of all of F. (This proof requires AC because the compactness theorem, and hence the completeness theorem, is applied to the uncountable theory T'.) ∎

Definition. Any model $^*\mathfrak{M}$ of T' is called a **nonstandard extension** of $\mathfrak{M}$.

We can view $\mathfrak{M}$ and $^*\mathfrak{M}$ as $\mathcal{L}_0$-structures or $\mathcal{L}$-structures, whichever is more convenient at the moment.

Notation. Let $^*\mathfrak{M}$ be a nonstandard extension of $\mathfrak{M}$. For each element x of $\mathfrak{M}$ (real number or set), there is a constant C_x representing it in $\mathcal{L}$. We let $^*x$ denote the interpretation of C_x in $^*\mathfrak{M}$. Think of $^*x$ as the **nonstandard version** of x.

In order to make it easier to think about $^*\mathfrak{M}$, we will make the following harmless assumptions about it:

(1) We assume that * is the identity on $\mathbb{R}$, so that $^*3 = 3$, etc. (So we only write * before *sets* of $\mathfrak{M}$.) Thus $\mathbb{R}$ is a subset of $^*\mathbb{R}$, the set of **hyperreal** numbers. Furthermore, we assume that all elements of $^*\mathbb{R}$ are urelements.

(2) We assume that the interpretation of the symbol $\in$ in $^*\mathfrak{M}$ is actually the relation $\in$, and the universe of $^*\mathfrak{M}$ (with the two sorts, numbers and sets, combined) is some transitive subset of the superstructure $V(^*\mathbb{R})$. As we will soon see, there is no way that it could be all of $V(^*\mathbb{R})$. Still, this assumption is quite comforting. It guarantees that if A is any set in $\mathfrak{M}$, then the object $^*A$ in $^*\mathfrak{M}$ is also a set, all of whose members are also in $^*\mathfrak{M}$. Furthermore, the rank of $^*A$ in $^*\mathfrak{M}$ will always be the same as the rank of A in $\mathfrak{M}$.

For notational simplicity, we will also usually omit the * before the symbols $+$, $\cdot$ and $<$, even though this is technically inaccurate: $^*+$ is a proper extension of the standard operation $+$, etc.

Lemma 7.5 (Main Lemma). *The mapping * from $\mathfrak{M}$ to $^*\mathfrak{M}$ is an elementary embedding (of $\mathcal{L}$-structures, and therefore also of $\mathcal{L}_0$-structures), with respect to restricted formulas.*

Proof. In spite of the importance of this lemma, there is almost nothing to prove, because $\mathcal{L}$ has a constant for each member of $\mathfrak{M}$, and both $\mathfrak{M}$ and $^*\mathfrak{M}$ are models of T, which includes every true statement about $\mathfrak{M}$ involving these constants.

So let P be any restricted formula, and let g be any $\mathfrak{M}$-assignment. Then form an $\mathcal{L}$-sentence P′ by replacing each free variable v_i of P by the constant $C_{g(v_i)}$. The point is that this replacement is equivalent to the interpretation of the free variables of P under g. So $\mathfrak{M} \models \mathrm{P}[g]$ iff $\mathfrak{M} \models \mathrm{P}'$ iff $\mathrm{P}' \in T$ iff $^*\mathfrak{M} \models \mathrm{P}'$ iff $^*\mathfrak{M} \models \mathrm{P}[* \circ g]$. ■

Corollary 7.6 (Transfer Principle). *Any restricted $\mathcal{L}$-statement that is true in $\mathfrak{M}$ remains true in $^*\mathfrak{M}$.*

The transfer principle is sometimes called **Leibniz's principle**, because Leibniz enunciated more or less the same point regarding the use of infinitesimals, as indicated in Section 7.1. To use the transfer principle, it is important to remember that for any object x in $\mathfrak{M}$, the constant C_x of $\mathcal{L}$ is interpreted as x in $\mathfrak{M}$ and $^*x$ in $^*\mathfrak{M}$.

With a bit of practice, one begins to learn just how powerful the transfer principle is. Here is a tiny sampler of its consequences: Since $3 < 4$ and $3 + 4 = 7$ are true in $\mathfrak{M}$, these statements are also true in $^*\mathfrak{M}$. Since $(\mathbb{R}, +, \cdot, <)$, an object in $\mathfrak{M}$, is an ordered field, so is $^*(\mathbb{R}, +, \cdot, <)$, an object in $^*\mathfrak{M}$. Since $\mathbb{Q}$ is a dense subset of $\mathbb{R}$, $^*\mathbb{Q}$ must be a dense subset of $^*\mathbb{R}$. Since the usual exponential function exp is an increasing function from $\mathbb{R}$ onto $\mathbb{R}^+$, the object $^*\exp$ in $^*\mathfrak{M}$ is an increasing function from $^*\mathbb{R}$ onto $^*\mathbb{R}^+$.

Let's show one of the claims in the previous paragraph in some detail: to say that $\mathbb{Q}$ is a dense subset of $\mathbb{R}$ means

$$\forall x, y \in \mathbb{R}[x < y \to \exists z \in \mathbb{Q}(x < z < y)].$$

Therefore, the following $\mathcal{L}$-sentence is true in $\mathfrak{M}$:

$$\forall x, y \in C_{\mathbb{R}}[x < y \to \exists z \in C_{\mathbb{Q}}(x < z < y)].$$

So this $\mathcal{L}$-sentence is in T, and hence it is true in $^*\mathfrak{M}$. But in $^*\mathfrak{M}$, the constants $C_{\mathbb{R}}$ and $C_{\mathbb{Q}}$ are interpreted as $^*\mathbb{R}$ and $^*\mathbb{Q}$, respectively. So $^*\mathbb{Q}$ must be a dense subset of $^*\mathbb{R}$.

More generally, since $\mathfrak{M}$ is a model of Zermelo's set theory (with urelements), all restricted theorems of this theory hold in $^*\mathfrak{M}$. In other words, for the most part, every true statement of ordinary mathematics (except for those based on replacement) holds in $^*\mathfrak{M}$, *when properly translated using the mapping* $*$.

Of course, the transfer principle would be a triviality if $\mathfrak{M}$ and $^*\mathfrak{M}$ were isomorphic. This possibility is eliminated by the construction of $^*\mathfrak{M}$: clearly, the element that interprets the constant h is a positive small number in $^*\mathbb{R}$, so $^*\mathbb{R}$ is nonarchimedean.

At this point, we seem to be confronted with a blatant paradox: if the transfer principle asserts that every statement that holds in $\mathfrak{M}$ also holds in $^*\mathfrak{M}$, how can $\mathbb{R}$ be Archimedean while $^*\mathbb{R}$ is not? The resolution of this paradox is based on the fact that the interpretations of power sets in $^*\mathfrak{M}$ must be "deficient." In other words, while $\mathfrak{M}$ is closed under subsets (if $A \subseteq B \in V(S)$, then $A \in V(S)$), $^*\mathfrak{M}$ cannot be. Thus we have another situation that is similar to Skolem's paradox.

Example 2. Let us illustrate this situation by considering the completeness property of $\mathbb{R}$. The usual definition of this property is that every nonempty subset of $\mathbb{R}$ with an upper bound has a least upper bound. To apply the transfer principle, we must rephrase this as: for every nonempty $A \in \mathcal{P}(\mathbb{R})$, if A has an upper bound then it has a least upper bound. Thus $^*\mathfrak{M}$ must satisfy the same statement with $\mathcal{P}(\mathbb{R})$ replaced by $^*\mathcal{P}(\mathbb{R})$. *If* $^*\mathcal{P}(\mathbb{R})$ *were the full power set of* $^*\mathbb{R}$*, this would make* $^*\mathbb{R}$ *complete, and thus Archimedean.* But in fact $^*\mathcal{P}(\mathbb{R})$ is a proper subset of $\mathcal{P}(^*\mathbb{R})$; in order for the transfer principle to work, $^*\mathcal{P}(\mathbb{R})$ must be missing all sorts of sets. For example, $\mathbb{R}$, $\mathbb{N}$ and $\mathbb{Z}$ must all be missing from $^*\mathcal{P}(\mathbb{R})$, because they are bounded in $^*\mathbb{R}$ (any large number is an upper bound for them), but they don't have least upper bounds. That is, even though these are subsets of $^*\mathbb{R}$, $^*\mathfrak{M}$ doesn't "recognize" them as such. It can be shown that if A is *any* infinite subset of $\mathbb{R}$, then $^*A$ is a proper extension of A, and A itself is not in $^*\mathcal{P}(\mathbb{R})$; furthermore, this implies that A is not a member of $^*\mathfrak{M}$. It also follows from this reasoning that $\mathcal{P}(^*\mathbb{R})$ is not in $^*\mathfrak{M}$.

More generally, for any set A in $\mathfrak{M}$, let $^\sigma A = \{^*x \mid x \in A\}$. (The use of the symbol σ here derives from the word "standard.") By transfer, whenever $x \in A$ holds in $\mathfrak{M}$, $^*x \in {}^*A$ must hold in $^*\mathfrak{M}$. In other words ${}^{\sigma}A \subseteq {}^{*}A$. But these sets are equal if and only if A is finite. If A is infinite, then ${}^{\sigma}A$ cannot be a member of ${}^{*}\mathfrak{M}$.

Exercise 7. Prove the claims made in the previous two sentences.

A succinct way of putting the main lesson of the previous example is that the operators * and $\mathcal{P}$ cannot commute. They do commute in this limited sense: for any set A in $\mathfrak{M}$, ${}^{*}\mathcal{P}(A)$ is the set of all subsets of ${}^{*}A$ *that are in* ${}^{*}\mathfrak{M}$. On the other hand, the transfer principle implies that * does commute literally with many other operators. Here are just a few examples:

Exercise 8. Prove that * commutes with:

(a) interval notation. That is, whenever $[a, b]$ is a closed interval in $\mathbb{R}$, ${}^{*}[a, b]$ is the set of all hyperreal numbers between a and b.

(b) ordered pairs: ${}^{*}(a, b) = ({}^{*}a, {}^{*}b)$, for any $a, b \in {}^{*}\mathfrak{M}$.

(c) function application: ${}^{*}(f(a))$ always equals ${}^{*}f({}^{*}a)$.

(d) Cartesian product: ${}^{*}(A \times B) = {}^{*}A \times {}^{*}B$, for any sets A and B in $\mathfrak{M}$.

This situation is strongly related to something we encountered in Section 6.2: many basic notions of set theory are absolute, but not the power set operation.

Exercise 9. For another nice example of the strangeness of ${}^{*}\mathfrak{M}$, try applying the transfer principle to the well-ordering property of $\mathbb{N}$. ${}^{*}\mathbb{N}$ is not really well ordered under ${}^{*}<$, but ${}^{*}\mathfrak{M}$ must "think" that it is. Draw some conclusions about ${}^{*}\mathbb{N}$ and ${}^{*}\mathcal{P}(\mathbb{N})$.

The ideas we have just introduced are important enough to deserve words:

Definition. Let B be a set in ${}^{*}\mathfrak{M}$. A subset of B is called **internal** or **external** according to whether it is or is not a member of the universe of sets in ${}^{*}\mathfrak{M}$.

From the above example about the completeness property, it follows that a subset of $\mathbb{R}$ (viewed as a subset of ${}^{*}\mathbb{R}$) is internal if and only if it is finite. Other important external subsets of $^*\mathbb{R}$ are the set of all large numbers, the set of all medium numbers, and the monad of any number. Assuming that any of these sets is in $^*\mathcal{P}(\mathbb{R})$ quickly leads to a contradiction.

Exercise 10. Show that $\mu(0)$, the set of all small numbers in $^*\mathbb{R}$, is external. From this, deduce that every monad is external.

A common reaction to this strange situation is that there is something mysterious, even a bit "fishy," about the structure $^*\mathfrak{M}$. But there is no voodoo here. It's just that power sets in $^*\mathfrak{M}$ are *rigged* to make all sorts of things work that would not work with true power sets. It took the brilliance of Robinson to see that an ordered field could be "externally" (that is, actually) nonarchimedean and at the same time "internally" (that is, with respect to its limited perspective) look just like $\mathbb{R}$.

We also need some guidelines for establishing that various sets are *internal*. Of course, every set of the form $^*A$ is internal, but there are all sorts of other sets that must be internal. Here is the most important method for establishing internality:

Lemma 7.7 (Internal Definition Principle). *If A is a subset of a set in $^*\mathfrak{M}$ and A is definable in $^*\mathfrak{M}$ (with parameters), then A is internal.*

Proof. We mentioned previously that all the axioms of Zermelo's set theory hold in $^*\mathfrak{M}$, by transfer. These axioms include separation, and the internal definition principle says precisely that separation holds in $^*\mathfrak{M}$. ■

Exercise 11. Let h be a nonzero small number in $^*\mathbb{R}$. Is $\{h\}$ internal? Is the interval $[h, 3]$ internal?

The converse of the internal definition principle holds trivially, so the internal sets are precisely the definable subsets of sets in $^*\mathfrak{M}$. The internal sets are also those sets that are elements of elements of $^*\mathfrak{M}$.

Sometimes, the fact that $^*\mathfrak{M}$ is a nonstandard extension of $\mathfrak{M}$ is not enough to prove some result that clearly seems correct. After several years during which the early practitioners of nonstandard analysis struggled with this difficulty, Robinson, Davis, and W. A. J. Luxemburg identified the right sort of additional condition that could be required of nonstandard models. Recall the definition of κ-saturation in Chapter 5. Nonstandard analysis uses the following slightly weaker version of this notion:

Definition. Let κ be an uncountable cardinal. A nonstandard model is called **κ-saturated** if every collection of fewer than κ *internal* sets is resilient.

For most purposes, it suffices to assume that $^*\mathfrak{M}$ is $\aleph_1$-saturated, but occasionally, more saturation is needed. If $^*\mathfrak{M}$ is κ-saturated for some κ greater than the cardinality of the universe of $\mathfrak{M}$ (not $^*\mathfrak{M}$!), then $^*\mathfrak{M}$ is called **polysaturated**, a term created by the witty mind of Keith Stroyan. So if $\mathfrak{M}$ is the usual standard model, polysaturation means $(\beth_\omega)^+$-saturation. Polysaturation is sufficient for all practical purposes. Our compactness theorem approach does not guarantee even $\aleph_1$-saturation of $^*\mathfrak{M}$. But with the ultrapower approach, it is possible to prove (in ZFC) the existence of polysaturated nonstandard models. Polysaturation is still a much weaker assumption than full saturation.

Convention. From now on, we always assume that $^*\mathfrak{M}$ has sufficient saturation for whatever purpose we have in mind, with polysaturation being the maximum amount needed.

The next three exercises, especially the last two, are intended to give some feel for the power of various saturation assumptions.

Exercise 12. Prove that in any $\aleph_1$-saturated model of the theory T defined earlier in this section, $^*\mathbb{R}$ is nonarchimedean.

Exercise 13. Prove that if $^*\mathfrak{M}$ is κ-saturated, then every infinite internal set of $^*\mathfrak{M}$ has cardinality at least κ.

This exercise applies even to infinite internal sets that are **hyperfinite**, meaning that $^*\mathfrak{M}$ "thinks" they are finite. A typical example of an infinite hyperfinite set is $\{1, 2, \ldots, n\}$, where n is a large natural number.

Exercise 14. Assume that $^*\mathfrak{M}$ is polysaturated. Prove that for every set A in $\mathfrak{M}$ there is a hyperfinite set B in $^*\mathfrak{M}$ such that $^\sigma A \subseteq B$.

By modifying the construction of $^*\mathfrak{M}$, we could even guarantee the existence of a single hyperfinite set that contains $^*x$ for every object x in $\mathfrak{M}$. Thus a hyperfinite set can have very large cardinality "in real life." We are almost ready to use $^*\mathfrak{M}$ to do some mathematics. First, here is a useful technical fact:

Proposition 7.8. *Let $x \in {}^*\mathbb{R}$ be limited. Then there is a unique real number y such that $x \approx y$.*

Proof. Let x be limited. Then there exists $n \in \mathbb{N}$ such that $x < n$. So the set $\{u \in \mathbb{R} \mid u < x\}$ is bounded, and therefore has a LUB y in $\mathbb{R}$. We claim that $x \approx y$. For suppose $n \in \mathbb{Z}^+$. If $y \leq x - 1/n$, then $y < y + 1/2n < x$, contradicting the definition of y. Similarly, if $y \geq x + 1/n$, then $x < y - 1/2n < x$, again contradicting the definition of y.

Uniqueness of y follows easily: if $x \approx y_1$ and $x \approx y_2$, then $y_1 \approx y_2$. But when y_1 and y_2 are real, this implies that $y_1 = y_2$. ■

Definition. For any limited $x \in {}^*\mathbb{R}$, the real number described in this proposition is called the **standard part** of x, denoted $St(x)$.

Figure 7.2 provides a helpful image of the set of limited numbers in $^*\mathbb{R}$. This set looks a lot like $\mathbb{R}$, except that each real is now replaced by its monad, a set of hyperreal numbers. I like to visualize the limited

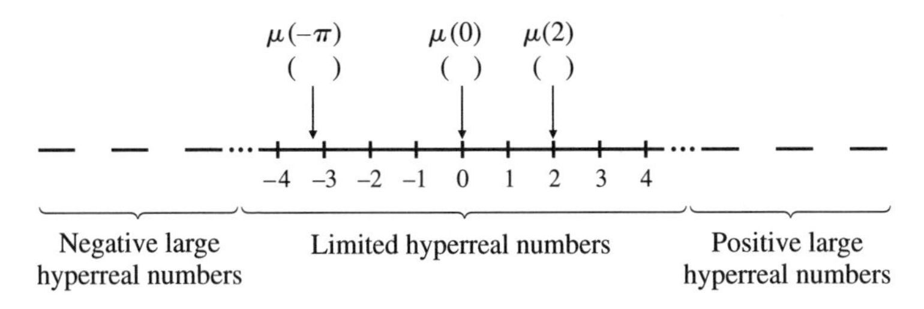

Figure 7.2. Hyperreal numbers, limited numbers, and monads

numbers as a line in which each point is somehow embedded in a blurry little "cloud." Within each of these clouds there is a unique real number, which is the standard part of every hyperreal number in that monad.

It is much harder to visualize the entire set ${}^*\mathbb{R}$, with its vast stretches of positive and negative large numbers. One simple observation is that, for each $x \in {}^*\mathbb{R}$, we can define the **galaxy** of x, meaning the set of all hyperreals whose distance from x is limited. Thus ${}^*\mathbb{R}$ is partitioned into (disjoint) galaxies, each of which looks just like the set of limited numbers. Of course, this is a much coarser partition of ${}^*\mathbb{R}$ than the partition created by the monads.

7.4 Nonstandard methods in mathematics

Now we can use nonstandard models to prove that the standard "Weierstrassian" (epsilon-delta) definitions of analysis and topology are equivalent to the infinitesimal-based definitions that Leibniz and Euler favored. These equivalences demonstrate that both approaches are valid and can be mixed at will. Combining them gives mathematicians more power than using either one exclusively. As one might expect, many nonstandard proofs are shorter and more elegant than the usual classical arguments.

Here are a few of these important equivalences. It is important to realize that these are *theorems*, not definitions. That is, the left side of each equivalence refers to the standard definition of a concept. The right side refers to any nonstandard model ${}^*\mathfrak{M}$.

Proposition 7.9. *Let $f : U \to \mathbb{R}$, where $U \subseteq \mathbb{R}$ is open, and let $a \in U$. Then f is continuous at a if and only if, whenever $x \approx a$, $({}^*f)(x) \approx f(a)$.*

Proof. For the forward direction, assume that f is continuous at a. In other words, this statement is true in $\mathfrak{M}$. Let ϵ be any positive real number. By the continuity assumption, we can choose a positive real δ such that

$$\forall x \in \mathbb{R}\,(|x - a| < \delta \to |f(x) - f(a)| < \epsilon)\,.$$

Since this statement is true in $\mathfrak{M}$, transfer tells us that the following statement is true in $^*\mathfrak{M}$ for the same ϵ and δ:

$$\forall x \in {}^*\mathbb{R}\,(|x-a| < \delta \rightarrow |({}^*f)(x) - ({}^*f)(a)| < \epsilon)\,. \tag{7.1}$$

So if $x \approx a$, we obtain $|({}^*f)(x) - ({}^*f)(a)| < \epsilon$. Since this holds for every positive real ϵ, we have $({}^*f)(x) - ({}^*f)(a) \approx 0$, or $({}^*f)(x) \approx ({}^*f)(a)$. And $({}^*f)(a) = f(a)$ by transfer.

For the reverse direction, assume that whenever $x \approx a$, $({}^*f)(x) \approx f(a) = ({}^*f)(a)$. Now let ϵ be any given positive real number. Then if δ is any positive infinitesimal, statement 7.1 is true. Therefore, the following statement holds in $^*\mathfrak{M}$:

$$\exists \delta \in {}^*\mathbb{R}\,[\delta > 0 \wedge \forall x \in {}^*\mathbb{R}\,(|x-a| < \delta$$
$$\rightarrow |({}^*f)(x) - ({}^*f)(a)| < \epsilon)]\,.$$

We can therefore apply transfer to this statement, so the same statement with all the asterisks removed is true in $\mathfrak{M}$. Since this holds for an arbitrary positive real ϵ, f is continuous at a. ■

If you examine this proof, you will see how it depends on clever choices of which quantifiers to include or omit when transfer is applied. In the forward direction, both ϵ and δ are left unquantified. In the reverse direction, it is crucial to quantify δ but not ϵ.

Give some thought to the content of this proposition. The standard ϵ-δ definition of continuity is rigorous but somewhat "clunky." Intuitively, a function is continuous at a number if a small change in the input creates only a small change in the output of the function. This is exactly what the right side of this proposition says!

Corollary 7.10. *Let $f : \mathbb{R} \rightarrow \mathbb{R}$. Then f is continuous if and only if, for every limited x in $^*\mathbb{R}$, $St({}^*f(x)) = f(St(x))$.*

Since the basic arithmetical operations are continuous (viewed as functions of one or two variables), this corollary tells us that the standard part mapping commutes with addition, subtraction, multiplication, and division.

Proposition 7.11. *Let f be as in Proposition 7.9. Then f is uniformly continuous if and only if, for any $u, v \in Dom(^*f)$, $u \approx v$ implies $(^*f)(u) \approx (^*f)(v)$.*

Exercise 15. To understand the subtle difference between the two previous propositions, consider the function $f(x) = 1/x$, with domain $\mathbb{R} - \{0\}$. Explain why this function is continuous at every point in its domain under the nonstandard criterion of Proposition 7.9, but is not uniformly continuous under the nonstandard criterion of Proposition 7.11.

Proposition 7.12. *Let (a_n) or simply a be any sequence of real numbers (that is, a function from $\mathbb{N}$ to $\mathbb{R}$), and let $L \in \mathbb{R}$. Then $\lim(a_n) = L$ if and only if, for every large natural number n, $(^*a)_n \approx L$.*

Exercise 16. Prove this proposition. You may use an argument that is very similar to the proof of Proposition 7.9.

Proposition 7.13. *Let f be as in Proposition 7.9, and let $a \in U$ and $L \in \mathbb{R}$. Then $f'(a) = L$ if and only if, whenever $x \approx a$ and $x \neq a$,*

$$\frac{(^*f)(x) - f(a)}{x - a} \approx L.$$

This proposition allows us to define the derivative of a function almost exactly as Leibniz did:

$$f'(x) = St\left[\frac{(^*f)(x+h)] - f(x)}{h}\right],$$

where h is a nonzero small number, provided that the value of the right-hand side is the same for all such h. The application of the standard part function on the right-hand side corresponds to discarding "negligible" (essentially, infinitesimal) terms, a practice used by many seventeenth century mathematicians such as Fermat and Leibniz. In modern parlance, we say instead that we let h or Δx "approach" 0 or "go to" 0.

Example 3. Let $f(x) = x^2$. Then we have $f(x+h) = x^2 + 2xh + h^2$, so $f(x+h) - f(x) = 2xh + h^2$ and $[f(x+h) - f(x)]/h = 2x + h$, for any nonzero h. By transfer, all of this holds in $^*\mathfrak{M}$ with f replaced by $^*f$. So if $x \in \mathbb{R}$ and h is small but not zero, we obtain

$$f'(x) = St(2x + h) = 2x.$$

Do not confuse this procedure, in which we use $^*\mathfrak{M}$ and $^*f$ to compute the derivative of a standard function f at a standard real number x, with the process of computing the derivative of $^*f$. Of course, since f in this example is the squaring function, the transfer principle implies that $^*f$ is the squaring function in $^*\mathfrak{M}$, and so its derivative is also $2x$.

Proposition 7.14. *Let f be piecewise continuous on $[a, b]$, and let $L \in \mathbb{R}$. Then $\int_a^b f(x)\,dx = L$ if and only if, for every hyperfinite Riemann sum S of $^*f$ over $^*[a, b]$ such that the width of every subinterval is small, $S \approx L$.*

The next four results give the nonstandard versions of some of the basic notions of topology. For simplicity, we state them in the context of $\mathbb{R}$.

Proposition 7.15. *Let $x \in A \subseteq \mathbb{R}$. Then x is in the interior of A if and only if $\mu(x) \subseteq {}^*A$.*

Corollary 7.16. *Let $A \subseteq \mathbb{R}$. Then A is open if and only if, for every x in A, $\mu(x) \subseteq {}^*A$.*

Exercise 17. Prove the previous proposition and corollary. (There is almost nothing to prove for the corollary.)

Proposition 7.17. *Let $A \subseteq \mathbb{R}$. Then A is closed if and only if, for every y in $^*A$ and every x in $\mathbb{R}$, $x \approx y$ implies $x \in A$.*

Proof. Recall that A is closed if and only if its complement A' is open. By the previous corollary, that means that for every $x \in A'$, $\mu(x) \subseteq {}^*(A')$. If we "unravel" this, it becomes: for every $x \in A'$ and every $y \in {}^*\mathbb{R}$, $y \approx x$ implies $y \in {}^*(A')$. It is easy to show that $^*(A') = ({}^*A)'$, and so we have that A is closed if and only if for every

$x \in \mathbb{R}$ and every $y \in {}^*\mathbb{R}$, $x \notin A$ and $y \approx x$ implies $y \notin {}^*A$. It is a simple exercise in propositional logic to show that this is equivalent to what we want. ■

Exercise 18. Complete this proof by first proving that ${}^*(A') = ({}^*A)'$, or more precisely ${}^*(\mathbb{R} - A) = {}^*\mathbb{R} - {}^*A$, and then verifying (perhaps by a truth table) the claim in its last sentence.

Proposition 7.18. *Let $A \subseteq \mathbb{R}$. Then A is compact if and only if for every $y \in {}^*A$ there is an $x \in A$ such that $x \approx y$.*

Proof. The simplest proof of this result for subsets of $\mathbb{R}$ or $\mathbb{R}^n$ is via the Heine–Borel theorem. (See the next exercise.) Instead, we give a proof that generalizes to all topological spaces. This will allow us to give a short proof of the Heine–Borel theorem via nonstandard analysis.

So, for the forward direction, assume that A is compact and we have $y \in {}^*A$ for which there is no $x \in A$ such that $x \approx y$. By definition, that means that every $x \in A$ has a neighborhood U_x such that $y \notin {}^*U_x$. Since the collection of open sets $\{U_x \mid x \in A\}$ covers A, there is a finite subcovering by compactness. That is, for some $x_1, x_2, \ldots, x_n \in A$, we have $A \subseteq U_{x_1} \cup U_{x_2} \cup \cdots \cup U_{x_n}$. By transfer, ${}^*A \subseteq {}^*U_{x_1} \cup {}^*U_{x_2} \cup \cdots \cup {}^*U_{x_n}$. But then y must be in one of the ${}^*U_{x_i}$'s, contradicting its definition.

For the reverse direction, assume that A is not compact. Then there is a collection $\{C_i \mid i \in I\}$ of closed subsets of $\mathbb{R}$ such that the collection $\{A \cap C_i \mid i \in I\}$ is not resilient. By transfer, the collection $\{{}^*A \cap {}^*C_i \mid i \in I\}$ also has the finite intersection property. So by saturation its intersection is nonempty. Let y be in this intersection, and assume that $x \in A$ and $x \approx y$. Since $y \in {}^*C_i$ for every $i \in I$ and ${}^*C_i$ is closed, it follows by Proposition 7.17 that $x \in C_i$ for every $i \in I$. But then $x \in \bigcap_{i \in I} A \cap C_i$, contradicting the fact that the intersection of this collection is empty. Thus the right-hand side of the equivalence is false. ■

Compare the last two propositions. The right-hand side of Proposition 7.17 says that if any given element of ${}^*A$ has a standard part, that standard part must be in A. The right-hand side of Proposition 7.18

says that every element of ${}^*A$ has a standard part, and that standard part must be in A.

Exercise 19. Recall the Heine–Borel theorem, which states that a subset of $\mathbb{R}^n$ is compact if and only if it is closed and bounded. Use this theorem to prove Proposition 7.18 (for subsets of $\mathbb{R}$ only).

Results 7.9 through 7.12 and 7.15 through 7.18 can easily be generalized to an arbitrary metric space (X, d) in place of $\mathbb{R}$. For $x, y \in {}^*X$, $x \approx y$ means ${}^*d(x, y) \approx 0$. From this, monads are defined more or less as before: for any $x \in {}^*X$, $\mu(x) = \{y \in {}^*X \mid y \approx x\}$. In particular, all of these results generalize to Euclidean space $\mathbb{R}^n$. These same results, except for Proposition 7.11, can be further generalized to arbitrary topological spaces, provided that we correctly redefine the notion of monads:

Definition. Let $(X, \mathcal{T})$ be a topological space in $\mathfrak{M}$. (That is, $\mathcal{T}$ is a basis for the open subsets of X.) For any $x \in X$, the **monad** of x is

$$\mu(x) = \bigcap\{{}^*U \mid x \in U \in \mathcal{T}\}.$$

An element y in ${}^*X$ is called **nearstandard** if $y \in \mu(x)$ for some $x \in X$.

In words, $\mu(x)$ is the intersection of the nonstandard versions of all basic open sets that contain x. It is a subset of ${}^*X$. Since $\mu(x)$ contains ${}^*x$ rather than x, you might think that $\mu({}^*x)$ would be a more sensible notation for this set. This is a reasonable argument, but $\mu(x)$ is the usual notation.

Exercise 20. Prove that the nearstandard members of ${}^*\mathbb{R}$ are precisely the limited ones. You may assume that $\mathcal{T}$ (for $\mathbb{R}$) consists of all open intervals in $\mathbb{R}$.

If x is an element of a metric space, then this topological definition of $\mu(x)$ coincides with the metric space definition of $\mu({}^*x)$: it consists of everything that is close to ${}^*x$. But in a general topological space, not every element of its nonstandard version is in a monad; only nearstandard elements are. Thus there is no natural equivalence relation $\approx$ on the nonstandard version of a general topological space. (But see part (b) of the next exercise.)

Proposition 7.19. *Let $(X, \mathcal{T})$ be a topological space in $\mathfrak{M}$, and let $x \in X$. Then there is a set $W \in {}^*\mathcal{T}$ such that ${}^*x \in W \subseteq \mu(x)$.*

Proof. For each open set $U \in \mathcal{T}$, let $C_U = \{V \in {}^*\mathcal{T} \mid {}^*x \in V \subseteq {}^*U\}$. By the internal definition principle, C_U is a nonempty internal set in ${}^*\mathfrak{M}$. Also, if $U_1, U_2, \ldots, U_n \in \mathcal{T}$, then $C_{U_1} \cap C_{U_2} \cap \cdots \cap C_{U_n} = C_{U_1 \cap U_2 \cap \cdots \cap U_n}$. Therefore, since $U_1 \cap U_2 \cap \cdots \cap U_n$ is also in $\mathcal{T}$, the collection $\{C_U \mid U \in \mathcal{T}\}$ has the finite intersection property. So by saturation, the intersection of all the C_U's is nonempty. (Here we need saturation in a cardinality greater than that of $\mathcal{T}$.) Let W be any set in the intersection of all the C_U's. ■

Exercise 21. Let $(X, \mathcal{T})$ be a topological space in $\mathfrak{M}$.

(a) Show that $(X, \mathcal{T})$ is Hausdorff if and only if, for any $x, y, \in X$, $\mu(x)$ and $\mu(y)$ are disjoint.

(b) If $(X, \mathcal{T})$ is Hausdorff, what is the reasonable way to define an equivalence relation $\approx$ on the set of all nearstandard elements of ${}^*X$? Show that your definition works.

(c) Explain why we need $(X, \mathcal{T})$ to be Hausdorff in order to define the standard part function on the set of all nearstandard elements of ${}^*X$.

For Hausdorff spaces, results 7.9, 7.12, and 7.15 through 7.18 generalize without change. For arbitrary spaces, they still generalize provided that they are stated in terms of monads instead of $\approx$. For example, Proposition 7.9 becomes: A function f is continuous at a point a if and only if ${}^*f[\mu(a)] \subseteq \mu(f(a))$. Proposition 7.18 has a particularly succinct statement: a topological space is compact if and only if every point in its nonstandard version is nearstandard.

What we have done so far is to give nonstandard equivalents of various concepts of analysis and topology. These are appealing and perhaps enlightening, but what makes these equivalences really useful is that we can use them to write proofs that are shorter and more intuitive than the usual proofs. It is essential to realize that theorems stated in the usual language of mathematics, but proved using these equivalences, are not "nonstandard results." They are results about standard mathematical objects with the same meaning that they normally have. They are just proved using a nonstandard model $^*\mathfrak{M}$. The rest of this section is devoted to such proofs.

Theorem 7.20 (Intermediate Value Theorem). *Every continuous real-valued function that is defined on a closed interval of real numbers, and takes values of opposite sign at the endpoints, has a zero.*

Proof. Let $f : [a, b] \to \mathbb{R}$, with $a < b$, be continuous. Assume $f(a) < 0 < f(b)$. (The case $f(a) > 0 > f(b)$ is handled similarly.) If n is any natural number and we partition the interval $[a, b]$ into n equal subintervals with "partition points" $a = p_0, p_1, \ldots, p_{n-1}, p_n = b$, there will be a largest p_k such that $f(p_k) < 0$. By transfer, this statement must hold for $^*f$ in $^*\mathfrak{M}$. So let n be a large natural number (that is, $n \in {}^*\mathbb{N} - \mathbb{N}$). Then partition $^*[a, b]$ as above and find the largest p_k such that $^*f(p_k) < 0$.

Since p_k is limited, $St(p_k)$ is defined; call this real number c. By definition, $c \approx p_k$. Also, since $p_{k+1} = p_k + (b - a)/n$, we have $p_{k+1} \approx p_k$ and thus $c \approx p_{k+1}$. So, by Proposition 7.9, $^*f(p_k) \approx {}^*f(c) \approx {}^*f(p_{k+1})$. But $^*f(p_k) < 0$ while $^*f(p_{k+1}) \geq 0$. These facts easily yield that $^*f(c) \approx 0$. But since c is real, transfer guarantees that $^*f(c) = f(c)$, a real number. So $f(c) = 0$. ∎

Exercise 22. Prove the extreme value theorem: every continuous real-valued function that is defined on a closed interval of real numbers achieves an absolute maximum and minimum. The proof can be very similar to the previous one.

Theorem 7.21 (Chain Rule). *Let f and g be as in Proposition 7.9, and $a \in \mathbb{R}$. If f is differentiable at a and g is differentiable at $f(a)$, then $g \circ f$ is differentiable at a and $(g \circ f)'(a) = f'(a) \cdot g'(f(a))$.*

Proof. We use Proposition 7.13. Assume all the givens and that $x \approx a$, $x \neq a$. For clarity, let's use Leibniz's notation: let $dx = x - a$,

$du = {}^*f(x) - {}^*f(a)$, and $dy = {}^*g({}^*f(x)) - {}^*g({}^*f(a))$. We must show that $St(dy/dx) = f'(a) \cdot g'(f(a))$. We consider two cases:

If $du = 0$, then $f'(a) = St(du/dx) = St(0) = 0$. But $du = 0$ also implies $dy = 0$, so both sides of the equation we want to establish are 0.

If, on the other hand, $du \neq 0$, then we have

$$St(dy/dx) = St(du/dx) \cdot St(dy/du)$$

by the remarks following Corollary 7.10

$$= f'(a) \cdot g'(f(a)), \quad \text{as desired.}$$

■

As Lindstrøm writes [Cut, page 19], "you can now prove the chain rule the way you always wanted to." This is also the way that Leibniz and many other mathematicians of the seventeenth and eighteenth centuries would have proved the chain rule. Of course, the notations don't quite match: for Leibniz the derivative was dy/dx, whereas we have to take $St(dy/dx)$. But Leibniz in essence applied the standard part operation even if he omitted it from his notation.

The nonstandard characterization of compactness that we established in Proposition 7.18 is particularly useful for simplifying proofs. From the proof of that proposition, you can see that the usual description of compactness in terms of *closed* sets is closely related to saturation. We now present several elegant proofs that make use of Proposition 7.18. We assume results 7.15 through 7.18 for arbitrary topological spaces.

Theorem 7.22 (Heine–Borel). *A subset of* $\mathbb{R}^n$ *is compact if and only if it is closed and bounded.*

Proof. Assume that A is a compact subset of $\mathbb{R}^n$. The fact that A must be closed follows immediately from the remarks following the proof of Proposition 7.18. Proposition 7.8 is implicitly involved here too.

To see that A is bounded, note that Proposition 7.18 implies that the absolute value of every member of ${}^*A$ is limited. So if K is any large positive number in ${}^*\mathbb{R}$, then $|x| < K$ for every x in ${}^*A$. Thus the statement

$$\exists u \in {}^*\mathbb{R} \forall x \in {}^*A(|x| < u)$$

holds in $^*\mathfrak{M}$. By transfer, the same statement holds in $\mathfrak{M}$ with the two stars removed. So there is a real number u such that A is contained in the ball of radius u centered at the origin, making A bounded.

For the reverse direction, assume A is closed and bounded. Since A is bounded, we can find a real number k such that every element of A has absolute value less than k. By transfer, the same statement holds for $^*A$, with the same k. So every element of A is limited, and again from the remarks following the proof of Proposition 7.18, the compactness of A follows from the fact that it's closed. ∎

Exercise 23. Using nonstandard methods, prove the forward direction of the Heine–Borel theorem for arbitrary metric spaces. (The reverse direction fails.) A subset of a metric space is called bounded if the set of distances between pairs of points in it is bounded.

Exercise 24. Using nonstandard methods, prove that a continuous function whose domain is a compact subset of $\mathbb{R}^n$ must be uniformly continuous.

Theorem 7.23. *Let f be a continuous function whose domain is compact. Then the range of f is also compact.*

Proof. Let A and B be the domain and range of f, respectively. Let y be any member of $^*B$. We need to show that y is nearstandard in $^*B$. By transfer, $^*f$ has domain $^*A$ and range $^*B$. So there is an x in $^*A$ such that $^*f(x) = y$. Since A is compact, $x \in \mu(a)$ for some $a \in A$. Since f is continuous, Proposition 7.9 (modified as in the remarks following Exercise 21) implies that $y \in \mu(f(a))$. So y is nearstandard. ∎

Recall the definition of the Cartesian product of topological spaces. If $\{(X_i, T_i) \mid i \in I\}$ is a collection of topological spaces, their product $\prod_{i \in I} X_i$ is topologized by taking as basic open sets all subsets of the form $\prod_{i \in I} U_i$, where each U_i is open in X_i and $U_i = X_i$ for all but a finite number of i's in I.

The next result is not easy to prove by standard methods. Most proofs of it are several pages long and quite complex. The axiom of choice is always required—indeed, this theorem is equivalent to AC. The shortness and elegance of the proof we will give is a perfect testament to the power of nonstandard methods.

Theorem 7.24 (Tychonoff). *Any Cartesian product of compact spaces is compact.*

Proof. Let $\{(X_i, T_i) \mid i \in I\}$ be a collection of compact spaces, and let $X = \prod_{i \in I} X_i$. For notational simplicity, we make the harmless assumption that $*$ is the identity on I. We must show that an arbitrary $f \in {}^*X$ is nearstandard. By definition of Cartesian products and transfer, f is a function with domain ${}^*I$ such that $f(i) \in {}^*X_i$ for each $i \in {}^*I$. For each $i \in I$, the compactness of X_i implies that $f(i)$ is nearstandard. So let $g \in X$ be such that $f(i) \in \mu(g(i))$ for every $i \in I$.

We claim that $f \in \mu(g)$. To prove this, we must show that $f \in {}^*V$ for every neighborhood V of g. Given such a V, there must be a basic open set U such that $g \in U \subseteq V$, and it suffices to prove that $f \in {}^*U$. By definition of the product topology on X, we know that

$$U = \{h \in X \mid h(i_1) \in U_{i_1} \wedge \cdots \wedge h(i_n) \in U_{i_n}\},$$

where $i_1, \ldots, i_n \in I$ and U_{i_k} is a neighborhood of $g(i_k)$ in X_{i_k} for $k = 1, 2, \ldots, n$. Therefore, by transfer,

$${}^*U = \{h \in {}^*X \mid h(i_1) \in {}^*U_{i_1} \wedge \cdots \wedge h(i_n) \in {}^*U_{i_n}\}.$$

Since $f(i) \in \mu(g(i))$ for every $i \in I$, we certainly have $f(i_k) \in {}^*U_{i_k}$ for $k = 1, 2, \ldots, n$. So $f \in {}^*U$, as desired. (The point of this argument is that if U is a basic open set in X, then the "restricted coordinates" of ${}^*U$ are the same as those of U. And these coordinates are all in I, not in ${}^*I - I$.) ∎

To make sure that we do not give the impression that nonstandard methods are useful only in calculus, analysis, and topology, we conclude with a nonstandard proof of a result in combinatorics—de Bruijn's theorem, which was stated in Section 5.3 and is also provable via the compactness theorem.

Proof. Assume that every finite subgraph of (G, R) has a k-coloring. For each $x \in G$, let B_x be the set of all k-colorings in $^*\mathfrak{M}$ on subgraphs of $(^*G, {}^*R)$ whose domain contains $^*x$. It is clear that B_x is internal. Also, if H is any finite subset of G, then the subgraph of (G, R) with domain H has a k-coloring by assumption, and such a coloring is a k-coloring of the subgraph of $(^*G, {}^*R)$ with domain $^*H$, by transfer. In other words, the collection of all the B_x's has the finite intersection property. Therefore, by saturation, the intersection of all the B_x's is nonempty. An element of this intersection is a k-coloring g of a subgraph of $(^*G, {}^*R)$ whose domain contains all of G^*. From this we get a k-coloring f of (G, R) by letting $f(x) = g(^*x)$ for every $x \in G$.

■

This proof works because the definition of a k-coloring is very simple, so simple that there is essentially no difference between k-colorings in $\mathfrak{M}$ and k-colorings in $^*\mathfrak{M}$.

Exercise 25. Try to mimic the proof of de Bruijn's theorem to show that if $A \subseteq \mathbb{R}$ and every finite subset of A has an upper bound, then A has an upper bound. What goes wrong?

We have seen many appealing nonstandard proofs of known theorems. But can we prove anything "new" with these methods? This is an important question that must be answered carefully. In a theoretical sense, the answer is no. More precisely, it is not hard to see (by transfer, essentially) that these nonstandard methods are conservative over standard mathematics. That is, any statement of ordinary mathematics that can be proved using nonstandard methods can also be proved without them. While it would be quite interesting if nonstandard analysis could actually increase the scope of "doable" mathematics, it is probably more comforting to know that this cannot occur.

But this theoretical answer does not mean that nonstandard methods are not useful for obtaining new results in mathematics. Quite a few theorems in a variety of areas have been proved first by nonstandard methods. The first and best known of these is the **Bernstein–Robinson theorem**, an important result about Hilbert space that had been open for about thirty years when it was proved in 1966. After the original non-standard proof was published, standard proofs of the theorem were also discovered. Since then, there have even been friendly rivalries between groups of mathematicians, one group working with standard methods and the other with nonstandard ones, to see which group could solve various problems more quickly.

CHAPTER 8
Constructive Mathematics

8.1 Introduction

Why should students of mathematics want to know something about constructive mathematics? Here is one answer: the constructivist school of mathematics shows us that there is a whole spectrum of mathematical activity, from arithmetical operations and other purely computational processes at one end, to abstract reasoning about very "remote" objects at the other end. Being aware of this spectrum, and understanding how to identify the "constructive" (or "computational" or "numerical") content of a given argument or definition, can make one a better mathematician.

Here is another, rather simple answer to this question: computers are becoming more and more important as a tool in mathematics, even in relatively "pure" mathematics, and constructive mathematics offers a good approximation to the part of classical (ordinary) mathematics that can be carried out computationally. In 1976, Kenneth Appel and Wolfgang Haken caught the world off guard when they made essential use of a computer program to prove the four-color theorem, one of the two or three most famous unsolved problems in mathematics at the time. Since then, the use of the computer as a tool in mathematics has accelerated substantially, and mathematicians of the twenty-first century will need to be aware of what can and cannot be done with this tool.

There is a tendency among mathematicians to identify constructive mathematics with the more specific movement known as **intuitionism**, which was started by Brouwer in the early 1900s. However, it would be erroneous to think of constructive mathematics as a modern invention. Throughout history, from the ancient Egyptians and Babylonians, through the Golden Age of Greece, and well into the nineteenth century, almost all mathematical activity was essentially constructive. It was only late in that century, with the development of the modern theory of real numbers, functions, and sets, that abstract definitions and reasoning became commonplace in mathematics. We mentioned earlier that intuitionism was to some extent a response to the paradoxes of set theory, but in fact it was more of a reaction to the overall trend toward abstraction in mathematics, and most specifically to Zermelo's introduction of the axiom of choice as a means to prove, highly nonconstructively, that $\mathbb{R}$ can be well ordered.

Let us now give some illustrative examples of the distinction between constructive and nonconstructive mathematics. As we do this, keep in mind that there are several distinct movements or philosophies under the loose label of constructivism. For instance, in Chapter 2 we mentioned Kronecker's objections to the use of infinite sets and processes in mathematics. This **finitist** position is one aspect of constructivism, but it is not the central viewpoint of constructive mathematics. We will confine ourselves in this section to examples that illustrate the constructive point of view in a general way.

Constructivists object to abstract reasoning in definitions as well as in proofs. In contemporary mathematics, it is perfectly acceptable to define an integer n by a phrase such as:

$$n = \begin{cases} 5, & \text{if the continuum hypothesis is true,} \\ 17, & \text{otherwise.} \end{cases}$$

To constructive mathematicians, this simply isn't a definition. To them, the definition of an integer (especially!) should give a clear, concrete procedure for computing that integer, and these words do nothing of the sort. The definition would not be improved by replacing the continuum hypothesis with some nonmathematical statement whose truth cannot presently be determined, such as "There is life in other galaxies" or "God exists." Nor would it be fixed by replacing the continuum hypothesis with a more concrete mathematical statement whose truth cannot presently be determined, such as Goldbach's conjecture (that every even number greater than 2 is the sum of two primes). You might want to compare this discussion to the brief discussion in Section 1.2 about what constitutes a proposition.

Note that this dubious definition is a typical **definition by cases**. Rejecting such definitions requires rejecting the classical **law of the excluded middle**, the tautology $P \vee \sim P$ (for arbitrary P), as an axiom. In other words, if one accepts the assertion that "the continuum hypothesis is true or the continuum hypothesis is false," then one must accept the above definition. But, to a constructive mathematician, a statement of the form $P \vee Q$ is established only by proving P or proving Q. The classical justification of the law of the excluded middle, in essence a truth table, is not accepted by constructivists. Many mathematicians who have heard a bit about intuitionism or constructivism think that its main tenet is the rejection of the law of the excluded middle. This is an oversimplification, but there is no question that the rejection of this tautology, and of Aristotelian (classical) logic in general as a valid basis for mathematical reasoning, is an important part of the constructivist viewpoint.

Here is a somewhat more relevant definition by cases: let f be the characteristic function of $\mathbb{Q}$ (in $\mathbb{R}$), that is, the function with domain $\mathbb{R}$ defined by $f(x) = 1$ if $x \in \mathbb{Q}$, and $f(x) = 0$ if $x \in \mathbb{R} - \mathbb{Q}$. This "pathological" function is now the standard example of a nowhere continuous function, but such definitions of functions (and, therefore, such functions) were not generally accepted until the late 1800s.

Is this an acceptable constructive definition of a function with domain $\mathbb{R}$? Almost certainly not. The first step would be to clarify what a real number is. One acceptable way to specify a real number, in classical or constructive mathematics, is to give a clear procedure for computing any requested finite number of decimal places of that real number. Clearly, there is no way to determine, from a finite part of the decimal expansion of a real number, whether that number is rational.

Also, there is no effective way to make that determination from a computer program that generates the digits of a real number. (This claim is less obvious than the previous one. See the next example.) So, under most constructive definitions of what a real number is, f is not a function with domain $\mathbb{R}$. As we will see, there are plausible constructive interpretations of real numbers and functions under which all functions from $\mathbb{R}$ to $\mathbb{R}$ are continuous.

Example 1. Suppose that a computer program (a Turing machine, say) computes a total recursive function g such that $g(n)$ is a digit for every n. Then we can say that this program defines a real number between 0 and 1, since g determines an infinite sequence of digits.

Now consider the function g with domain $\mathbb{N}$ defined by

$$g(n) = \begin{cases} 0, \text{ if Goldbach's conjecture holds for every even} \\ \text{number} \leq n, \\ \\ \text{the } n\text{th digit in the decimal expansion of } \pi, \text{ otherwise.} \end{cases}$$

Clearly, g is recursive, and every school of constructive (or classical) mathematics allows a Turing machine that computes g as a valid way of defining a real number. There is no problem with this definition by cases because a finite amount of arithmetical computation suffices to determine which case holds. But, as long as Goldbach's conjecture remains unsolved, there is no effective procedure for determining whether the number defined by g is rational. Therefore, there is no effective procedure for evaluating the above function f on this real number.

Now we can say more about the above comment that, in constructive mathematics, it is plausible to claim that all functions from $\mathbb{R}$ to $\mathbb{R}$ are continuous. In ordinary mathematics, we can define much simpler discontinuous functions than the characteristic function of the rationals. For instance, we can let $h(x) = 0$ if $x \leq 0$, and $h(x) = 1$ if $x > 0$. But this definition has the same defect as the definition of the function f: we have no way of evaluating h on the real number determined by the Turing machine described above. Constructively, the definition of h would be acceptable, but its domain would not be all reals. If it seems strange to imagine doing mathematics without discontinuous functions, note that all the elementary functions of mathematics (algebraic, trigonometric, exponential, logarithmic, and combinations thereof) are continuous at all points in their domains. The only discontinuous functions one encounters (in freshman calculus, say) are step functions and functions defined by cases.

Proofs by cases based on the unrestricted law of the excluded middle are also not, in general, acceptable to constructivists. Here is a well-known example:

Proposition 8.1. *There exist irrational numbers x and y such that x^y is rational.*

Proof. Let $r = \sqrt{2}^{\sqrt{2}}$. If r is rational, let $x = y = \sqrt{2}$, and we're done. Otherwise, let $x = r$ and $y = \sqrt{2}$. Then

$$x^y = [\sqrt{2}^{\sqrt{2}}]^{\sqrt{2}} = \sqrt{2}^2 = 2,$$

a rational number. ∎

Note that this existence proof provides no way to compute x. We know (from the two cases of the proof), that x is between 1 and 2, but unless we can determine whether r is rational, we have no idea whether x is between 1.4 and 1.5 or between 1.6 and 1.7. Thus, a constructive mathematician cannot accept this proof. (The theorem can be proved constructively, using a more sophisticated argument.) This brings us to what is perhaps the most important single tenet of constructivism: in order to prove that some object exists, one should provide an explicit "construction" of that object. Among the usual connectives and quantifiers, it is the existential quantifier that probably receives the strictest reinterpretation in constructive mathematics, with disjunction a close second.

Next we will give another nonconstructive existence proof, followed by a simple procedure for modifying the proof so that it becomes constructive. This brings up a very important point: there is a tendency to think of constructive mathematics as a primarily "negative" activity that rejects various definitions and proofs of mainstream mathematics. Regardless of whether there is some validity to this belief as an *historical* observation, there is no reason that this *should be* the main activity of constructive mathematics. Rather, constructive mathematics can play the very positive role of finding more concrete and informative proofs of classical mathematical theorems. Usually, if this cannot be done for the usual, general version of a theorem, it can at least be done for some slight modification of the theorem; and this type of modification also provides useful information. As we will see, Erret Bishop made particularly effective use of this approach. In cases where no constructive modification of a result is known, there is often little or no significance to the result anyway.

Theorem 8.2. *There exist transcendental real numbers. In fact, "most" reals (in the sense of cardinality) are transcendental.*

Proof. The function B defined in Appendix C shows that $\mathbb{N}^{<\omega}$ is denumerable. From this, it follows easily that the set of polynomials in one variable with integer coefficients is denumerable. We also know that each nonconstant polynomial has a finite number of solutions. Therefore, the set of all real solutions to such polynomials, which is the set of real algebraic numbers, is denumerable. But $\mathbb{R}$ is uncountable, by Cantor's Theorem. Therefore, the set of transcendental (nonalgebraic) reals is also uncountable, since the union of two denumerable sets is again denumerable. ■

This theorem and proof were published by Cantor in 1874, but it was not the first proof of the existence of transcendental numbers. That accomplishment belongs to Joseph Liouville, who discovered a large category of transcendental numbers in the 1840s. (In contrast, the fact that there are many *irrational* numbers, such as $\sqrt{2}$ and $\sqrt[3]{17}$, was known to the Pythagoreans about 2500 years ago.) For instance, he proved that

$$0.110001000000000000000001\ldots$$

is transcendental, where the nth 1 occurs in the $n!$th decimal place. So is any similar number obtained by replacing all the 1's with any sequence of nonzero digits. There was no theory of sets and cardinality at the time, but varying the nonzero digits in this manner would clearly produce an uncountable set of transcendental numbers.

Many mathematicians did not accept Cantor's proof of the existence of transcendentals, because of its highly nonconstructive character: it gives no clue as to how to compute even one transcendental number. Naturally, Kronecker was prominent among those who did not accept this theorem. Kronecker did not even believe Liouville's result or the relatively concrete proofs establishing the transcendence of e (Charles Hermite, 1873) and π (Leopold Lindemann, 1882). Kronecker supposedly told Lindemann "Of what use is your beautiful investigation regarding π? Why study such problems when irrational numbers do not exist?" Kronecker was not alone in doubting the existence of transcendental numbers. But it seems strange that a mathematician of his caliber did not believe in *irrational* numbers.

Here is how to modify Cantor's argument to prove the existence of transcendental numbers constructively: Cantor's proof of the uncountability of $\mathbb{R}$ (Proposition 2.3) is constructive. That is, the diagonalization procedure shows how to compute a real number that differs from every number in any given countable list of reals. Now, it is easy to define an infinite sequence consisting of all the polynomials with integer coefficients. Also, using standard techniques such as Newton's method, we can compute all of the real roots of any such polynomial to any given number of decimal places. In other words, we can define an infinite sequence $\{a_k \mid k \in \mathbb{N}\}$ of all the real algebraic numbers such that the function g defined by

$$g(k, n) = \text{the } n\text{th digit (after the decimal point) of } a_k$$

is recursive. So, by applying the diagonalization procedure to this list $\{a_k\}$, we get an effective procedure for computing a transcendental number. Furthermore, since the diagonalization allows more than one choice at each decimal place, this argument produces a set of transcendental numbers that has the same cardinality as $\mathbb{R}$.

Here is one more example of nonconstructive and constructive proofs of the same result. This theorem was one of Gauss's major achievements, and he was so interested in it that he published five different proofs. However, the first proof we will provide, also found in most modern textbooks, is not due to Gauss. It is quite short and simple, but not constructive.

Theorem 8.3 (Fundamental Theorem of Algebra). *Every nonconstant polynomial with coefficients in $\mathbb{C}$ has a root in $\mathbb{C}$.*

Proof. Assume that $f(z)$ is a polynomial with coefficients in $\mathbb{C}$ but no root in $\mathbb{C}$. Then the function $1/f$ is defined for all complex numbers. Also, since polynomials are differentiable, so is $1/f$. Finally, since every nonconstant polynomial approaches ∞ as its variable approaches ∞, in $\mathbb{C}$ as well as in $\mathbb{R}$, it follows that $1/f(z) \to 0$ as $z \to \infty$. From this it is easy to show that the range of $1/f$ is bounded in absolute value. But an important result of complex analysis known as Liouville's theorem states that every differentiable, bounded function from $\mathbb{C}$ to $\mathbb{C}$ is constant. Thus $1/f$ is constant, which implies that f is also constant. ∎

This proof is elegant, but it is highly nonconstructive since it gives no way of finding a root of a given polynomial. Note that this theorem has the very common "$\forall\exists\ldots$" quantifier structure, which means that it essentially asserts the existence of a certain function (from polynomials to complex numbers). A constructive proof should provide a rule for evaluating the function. Gauss understood the importance of numerical content in mathematics, and provided at least one constructive proof of the fundamental theorem. Here is the basic idea of a procedure for locating a root of a given polynomial: first, from the coefficients of a polynomial f, it is not too hard to compute a number M such that all roots of f must have absolute value less than M. (For instance, if $f(z) = z^3 + 7z^2 - 5z + 22$, we can let $M = 35$. Why?) Then, once we have the roots of f bounded in this way, we can compute one to any desired accuracy by dividing up the bounded disk $z < M$ into a sufficiently fine grid.

It is now time to present a more systematic treatment of the constructive approach to mathematics. We will concentrate on the work of two particularly important advocates of this viewpoint, Brouwer and Bishop. Some good sources for constructive mathematics and its history are [Hey], [Tro], [Bee], [TvD], and [KV]. Section 8.3 is devoted to [BB].

8.2 Brouwer's intuitionism

The intuitionist movement began in 1907 with the publication of Brouwer's thesis. The old adage that "there's nothing new under the sun" can be invoked here, with some validity. We have already discussed the important role of Kronecker and the finitist viewpoint as a predecessor to Brouwer's work. Brouwer was also influenced by a group of contemporary French mathematicians, sometimes referred to as **semi-intuitionists**, that included Poincaré, Borel, Baire, and Lebesgue. They mistrusted the use of infinite sets and abstract reasoning, including nonconstructive existence proofs.

Even though Hilbert was the world's most influential mathematician during the early part of the twentieth century, he considered Brouwer to be a formidable opponent. Hilbert insisted on a strict finitist approach for metamathematics—the discipline that was intended to prove the consistency and completeness of the formal systems used in mathematics. But as to the practice of mathematics itself, Hilbert was adamant in his support for both abstract set theory and the full power of the nonconstructive reasoning allowed by classical logic. His most well-known statements on the subject were "No one shall evict us from the paradise that Cantor has built for us," and "Depriving a mathematician of the principle of proof by contradiction [of an object's existence] is like depriving a boxer of his fists."

Brouwer's most original idea was to attempt to define the subject matter of mathematics in terms of the *activity* of mathematics. He believed that what mathematicians do (or *should* do) is carry out "intuitive mental constructions." These constructions are not based on

L. E. J. Brouwer (1882–1966) graduated from high school in Hoorn, the Netherlands, at the age of fourteen, but then he had to spend two years learning Latin and Greek to qualify for university admission. While still an undergraduate, he proved original results in dynamics. Brouwer was one of the founders of modern topology and proved many important theorems in the subject by age thirty. Ironically, he later rejected most of these results because their proofs were not intuitionistically acceptable.

Brouwer's work in foundations was closely connected with his strong interest in philosophy and mysticism. His doctoral dissertation, "The Unreliability of Logical Principles," was a very radical document. Brouwer's advisor urged him to stick to more "respectable" mathematics, to no avail. His ideas attracted much attention, and during the 1920s he was the most influential critic of Hilbert and the formalist program. Berlin was the largest and most important city in Germany, but had no university as prestigious as Göttingen's, where Hilbert presided. Brouwer became a popular figure in Berlin, especially to those who wanted to challenge the supremacy of Göttingen. Unfortunately, their rivalry became so intense that Hilbert felt compelled to use heavy-handed tactics to get Brouwer removed from the editorial board of *Mathematische Annalen* in 1928. Brouwer was understandably hurt and discouraged by this action.

During World War II, Brouwer actively helped both the Dutch resistance and Jewish students, but in 1943 he encouraged his students to sign a declaration of loyalty to Germany that the Nazi occupiers were requiring. This was undoubtedly a pragmatic move on his part, but after the war he was suspended for several months because of it. Brouwer was once again very upset and even considered emigrating, but ultimately he stayed in Holland.

language, logic, formal axioms, manipulation of symbols, or abstract concepts. Language and logic are useful to mathematicians, but they are tools rather than essential building blocks. To Brouwer, the mental constructions of the mathematician are more basic than concepts based on language, logic, or truth. In some sense, these constructions are subjective: there is no guarantee that one mathematician's constructions will be the same as, or even consistent with, another mathematician's constructions. Yet Brouwer was confident that mathematical intuition was universal and that, as a result, mathematicians would be able to communicate their constructions with each other and would be able to agree about which constructions were correct and which were not.

Like Kronecker and almost all constructivists, Brouwer accepted the natural numbers as the most basic and universal part of mathematics. In his view, the conception of a natural number and of the entire sequence of natural numbers, elementary arithmetical operations such as addition and multiplication, and the principle of mathematical induction are all clearly within the realm of intuitive mental constructions. In other words, Brouwer accepted all the *proper* axioms of Peano arithmetic as a valid part of constructive mathematics. Brouwer was therefore not a strict finitist, since he had no objection to the notion of the infinite set $\mathbb{N}$.

Even though Brouwer did not believe that logic should be the basis of mathematical reasoning, he realized that logic is important for the expression of mathematical ideas. Therefore, he carefully explained the intuitionistic interpretation of the logical connectives and quantifiers. Essentially, he gave inductive criteria, with respect to the logical structure of a mathematical statement, for what should constitute a proof or construction of that statement. Here is a brief summary of his analysis:

(1) A proof or construction of a conjunction should consist of proofs of both conjuncts. This is the same as in classical mathematics.

(2) A proof of a disjunction should consist of a proof of one of the disjuncts. As we have already pointed out, this is *much stronger* than the classical criterion.

(3) A proof of an implication $P \to Q$ should consist of a construction that explicitly shows how to transform any given proof of P into a proof of Q. This is again stronger than the classical criterion. Under this definition, however, it might be easier to prove $P \to Q$ than its classical equivalent $\sim P \vee Q$.

(4) Intuitionistically, $P \leftrightarrow Q$ is taken to be an abbreviation for $(P \to Q) \wedge (Q \to P)$.

(5) $\sim P$ is viewed as an abbreviation for the statement that P leads to a contradiction, usually (for specificity) $P \to (0 = 1)$. So a proof of $\sim P$ should be a construction that shows how to transform any given proof of P into a proof of $0 = 1$. Essentially, this means that we have a constructive proof that there can be no proof of P.

(6) A proof of $\forall x P(x)$ should be a construction that shows how to transform any construction of an object a, together with a proof that a is in the intended domain of the variable x, into a proof of $P(a)$.

(7) Finally, a proof of $\exists x P(x)$ should consist of three parts: a construction of an object a, a proof that a is in the intended domain of the variable x, and a proof of $P(a)$. As noted earlier, this criterion for "constructive existence" is very strong. We will see that most of the important existence theorems of basic analysis—the intermediate value theorem, the mean value theorem, etc.—must be weakened somewhat in order to be proved constructively.

So far, nothing we have indicated of Brouwer's work is particularly strange. His interpretation of $\mathbb{N}$ and its theory is functionally the same as in classical mathematics, while his interpretation of the logical connectives and the notion of proof is more restrictive, especially with respect to the existential quantifier, disjunction, and implication. The point at which intuitionism became much more involved was in its definition and use of real numbers. Brouwer accepted the notion that Cauchy sequences of rationals could be used as the basis for defining real numbers. For now, let's not worry about whether it is necessary to define a real number as an equivalence class of Cauchy sequences rather than a single sequence.

Brouwer found himself wrestling with the problem of whether there are "enough" constructive real numbers. A constructive Cauchy sequence of rationals should be given by an effective procedure. How many effective procedures are there? If an effective procedure must be given by a Turing machine, or a finite set of words or instructions in any reasonable (natural or formal) language, then there are only a countable number of them. In any event, it is hard to imagine that there could be an uncountable number of effective procedures; yet Cantor's theorem asserts that the set of reals is uncountable.

To deal with this difficulty, Brouwer developed a complicated theory of **lawlike** sequences and **lawless** sequences. A lawlike sequence is one that is given by an effective procedure in the usual sense. A lawless sequence could be determined by a process such as an infinite sequence of coin flips, or by an infinite sequence of arbitrary choices by a person. Brouwer argued that both of these types of sequences should be allowed in constructive mathematics, and used the term **choice sequence** or **infinitely proceeding sequence** to include both types. Later in his career, Brouwer considered even more esoteric ways of defining sequences, such as sequences determined by the activity of an "ideal mathematician" attempting to prove or disprove a certain theorem over time.

Unfortunately, Brouwer and his followers were not able to come up with a clear, satisfactory theory of the continuum. First of all, it is difficult to give a mathematically acceptable definition of lawless sequences and choice sequences. Then, depending on what postulates one accepts for these various types of sequences, one gets quite different mathematical consequences. For example, if real numbers must be defined by lawlike sequences, then one cannot prove that all functions from $\mathbb{R}$ to $\mathbb{R}$ are continuous. But under some sophisticated but plausible postulates about choice sequences, Brouwer was able to prove this continuity result if real numbers are defined by arbitrary choice sequences.

It's too bad that Brouwer's intuitionism predated the development of both recursion theory and model theory. Mathematically precise definitions of notions like "algorithm" and "effective procedure" were not given for almost three decades after Brouwer's thesis. This is not to suggest that Brouwer would have accepted "recursive" as a synonym for "constructive," with respect to functions on $\mathbb{N}$. Brouwer never endorsed this approach after the invention of recursion theory, although it was used with some success by the Russian school of constructivists in the 1950s, as we will see in the next section. But recursive objects could have been used at least as an illustration and model of, and perhaps a close approximation to, what Brouwer had in mind early in his career. Perhaps Brouwer's ideas would have seemed less alien to the mathematical community if this interpretation had been available. Furthermore, if one uses recursive functions to interpret constructive real numbers and functions, Cantor's theorem can still turn out to be true in such a model: the set of constructive real numbers can fail to be constructively countable, even if it is "really" countable. This is almost the same situation as Skolem's paradox involving countable models of set theory, which no longer troubles anyone.

The unappealing nature of Brouwer's theory of real numbers is one of the major reasons that intuitionism did not gain a wider following. Another important reason is that, until Bishop came along, it was generally believed (in particular, by both Brouwer and Hilbert) that doing mathematics constructively meant giving up most of the important results of modern mathematics. One can give other reasons, related to Brouwer himself: his unwillingness to make compromises or to be "pinned down" as to the exact meaning of various concepts, and his eagerness to prove theorems that contradicted classical mathematics. This is a shame, because Brouwer had great insight into the weaknesses of abstract mathematics and presented a philosophy and a system under which mathematics could be made more concrete, and therefore more meaningful.

8.3 Bishop's constructive analysis

In 1967, Bishop published a remarkable book entitled *Foundations of Constructive Analysis*, revised in collaboration with Douglas Bridges as [BB]. Bishop's main achievement was to present, for the first time, a constructive development of the major part of real and complex analysis. In order to accomplish this, he wisely sidestepped the philosophical issues that had embroiled Brouwer and most of his followers. In the prolog of [BB], Bishop states, "This development is carried out with an absolute minimum of philosophical prejudice concerning the nature of constructive mathematics. There are no dogmas to which we must conform. Our program is simple: to give numerical meaning to as much as possible of classical abstract analysis."

In particular, Bishop completely avoided the struggles of the intuitionists to make sure that the continuum would "look right." In the last section we pointed out that Brouwer's theory of the real numbers was quite involved and counterintuitive. Bishop calls Brouwer's theory of $\mathbb{R}$ "semimystical" and writes, rather harshly, "A bugaboo of both Brouwer and the logicians has been compulsive speculation about the nature of the continuum. ... In Brouwer's case there seems to have been a nagging suspicion that unless he personally intervened to prevent it, the continuum would turn out to be discrete." ([BB], page 9). Presumably, Bishop was using the word "discrete" playfully, and meant "countable."

Brouwer's approach also led to theorems that are false in classical mathematics. In contrast, what Bishop called his "straightforward realistic" approach to the reals and to mathematics leads to no results that are false or unknown in ordinary mathematics (which Bishop calls "idealistic mathematics"). For every important theorem of analysis, [BB] either proves it literally, or proves a weaker version that pinpoints the constructive content of the usual abstract version.

The remainder of this section is devoted to an outline of some of the basic material in [BB]. All definitions, theorems, and proofs in this discussion (but not remarks and exercises, for the most part) are taken from that source. The words "prove" and "show" are always assumed to refer to constructive proofs, including when used in the exercises.

The mathematical development in [BB] begins by assuming, without much comment, the usual properties of the natural numbers and the integers (presumably, including Peano's axioms). This is no different from the starting point of intuitionism. The first point of divergence from classical mathematics is that a **rational number** is defined to be an expression of the form p/q (essentially, an ordered pair of integers) with $q \neq 0$, rather than an equivalence class of ordered pairs. This is characteristic of Bishop's program; an infinite equivalence class is a rather abstract object, and constructive definitions should be as concrete as possible. The next step in [BB] is to define equality of rational numbers in the obvious way: $p/q = r/s$ if and only if $ps = qr$. In general, the definition of a **set** in [BB] consists of two parts: a (constructive) definition of what it takes to be a member of the set, and a (constructive) definition of the **equality relation** on that set, which must be an equivalence relation.

This notion of a set demands care in the definition of a function: an **operation** from a set A into a set B is a "finite routine" f that assigns an element $f(a)$ of B for each given element a of A. A **function** is then an operation such that $f(a) = f(b)$ whenever $a, b \in A$ and $a = b$. Equality of two functions from A into B is defined in the usual way and is certainly an equivalence relation, so the collection of all functions from A into B is again a set.

Note the phrase "finite routine." This is elaborated by saying that "this routine must afford an explicit, finite, mechanical reduction of the procedure for constructing $f(a)$ to the procedure for constructing a." But there is no precise definition of words such as "explicit," "finite," and "mechanical." In particular, [BB] takes no position on Church's thesis (and does not even mention it!). Certainly, Bishop would accept that every Turing machine determines a finite routine, but there is no discussion of any possible converse. This omission is clearly deliberate. Bishop realized that his program could only be hindered by efforts to pin down the exact meaning of the words "finite routine."

A **sequence** is a function whose domain is $\mathbb{Z}^+$, the set of positive integers. A sequence $\{x_n \mid n \in \mathbb{Z}^+\}$ may be denoted $\{x_n\}$ or simply x. The notion of a sequence leads to the definition of a real number:

Definition. A **real number** is a sequence $\{x_n\}$ of rational numbers that is **regular**, meaning that $|x_m - x_n| \leq \frac{1}{m} + \frac{1}{n}$, for all $m, n \in \mathbb{Z}^+$.

Two reals x and y are **equal** if $|x_n - y_n| \leq 2/n$, for all $n \in \mathbb{Z}^+$.

Exercise 1. Show that this equality relation on real numbers is an equivalence relation.

As with rational numbers, Bishop avoids defining a real number as an equivalence class. For the rest of this section, x, y, and z always stand for reals. The set of all reals is denoted $\mathbb{R}$.

Note that this concept of a regular sequence is much more specific than the concept of a Cauchy sequence used by Brouwer (and most mathematicians) to define real numbers. A Cauchy sequence can have a large finite number of terms that oscillate wildly before the terms "settle down," but it is clear that in a regular sequence, terms cannot differ by more than $3/2$. Bishop could have used Cauchy sequences to define reals, but various proofs are simplified by using regular sequences. On the other hand, it doesn't work out constructively to define a real number as a decimal, as we will soon see. This would be too restrictive.

As usual, each rational number r is "identified" with the real number x such that $x_n = r$ for every $n \in \mathbb{Z}^+$.

Definitions. The reals $-x$, $\max\{x, y\}$, and $\min\{x, y\}$ are defined in the obvious componentwise fashion. Also, we let $|x| = \max\{x, -x\}$.

Because of the extra specificity in the notion of a regular sequence, we must be careful how we define addition and multiplication of reals. Before multiplication can be defined, an auxiliary notion is needed:

Definition. For any real x, its **canonical bound** K_x is the least integer that is greater than $|x_1| + 2$. It is clear that $K_x > |x_n|$ for every n.

Definitions.

(a) $x + y = \{x_{2n} + y_{2n} \mid n \in \mathbb{Z}^+\}$,

(b) $xy = \{x_{2kn} \cdot y_{2kn} \mid n \in \mathbb{Z}^+\}$, where $k = \max\{K_x, K_y\}$.

Exercise 2.

(a) Show that $\max\{x, y\}$, $x + y$, and xy are always real numbers.

(b) Show by counterexamples that the "obvious" definitions of $x + y$ and xy (obtained by omitting "2" and "$2k$" from the subscripts) do not always define real numbers.

Using these definitions of $x + y$, xy, and $-x$, it is easy to show that the set of reals becomes a commutative ring with unity.

Definitions. We can now define various ordering relations on $\mathbb{R}$:

(a) x is **positive** (or $x \in \mathbb{R}^+$) if $x_n > 1/n$ for some $n \in \mathbb{Z}^+$.

(b) x is **nonnegative** (or $x \in \mathbb{R}^{0+}$) if $x_n \geq -1/n$ for every $n \in \mathbb{Z}^+$.

(c) $x > y$ (or $y < x$) if $x - y$ is positive.

(d) $x \geq y$ (or $y \leq x$) if $x - y$ is nonnegative.

(e) $x \# y$ ("x is **apart from** y") if $x < y$ or $y < x$.

The last definition is worth thinking about. It illustrates an important part of Bishop's program, namely to "make every concept affirmative"; that is, to avoid defining important notions as the negations of other notions. (In [BB], the symbol $\neq$ is used rather than # for apartness.)

Exercise 3. Prove that $x \# y$ implies $x \neq y$ (that is, $\sim (x = y)$).

In ordinary mathematics (but still using Bishop's definitions of $x = y$ and $x \# y$), this implication and its converse are both trivial. Does the converse hold in constructive mathematics? It turns out that this converse is one version of a postulate known as **Markov's principle**, which was accepted as an axiom by the Russian constructivist movement led by A. A. Markov. However, most constructive mathematicians do not accept this axiom. Brouwer proposed a counterexample to Markov's principle [Hey, Section 8.1.1], but it is based on a more esoteric definition of real numbers and is also not widely accepted. The position taken by Bishop is simply, "There is no known [constructive] proof or counterexample for Markov's principle" [BB, page 63].

If you try to prove Markov's Principle constructively, you will see that $x \neq y$ implies $\sim\sim (x \# y)$, but there is no apparent way to eliminate this double negation.

The Russian constructivist school was unique in another respect: among all the major branches of constructive mathematics, it was the only one that accepted Church's thesis. Therefore, this approach is also known as **constructive recursive mathematics**.

Exercise 4. Try to prove:

(a) $x = y$ implies $\sim (x \# y)$.

(b) the converse of (a).

Lemma 8.4. *A real x is positive if and only if there exists $N \in \mathbb{Z}^+$ such that $x_m \geq 1/N$ whenever $m \geq N$.*

Proof. Assume x is positive. Find an n such that $x_n > 1/n$ and then choose N such that $2/N \leq x_n - 1/n$. Then

$$x_m \geq x_n - |x_m - x_n| \geq x_m - 1/m - 1/n \geq x_n - 1/n - 1/N > 1/N$$

whenever $m \geq N$.

Conversely, if such an N exists, let $n = N + 1$. Then $x_n > 1/n$.

■

With this lemma, we can define the reciprocal of any real x that is apart from 0:

Proposition 8.5. *Assume $x \# 0$, and let N be as in Lemma 8.4. Define a sequence y by: $y_n = 1/(x_{N^3})$ if $n < N$, and $y_n = 1/(x_{nN^2})$ if $n \geq N$. Then*

(a) *y is a real number.*

(b) $xy = 1$.

(c) *If x is positive (respectively, negative), then so is y.*

(d) *If $xz = 1$, then $z = y$.*

The real number y defined in this proposition is denoted x^{-1}. Note that we cannot quite say that our constructive $\mathbb{R}$ has now been shown to be a field: we've proved that every number that's apart from 0 has a reciprocal, but not that every nonzero number has a reciprocal.

Exercise 5. Explain the difficulty in trying to define x^{-1} (that is, a real such that $xx^{-1} = 1$) for any real x such that $x \neq 0$.

We mentioned that Bishop takes no position on Church's thesis. However, [BB] does prove Cantor's theorem, in the following strong form:

Theorem 8.6. *Given any reals x and y such that $x < y$ and any sequence (a_n) of reals, there exists a real z such that $x \leq z \leq y$ and $z \# a_n$ for all $n \in \mathbb{Z}^+$.*

The proof is similar to Cantor's original diagonalization argument. At first thought, it appears that this result implies that Church's thesis is false, since there are only countably many Turing machines and so Church's thesis (under which every real number would have to be given by a Turing machine) would make the reals countable. But this reasoning is faulty. In the context of [BB], Cantor's theorem only proves that there is no *constructive* sequence that lists all real numbers in an interval. And this does not contradict Church's thesis.

Example 2. We can again use an unsolved problem such as Goldbach's conjecture to construct examples that illustrate the difference between classical mathematics and Bishop's constructive mathematics. For instance, define a sequence u by

$$u_n = \begin{cases} 0, & \text{if Goldbach's conjecture holds} \\ & \text{for every even number} \leq n, \\ 2^{-k}, & \text{if } k \text{ is the least counterexample to} \\ & \text{Goldbach's conjecture, and } k \leq n \end{cases}$$

Then it is easy to show that u is a real number, and $u \geq 0$. But currently we cannot prove that $u > 0$ or that $u = 0$. Therefore, to a constructivist, the disjunction $u > 0$ or $u = 0$ is not true at this time. We say that u is a **weak counterexample** or a **constructive counterexample** to the statement $\forall x[x \geq 0 \to (x > 0 \vee x = 0)]$. This does not imply that this statement has been refuted or is false!

Note the words "at this time" above. In mainstream mathematics, it is very uncommon to mention time when speaking of the truth of a mathematical statement. That is because classical mathematics subscribes to something like the Platonist viewpoint, in which mathematical objects and statements have an absolute, permanent status; a statement cannot suddenly become true at a certain moment. But in constructive mathematics, the assertion that a statement is true means that we can carry out some construction or effective procedure. There can always be new constructions, which can create new truths. So if and when Goldbach's conjecture is ever settled, the disjunction $u > 0$ or $u = 0$ will become constructively true.

Exercise 6. Construct a weak counterexample w to the statement $\forall x \in \mathbb{R}(x \geq 0 \vee x \leq 0)$. (Hint: Split the second case of the previous definition of u into two subcases, based on whether the first counterexample to Goldbach's conjecture is a multiple of 4.)

Exercise 7.

(a) Let w be the real number just defined. Show that the number $1 - w$ is a weak counterexample to the statement that every real number has a decimal expansion.

(b) Does this also hold for the number $1 - u$, where u is as defined in the previous example?

Exercise 8.

(a) Prove that if $x > y$ is false, then $x \leq y$.

(b) Prove that if $x \leq y$ is false, then $\sim\sim (x > y)$.

Once again, the double negation in part (b) cannot be eliminated. In fact, without the double negation, this implication is another (constructively equivalent) version of Markov's Principle.

One of the essential axioms of $\mathbb{R}$ in classical mathematics is the completeness property, that every set of real numbers that is bounded above has a least upper bound. This property is not constructively correct. In fact, it fails even for constructive sequences of integers:

Example 3. Define a sequence (a_n) by

$$a_n = \begin{cases} 0, & \text{if Goldbach's conjecture holds for every} \\ & \text{even number} \leq n, \\ 1, & \text{otherwise.} \end{cases}$$

Certainly, (a_n) is a constructive sequences of integers, and its range is a set of integers. This set is bounded above, by 2 or even by 1. But, constructively, $\{a_n\}$ has no least upper bound. There is no known effective procedure for calculating even the integer part of the least upper bound of this set.

There is perhaps no clearer example of the difference between abstract and constructive mathematics than this. Geometrically, it seems completely obvious that a sequence or set of 0's and 1's has a least upper bound. But it is also quite clear why this property is not constructive. The example defines a simple effective procedure (a recursive function, in fact) for defining such a sequence for which there is no least upper bound in a computationally meaningful sense.

It is important to emphasize that most constructive mathematicians do not contend that the completeness property is false, or that abstract mathematics is not worth pursuing. Rather, they would like mathematicians to be aware of the distinction between abstract mathematics and mathematics with numerical or computational content, and to appreciate the value of striving for computational content in their work.

Exercise 9. Clearly, the sequence (a_n) defined above is not necessarily a real number. That is, (a_n) is regular if and only Goldbach's conjecture is true. But what if we had defined a real number to be any Cauchy sequence. Can we prove (constructively) that (a_n) is Cauchy?

From our earlier remarks, the reader should expect that [BB] proves a constructive version of the completeness property. Indeed it does. First, we need constructive versions of some basic definitions:

Definition. Let A be a set of real numbers that is nonempty (in the constructive sense that we can compute an element of it). Then y is an

upper bound of A if $x \leq y$ for every $x \in A$. A **least upper bound** of A is an upper bound y of A such that, for every $\epsilon > 0$ there is an $x \in A$ such that $x > y - \epsilon$.

Lower bounds and greatest lower bounds are defined similarly.

Note the affirmative definition of a least upper bound, in contrast to the usual negative one: an upper bound such that *no smaller number* is an upper bound. The least upper bound and the greatest lower bound of a set are unique (up to equality), if they exist. We use the usual notation $\text{LUB}(A)$ and $\text{GLB}(A)$.

Theorem 8.7 (Constructive Completeness Property). *Let A be a nonempty set of real numbers that is bounded above. Then* $\text{LUB}(A)$ *exists if and only if for all $x, y \in \mathbb{R}$ with $x < y$, either y is an upper bound of A or there is number in A that is greater than x.*

In ordinary mathematics, the condition after the words "if and only if" is always true, so this theorem is equivalent to the usual completeness property. More to the point, this result guarantees that nonempty sets that are bounded above will "normally" have least upper bounds.

Exercise 10. Show that the condition after the words "if and only if" in the above theorem fails constructively for the range of the sequence (a_n) defined above using Goldbach's conjecture.

Functions and continuity

Definitions. Let $a, b \in \mathbb{R}$. We define the **bounded intervals** $[a, b]$, (a, b), $[a, b)$ and $(a, b]$ in the usual way. For example, $(a, b]$ means $\{x \in \mathbb{R} \mid a < x \leq b\}$. Such an interval is called **proper** if $a < b$. A nonempty, closed, bounded interval is called a **compact interval**. We also define **unbounded intervals** or **rays** such as $[a, \infty)$ in the usual way.

Definition. A real-valued function f defined on a *compact interval I* is said to be **continuous** on I if there is a function ω defined on positive reals such that whenever $\epsilon > 0$, $x, y \in I$, and $|x - y| \leq \omega(\epsilon)$, we have $|f(x) - f(y)| \leq \epsilon$.

Definition. A real-valued function defined on an *arbitrary interval* J is said to be **continuous** on J if it is continuous on every compact subinterval of J.

The reader is encouraged to compare these definitions with the classical definition of continuity. The explicit mention of the modulus of continuity ω is not an essential difference. The number $\omega(\epsilon)$ is usually called δ, and the classical definition requires that δ is a function of ϵ. So the constructive definition merely emphasizes something that is implicit in the classical definition.

What is more significant is that in the constructive definition, $\omega(\epsilon)$ or δ can depend on ϵ but not on x. Recall that this stronger requirement is normally called **uniform continuity**. But in classical mathematics, a continuous function defined on a closed interval must be uniformly continuous, and so a continuous function defined on an arbitrary interval is uniformly continuous on every closed subinterval of the domain.

In other words, Bishop's definition of continuity *on an interval* is equivalent to the classical one. But [BB] simply ignores the notion of **pointwise continuity**: there is no terminology for a function that is continuous at a point without being continuous on some interval containing that point. Bishop addresses this specific issue in his prolog, stating that one of his guiding principles is to "avoid definitions that are not relevant. (The concept of a pointwise continuous function is not relevant; a continuous function is one that is uniformly continuous on compact intervals.)" Indeed, functions that are pointwise continuous (somewhere) but not continuous in Bishop's sense are rare and of limited interest.

Example 4. The most common example of a function that is continuous at just one point (namely, at 0) is

$$f(x) = \begin{cases} 0, & \text{if } x \text{ is rational,} \\ x, & \text{otherwise.} \end{cases}$$

Obviously, this function is "pathological"; it has no importance in mathematics except as a counterexample. Constructively, it is a function but we do not expect its domain to contain any proper interval.

As we mentioned in Section 8.1, every known definition of a discontinuous function is also nonconstructive. Bishop takes no position on the existence of discontinuous functions, except to write "It has been asserted by Brouwer that all functions [from $\mathbb{R}$ to $\mathbb{R}$] are continuous, but no acceptable proof of this assertion is known [BB, page 67].

Exercise 11.

(a) Consider a typical classical discontinuous function, such as the function h defined in Section 8.1: $h(x) = 0$ if $x \leq 0$, and $h(x) = 1$ if $x > 0$. Is h a function with domain $\mathbb{R}$ in the sense of [BB]? Why or why not?

(b) Now consider the function f defined classically by: $f(x) = 0$ if $x \leq 0$, and $f(x) = x$ if $x > 0$. In form, this is very similar to the definition of h, but f is classically continuous. Is f a function with domain $\mathbb{R}$ in the sense of [BB]? Prove your answer.

(c) State (but do not prove) a generalization of your answer to (b), whether positive or negative, to continuous functions defined by cases.

Much of the basic theory of continuous functions goes through constructively. For example, all polynomials are continuous; the absolute value function is continuous; the sum, difference, product, maximum, and minimum of any two continuous functions on the same interval J are again continuous on J; the reciprocal of any function that is bounded away from zero on each compact subinterval of its domain is continuous; and the composition $g \circ f$ of two continuous functions is continuous, provided that f maps compact subintervals of its domain to compact subintervals of the domain of g.

In classical mathematics, this last condition on f is always true; in fact, it is obviously equivalent to the intermediate value theorem (stated in Section 1.3, Example 2). This result is a typical example of an existence theorem that does not give an explicit method for calculating the object whose existence is asserted. Furthermore, its proof relies heavily on the completeness property of $\mathbb{R}$. So it should come as no surprise that the usual version of this theorem is not constructively valid. Here is Bishop's constructive modification:

Theorem 8.8. *Let f be a continuous function on an interval I, and let a and b be points of I with $f(a) < f(b)$. Then for every y in $[f(a), f(b)]$ and every $\epsilon > 0$, there exists x in $[\min\{a, b\}, \max\{a, b\}]$ such that $|f(x) - y| < \epsilon$.*

Thus the classical version of the theorem, which asserts the existence of an exact point on a graph at a given y-value, must be replaced by the existence of arbitrary close approximations. However, [BB] does include several results that assert exactness (that is, $f(x) = y$) under additional assumptions about f. In particular, exactness follows if f is a polynomial or f is strictly increasing on the interval $[\min\{a, b\}, \max\{a, b\}]$.

Theorem 8.8 is an "equal hypothesis" modification of the classical intermediate value theorem. The constructive results referred to in the previous paragraph, which achieve the classical conclusion under extra hypotheses, are "equal conclusion" modifications. Most mathematicians will find the latter type of modification more satisfactory, provided that the extra hypotheses are not too restrictive.

Exercise 12. Here is an example of a constructive counterexample to the classical intermediate value theorem.

(a) First, recall the number w defined in Exercise 6. Note that (classically!) $w = 0$, $w = 2^{-k}$, or $w = -2^{-k}$ for some k in $\mathbb{Z}^+$.

(b) Now consider this function f on the interval $[0, 3]$: the graph of f consists of a straight line between the points $(0, -1)$ and $(1, w)$, a straight line between $(1, w)$ and $(2, w)$, and a straight line between $(2, w)$ and $(3, 1)$. Sketch the possible graphs of f, and prove that f is a constructively acceptable function.

(c) Show that the existence of an x such that $f(x) = 0$ does not hold constructively. In fact, such an x cannot be computed to even one decimal place.

(d) Show that f does, however, satisfy Theorem 8.8 with $y = 0$. That is, for every $\epsilon > 0$, we can exactly determine an x such that $|f(x)| < \epsilon$.

Exercise 13. Another important classical existence theorem is the **extreme value theorem**: a continuous function defined on a compact interval has a maximum and a minimum. Define a constructive counterexample to this theorem. (Hint: You can use the same w again.)

Differentiation

Definition. Let f and g be continuous functions on a proper compact interval I, and let δ be an operation from $\mathbb{R}^+$ to $\mathbb{R}^+$ such that

$$|f(y) - f(x) - (y - x)g(x)| \leq \epsilon|y - x|$$

whenever $x, y \in I, \epsilon > 0$, and $|y - x| \leq \delta(\epsilon)$. Then f is said to be **differentiable** on I, g is called a **derivative** of f on I, and δ is a **modulus of differentiability** for f on I.

These definitions are extended to functions defined on an arbitrary proper interval in the same way that the definition of continuity was extended.

The derivative of a function on an interval is unique, up to equality.

As with continuity, this definition excludes various unusual or pathological functions from being considered differentiable. For Bishop, a differentiable function must have a (uniformly) continuous derivative on an interval.

Most of the basic derivative formulas hold constructively, including the sum, product, and quotient rules, and the usual formula for differentiating polynomials. The chain rule for the derivative of $g \circ f$ holds, under the same additional assumption needed to prove that the composition of continuous functions is continuous: that f maps compact subintervals of its domain to compact subintervals of the domain of g.

Not surprisingly, the two main existence theorems of basic differential calculus do not quite hold constructively in their usual forms. Here are their equal hypothesis modifications in [BB]:

Theorem 8.9 (Rolle's Theorem I). *Let f be a differentiable function on an interval $[a, b]$ such that $f(a) = f(b)$, and let $\epsilon > 0$. Then there exists x in $[a, b]$ such that $|f'(x)| < \epsilon$.*

Theorem 8.10 (Mean Value Theorem). *Let f be a differentiable function on an interval $[a, b]$, and let $\epsilon > 0$. Then there exists x in $[a, b]$ such that $|f(b) - f(a) - f'(x)(b - a)| < \epsilon$.*

There are also equal conclusion modifications of these results. Here is the one for Rolle's theorem:

Definition. A function f is called **locally nonconstant** on an interval J if, for every $x, y \in J$ with $x < y$, there is a z in (x, y) such that $f(z) \# f(x)$.

Theorem 8.11 (Rolle's Theorem II). *Let f be a locally nonconstant, differentiable function on an interval $[a, b]$ such that $f(a) = f(b)$. Then there exists x in $[a, b]$ such that $f'(x) = 0$.*

All of the usual functions encountered in elementary calculus, except for constants, are (constructively) locally nonconstant. So, for all practical purposes, the classical Rolle's theorem and mean value theorem, with their exact conclusions, hold constructively.

One of the most powerful applications of differential calculus is the ability to write the vast majority of important elementary functions as "infinite polynomials," that is, as Taylor series. Taylor's theorem in classical mathematics is an existence theorem, but it states an inequality involving a remainder term. In other words, it asserts an approximation about an unknown number rather than an equation. Therefore, it should come as no surprise that this theorem holds constructively.

The basic theory of sequences and series of real numbers and functions is developed constructively in [BB]. In particular, it is shown that a sequence of real numbers *or* a sequence of continuous functions converges if and only if it is a Cauchy sequence. Using these results and

Taylor's theorem, [BB] shows that the usual Taylor series of the exponential, logarithmic, trigonometric, and inverse trigonometric functions converge and obey all the common properties that make these functions so useful.

Integration

The development of the theory of integration in [BB] is very close to the classical approach, at least for continuous functions:

Definition. Let a and b be reals with $a < b$. A **partition** P of the interval $[a, b]$ is a finite sequence $(a_0, a_1, \dots, a_n)$ of reals such that $a = a_0 \leq a_1 \leq \cdots \leq a_n = b$. The **mesh** of P is $\max\{a_{i+1} - a_i \mid 0 \leq i \leq n-1\}$.

Notation. If f is a continuous function on an interval $[a, b]$ and P is a partition of $[a, b]$, then $S(f, P)$ denotes an arbitrary Riemann sum based on f and P. Note that $S(f, P)$ is not uniquely determined by f and P.

Theorem 8.12. *Let f be a continuous function on a compact interval $[a, b]$, with modulus of continuity ω. Then there is a unique number L such that, for any $\epsilon > 0$, any partition P of $[a, b]$ with mesh $\leq \omega(\epsilon)$, and any Riemann sum $S(f, P)$, we have*

$$|S(f, P) - L| \leq \epsilon(b - a).$$

Definition. The number L described in the previous theorem is called the **(definite) integral of f from a to b**, denoted $\int_a^b f(x)\,dx$.

This definition is extended in the usual way to integrals in which $b < a$. The standard properties of the definite integral, including the facts that the integral is a linear operator and that the interval of integration can be split up, hold constructively. Of course, the most important result in the subject is the fundamental theorem, and this holds without modification:

Theorem 8.13 (Fundamental Theorem of Calculus). *Let f be a continuous function on an interval J, and let a be any fixed point in J. For each x in J, define $g(x)$ to be $\int_a^x f(t)\,dt$. Then $g' = f$ on J. Also, if h is any function such that $h' = f$ on J, then $g - h$ is constant on J.*

The material in this section is a brief outline of the material in the first two chapters of [BB]. To give you some idea of the scope of Bishop and Bridges's work, here are the titles of the remaining seven chapters: Set Theory, Metric Spaces, Complex Analysis, Integration, Normed Linear Spaces, Locally Compact Abelian Groups, and Commutative Banach Algebras. Their work convincingly refutes the claim that the constructive approach to mathematics severely limits what one can accomplish. Indeed, one could argue that the constructive approach enriches the development of mathematics by requiring us to restrict our definitions and arguments to the realm of the numerically meaningful.

APPENDIX A

A Deductive System for First-order Logic

There are many correct ways to define the notion of deduction in first-order logic. Here we give one rather simple one, essentially as in [End]. All of the terminology used in this appendix is explained in Chapter 1.

Logical axioms

The logical axioms are all generalizations of all formulas of the following types.

(1) All tautologies.

(2) Universal specification: $\forall x \mathrm{P}(x) \to \mathrm{P}(t)$, where t is a term of the same sort as the variable x, $\mathrm{P}(t)$ is the result of replacing all free occurrences of x in $\mathrm{P}(x)$ by t, and no free variable of t becomes bound in $\mathrm{P}(t)$.

(3) $\forall x(\mathrm{P} \to \mathrm{Q}) \to (\forall x \mathrm{P} \to \forall x \mathrm{Q})$.

(4) $\mathrm{P} \to \forall x \mathrm{P}$, where x does not occur free in P.

(5) Definition of $\exists$: $\exists x \mathrm{P} \leftrightarrow \sim \forall x \sim \mathrm{P}$.

(6) $x = x$.

(7) Substitution of equals: $x = y \rightarrow [P(x) \leftrightarrow P(y)]$, where P is atomic and P(y) is the result of replacing one or more occurrences of x in P(x) by y.

Rule of inference

(1) Modus ponens: From two steps of the form P and P $\rightarrow$ Q, Q may be concluded.

APPENDIX B

Relations and Orderings

We begin this appendix with a very brief review of relations in general. The concept of the **ordered pair** (x, y) of any two objects x and y may be left as an undefined concept, or it may be defined rigorously, as in Section 2.3. Once we have ordered pairs at our disposal, we can iterate the process to define ordered triples, and ordered n-tuples in general. Specifically, (x, y, z) is usually defined to be $((x, y), z)$, although $(x, (y, z))$ would work just as well.

We can then define the **Cartesian product** of two sets A and B by $A \times B = \{(x, y) \mid x \in A \text{ and } y \in B\}$. We also have extended Cartesian products; for example, $A \times B \times C$ means $(A \times B) \times C$ or, equivalently, $\{(x, y, z) \mid x \in A \text{ and } y \in B \text{ and } z \in C\}$. One writes A^2 for $A \times A$, A^3 for $A \times A \times A$, etc. This notation is technically ambiguous, since A^n also means the set of all functions from n to A. But in many situations, the difference between the two possible sets denoted A^n is not significant.

An n-ary **relation** is simply a set of ordered n-tuples. The words "unary," "binary," and "ternary" mean 1-ary, 2-ary, and 3-ary, respectively. Note that a unary relation is just a set. Without any adjective, the word "relation" usually means a binary relation. An n-ary relation on a set A is defined to be any subset of A^n.

It is common in mathematics to use the word "relation" for a statement with free variables that is used to define a set of ordered n-tuples. For instance, we might say "Consider the relation $x < y$ on the set of real numbers," or simply "Consider the less-than relation on $\mathbb{R}$." Technically this refers to the relation $\{(x, y) \mid x, y \in \mathbb{R} \text{ and } x < y\}$. It is more precise to call $x < y$ a binary **predicate** that we are using to define a relation, but it is often not important to worry about this distinction.

Orderings

For the rest of this appendix, we assume that R is a binary relation. We write xRy as an abbreviation for $(x, y) \in R$. As usual, the **domain** of R is $Dom(R) = \{x \mid \exists y(xRy)\}$ and the **range** of R is $Rng(R) = \{y \mid \exists x(xRy)\}$.

Definitions. We say that R is:

reflexive (on A) if xRx, for all $x \in A$;
antisymmetric if, whenever xRy and yRx, then $x = y$;
transitive if, whenever xRy and yRz, then xRz;
a **preordering** (on A) if it is reflexive and transitive;
a **partial ordering** (on A) if it is an antisymmetric preordering;
a **total ordering** or a **linear ordering** or a **chain** (on A) if it is a partial ordering and also satisfies **trichotomy**: for any $x, y \in A$, either xRy, yRx, or $x = y$. This last condition may also be described by saying that any two elements of A are **comparable** under R.

The words "on A" are in parentheses in several of these definitions because they are often omitted in practice. When that occurs, the usual implication is that the unmentioned A is $Dom(R) \cup Rng(R)$.

If R is a partial ordering, it is common to write $x \leq y$ for xRy. We can then write $x \geq y$ for $y \leq x$, $x < y$ for $x \leq y \wedge x \neq y$, and $x > y$ for $y < x$. The relation $\geq$ defines a new partial ordering on A; it is simply R^{-1}. (Of course, it's permissible to reverse all of this by writing $x \geq y$ for xRy and $x \leq y$ for the inverse relation.)

On the other hand, the relations defined by $<$ and $>$ are not partial orderings as defined above. Rather, they are **irreflexive** partial orderings, meaning that they are transitive and **irreflexive**: $x < x$ is always false. From this follows **strong antisymmetry**: whenever $x < y$ holds, then $y \not< x$. Conversely, if S is an irreflexive partial ordering on A, then the relation on A defined by (xSy or $x = y$) is a (reflexive) partial ordering on A. Furthermore, a reflexive partial ordering is total if and only if the associated irreflexive ordering is total. In other words, it is easy to go "back and forth" between reflexive and irreflexive orderings, and we will freely do so.

Definitions. Let R be a partial ordering and $x \in B \subseteq Dom(R)$. Then x is called:

a **minimal** element of B if $\sim \exists y \in B(y < x)$;
a **maximal** element of B if $\sim \exists y \in B(y > x)$;
the **least** element of B if $x \leq y$, for all $y \in B$;
the **greatest** element of B if $x \geq y$, for all $y \in B$.

The following facts are elementary: if a subset of the domain of a partially ordered set has a least element or a greatest element, then that element is unique. A least element must be a minimal element, and a greatest element must be a maximal element; in a total ordering, the converses also hold.

Definitions. A partial ordering is called **well-founded** if every nonempty subset of its domain has a minimal element. (This definition can actually be made for binary relations in general.) A well-founded total ordering is called a **well-ordering**.

So a well-ordering is a total ordering in which every nonempty subset of the domain has a least element.

Definitions. If A is a partially ordered set, $B \subseteq A$, and $x \in A$, we say that x is an **upper bound** of B if $y \leq x$ for every $y \in B$. If B has a least upper bound, then it is of course unique and is denoted LUB(B) or Sup(B) (the **supremum** of B). In a totally ordered set, a subset with no upper bound is said to be **unbounded above** or **cofinal**.

Similarly, we define what is meant by a **lower bound** of B. If B has a greatest lower bound, then it is unique and is denoted GLB(B) or

Inf(B) (the **infimum** of B). In a totally ordered set, a subset with no lower bound is said to be **unbounded below** or **coinitial**.

Notation. In any total ordering, we can define the usual types of **bounded intervals** (a, b), $[a, b]$, $[a, b)$, and $(a, b]$. We can also define **rays**: the type of sets that, in $\mathbb{R}$, would be denoted $(-\infty, b)$, $[a, \infty)$, etc.

We will use the term **interval** to mean either a bounded interval or a ray. Also, the **initial segment** defined by an element a means the ray $\{x \mid x < a\}$. This term is usually applied only to well-orderings.

Definitions. Let a be an element of a totally ordered set. The **immediate successor** of a is the least element that is greater than a. The **immediate predecessor** of a is defined similarly.

The immediate successor and predecessor of an element are obviously unique, if they exist. Also, b is the immediate successor of a if and only if a is the immediate predecessor of b.

Definition. A total ordering is called **discrete** if every element that is not the greatest (respectively, least) element in the ordering has an immediate successor (respectively, predecessor).

Definition. Suppose that A is a totally ordered set and $B \subseteq A$. We say that B is a **dense** subset of A if, whenever $x, y \in A$ and $x < y$, there exists $z \in B$ such that $x < z < y$.

A total ordering is called dense if its domain has at least two members and is a dense subset of itself. The second conjunct simply means that no element has an immediate successor or an immediate predecessor.

Every subset of the domain of a discrete ordering is again discrete under the restriction of the original ordering. Every *interval* with more than one element (but not every subset) in a dense ordering is again dense under the restriction of the original ordering.

Discrete orderings and dense orderings may be thought of as opposite ends of a spectrum. No ordering is both discrete and dense.

Example 1. Every total ordering on a finite set is discrete. The usual ordering on $\mathbb{Z}$ is discrete; therefore, by the previous remark, so is the usual ordering on $\mathbb{N}$.

$\mathbb{Q}$ is a dense subset of $\mathbb{R}$. It follows that the usual orderings on $\mathbb{R}$ and $\mathbb{Q}$ are dense. Therefore, so are the orderings on all intervals in $\mathbb{R}$ and $\mathbb{Q}$ (except for intervals that are empty or include only one point).

There are several useful statements involving orderings that are equivalent (in ZF) to the axiom of choice. Here are the definitions of two of the most important ones:

Definition. **Zorn's lemma** states that a partial ordering in which every chain (totally ordered subset) has an upper bound must have a maximal element.

The second proof of Theorem 2.20 given in Section 2.5 illustrates the typical use of Zorn's lemma.

Definition. The **Hausdorff maximal principle** states that every chain in a partial ordering must be contained in some maximal chain.

Functions and equivalence relations

Orderings are one of the three most important types of binary relation used in mathematics. For the sake of completeness, here are the definitions of the other two.

Functions are also a type of relation. Specifically, as a set of ordered pairs, a **function** from A to B is simply a subset of $A \times B$ in which each element of A occurs in exactly one of the ordered pairs. This is the standard set-theoretic definition of a function. However, as noted at the beginning of Section 3.2, there are situations in which this definition is not quite satisfactory.

If f is a function, $f(x) = y$ is the usual way of writing $(x, y) \in f$. This is very handy because it makes it possible to write $f(x)$ as a term, and to substitute other terms for the variable x. Also, the notation $f : A \to B$ means that f is a function from A to B. Note that this implies that A is precisely the domain of f, but B can be any set that contains the range of f. Part of the reason for this convention is convenience. For instance, if f is the real-valued function defined by $f(x) = x^2 + \sin(x)$, we can write $f : \mathbb{R} \to \mathbb{R}$ without needing to take the trouble to determine the exact range of f.

The other very basic type of relation is equivalence relations. A binary relation R is said to be **symmetric** if $xRy \leftrightarrow yRx$ holds for all x and y. An **equivalence relation on** A is a relation with domain A that is reflexive, symmetric, and transitive. An equivalence relation normally expresses some way in which two objects are similar or alike. For example, the predicate "x and y were born in the same year" defines an equivalence relation on any set of people. Congruence and similarity define equivalence relations on any set of triangles or any other set of geometric shapes. The property of having the same integer part or the same decimal part defines an equivalence relation on $\mathbb{R}^+$.

If R is an equivalence relation on A and $x \in A$, the **equivalence class** of x, denoted $[x]_R$ or simply $[x]$ when there is no possibility of confusion, is the set $\{y \mid xRy\}$. Intuitively, $[x]_R$ is the set or "club" of objects that are similar to x, under R. For example, if R is the equivalence relation on people based on year of birth and Lucian was born in 1988, then $[\text{Lucian}]_R$ is the set of all people born in 1988. The most important mathematical fact about equivalence relations is that the equivalence classes must partition the domain. That is, the union of all the equivalence classes is the whole domain, and any two equivalence classes are either identical or disjoint. There can be no "partial overlap." Furthermore, this situation actually provides a one-to-one correspondence between the collection of all equivalence relations on any given set and the collection of partitions on that set.

APPENDIX C

Cardinal Arithmetic

This appendix is related to material in at least three sections of the text: 2.2, 2.5, and 3.2. Chapter 2 explains that the word "cardinal" can be defined in three different ways. Let's review these meanings briefly.

Definition. For any set x, its **cardinal** or **cardinality**, denoted $Card(x)$, is either:

(a) the class of all sets y such that $x \sim y$ (Frege cardinals; not a rigorous definition in ZF or ZFC),

(b) the set of all sets y of least rank such that $x \sim y$ (Scott's adaptation of Frege cardinals; rigorous in ZF or ZFC), or

(c) the least ordinal α such that $x \sim \alpha$ (von Neumann cardinals; rigorous in ZFC but not defined for all sets in ZF).

The material in this appendix is written in accordance with definitions (a) and (b), under which a cardinal is a collection of sets of the same size. It is not hard to rewrite this material to fit definition (c).

The letters κ, μ, and ν, possibly with subscripts, will denote cardinals.

Section 2.2 gives the definitions of the basic relations $x \preceq y$ and $x \prec y$ on *sets*. The relations $\kappa \leq \mu$ and $\kappa < \mu$ on cardinals are defined from these. For example, $\kappa < \mu$ means that $x \prec y$, where $x \in \kappa$, $y \in \mu$. It is very easy to show that this definition is well defined, meaning that it does not depend on the choice of x and y. All of our subsequent definitions involving cardinals are also well defined. Similarly, words such as "finite" and "uncountable" can be applied to cardinals without ambiguity.

Definitions (Cardinal Arithmetic). Let $Card(A_i) = \kappa_i$ $(i = 1, 2)$. Then:

(a) $\kappa_1 + \kappa_2 = Card[(A_1 \times \{1\}) \cup (A_2 \times \{2\})]$.

(b) $\kappa_1 \cdot \kappa_2 = Card(A_1 \times A_2)$.

(c) $\kappa_1^{\kappa_2} = Card(A_1^{A_2})$.

The set on the right-hand side of part (a) above is called the **formal disjoint union** of A_1 and A_2, denoted $A_1 \coprod A_2$. Clearly, we can't use $A_1 \cup A_2$ there, unless we already know that A_1 and A_2 are disjoint. The set on the right-hand side of (c) is, as usual, the set of all functions from A_2 to A_1.

We now list many of the basic properties of cardinal arithmetic, noting which ones require AC:

Proposition C.1.

(a) *Cardinal addition and multiplication are associative and commutative, and satisfy the distributive law.*

(b) *On finite cardinals, these three operations coincide with the usual operations of arithmetic (and therefore with ordinal arithmetic as well).*

(c) *(AC) If κ or ν is infinite, then $\kappa + \nu = Max(\kappa, \nu)$. If, in addition, neither κ nor ν is zero, then $\kappa \cdot \nu = Max(\kappa, \nu)$. ("Max" stands for "maximum.")*

(d) *(AC) If κ is infinite, then the union of κ sets of cardinality κ has cardinality κ.*

(e) *(AC) If A is infinite, then the set of all finite sequences of members of A, denoted $A^{<\omega}$, has the same cardinality as A.*

(f) *For any set x, $\mathcal{P}(x) \sim 2^x$. In other words, if $Card(x) = \kappa$, then $Card(\mathcal{P}(x)) = 2^\kappa$. (Here, 2 is the ordinal $\{0, 1\}$.)*

(g) **(Cantor's Theorem, restated using (f)).** *For every κ, $\kappa < 2^\kappa$.*

(h) $\kappa^\mu \cdot \kappa^\nu = \kappa^{\mu+\nu}$.

(i) $(\kappa^\mu)^\nu = \kappa^{\mu \cdot \nu}$.

(j) $(\kappa \cdot \mu)^\nu = \kappa^\nu \cdot \mu^\nu$.

Rather than prove any parts of this proposition here, we just illustrate a couple of useful special cases. Suppose we want to establish part (c) for $\kappa = \nu = \aleph_0$, the cardinality of $\mathbb{N}$. This amounts to showing that $\mathbb{N} \times \{1, 2\} \sim \mathbb{N}$ and $\mathbb{N} \times \mathbb{N} \sim \mathbb{N}$. A simple bijection f from $\mathbb{N} \times \{1, 2\}$ to $\mathbb{N}$ is given by

$$f(n, 1) = 2n + 1, \quad \text{and} \quad f(n, 2) = 2n.$$

(Recall that we are assuming that $0 \in \mathbb{N}$.) A simple bijection B_2 from $\mathbb{N} \times \mathbb{N}$ to $\mathbb{N}$ is given by $B_2(m, n) = 2^m(2n+1) - 1$. Note that the axiom of choice is not needed to define these bijections.

By iterating the function B_2, we can define a bijection B_k between $\mathbb{N}^k$ and $\mathbb{N}$ for each positive integer k. Specifically, let

$$B_k(a_1, a_2, \dots, a_k) = B_2[a_1, B_{k-1}(a_2, a_3, \dots, a_k)],$$

for any $k > 2$. We also define B_1 to be the identity on $\mathbb{N}$.

Similarly, (assuming AC), it follows that $\mu^k = \mu$ whenever μ is infinite and k is a nonzero finite cardinal. By the way, for the purposes of cardinal arithmetic, it doesn't matter whether we define the set A^k by iterating the operation $\times$ or as the set of all functions from k to A.

Part (e) of this proposition can also be proved without AC when $A = \mathbb{N}$. We can explicitly define a bijection B between $\mathbb{N}^{<\omega}$ and $\mathbb{N}$ by letting $B(\emptyset) = 0$, and

$$B(a_1, a_2, \dots, a_k) = 2^{a_1+1} \cdot 3^{a_2+1} \cdot \dots \cdot {p_k}^{a_k+1} - 1,$$

where p_k denotes the kth prime.

Note that parts (f), (h), (i) and (j) of Proposition C.1 arc a direct restatement of Proposition 2.18 in Section 2.5, stated in terms of cardinals rather than individual sets. It should also be clear that parts (h), (i), and (j) are completely analogous to the main laws of exponents in elementary algebra.

Infinitary cardinal operations

It is also fruitful to define **infinitary** arithmetical operations on cardinals. For the rest of this appendix, we always assume AC. Although several of the definitions we will give have versions that make sense without the axiom of choice, they become much more complex without it.

Definition. Let $\{\kappa_i \mid i \in I\}$ be an indexed family of cardinals. Choose $A_i \in \kappa_i$ for each $i \in I$. Then:

(a) $\sum_{i \in I} \kappa_i = Card[\bigcup_{i \in I}(A_i \times \{i\})]$.

(a) $\prod_{i \in I} \kappa_i = Card(\prod_{i \in I} A_i)$.

The set on the right-hand side of (b) is the usual **Cartesian product** of the indexed family $\{A_i \mid i \in I\}$, that is, the set of all functions f with domain I such that $f(i) \in A_i$ for every i.

Notation. The least cardinal that is greater than κ is denoted κ^+.

In the presence of AC, the existence of κ^+ follows directly from Hartogs's theorem. The following notation is defined in Section 2.5, in a slightly different way:

Definition. The cardinal $\aleph_\alpha$ is defined by induction on α as follows:

$\aleph_0 = Card(\mathbb{N})$.
$\aleph_{\alpha+1} = (\aleph_\alpha)^+$.
For limit ordinals λ, $\aleph_\lambda = \sum_{\alpha \in \lambda} \aleph_\alpha$.

When the von Neumann definition of cardinals is being used, $\aleph_\alpha$ is often written ω_α. The following notation is less common in mainstream mathematics but is often useful:

Definition. The cardinal $\beth_\alpha$ is defined by induction on α as follows ($\beth$ is the Hebrew letter beth):

$\beth_0 = Card(\mathbb{N})$.
$\beth_{\alpha+1} = 2^{\beth_\alpha}$.
For limit ordinals λ, $\beth_\lambda = \sum_{\alpha\in\lambda} \beth_\alpha$.

The notation $\beth_\alpha$ creates a concise way of stating CH: $\beth_1 = \aleph_1$. In ZFC (but not in ZF), GCH becomes $\forall\alpha(\beth_\alpha = \aleph_\alpha)$. So this notation is most useful when GCH is not being assumed. For example, it is clear that $Card(\mathcal{P}(\mathbb{R}))$ is $\beth_2$, but without GCH we cannot say where $\beth_2$, or even $\beth_1$, fits in the hierarchy of $\aleph$'s. Cantor's theorem implies trivially that $\beth_\alpha \geq \aleph_\alpha$ for every α, but not much else is obvious.

APPENDIX D

Groups, Rings, and Fields

At various points in this book, examples and results are given that pertain to the most important types of algebraic structures: groups, rings, and fields, as well as more specialized structures such as integral domains, ordered fields, etc. This appendix provides the definitions of some of the most important algebraic structures, plus a few examples and basic facts (related to topics discussed in the text, for the most part). It is intended as a reference for readers who are a bit rusty on these concepts. However, if you have never studied these structures (usually covered in courses called "abstract algebra" or "modern algebra"), you will probably need more than this appendix in order to understand the parts of the book that discuss them.

The one type of algebraic structure that is discussed in the text but not defined in this appendix is vector spaces. This decision has been made with the hope that most readers will have encountered vector spaces in relatively low-level courses in subjects such as matrix algebra or even calculus, if not in a linear algebra course.

Throughout this appendix, the symbols $*$, $\cdot$ and $+$ denote **binary operations** on a set A, functions from $A \times A$ to A. With these symbols, we always use "infix" notation rather than the usual function notation. That is, the result of applying the operation $*$ to the ordered pair (x, y) is written $x * y$, rather than the strange-looking $*(x, y)$. (The same thing is done with ordering relations—see Appendix B.) We may also abbreviate $x * y$ or $x \cdot y$ further to xy, as we do with ordinary multiplication.

Groups

Definition. A **group** is a set A together with a binary operation $*$ on A satisfying these conditions:

1. The **associative law** holds: $(x * y) * z = x * (y * z)$, for every x, y, and z in A.
2. There is a (two-sided) **identity** element, that is, an element e in A such that $x * e = e * x = x$, for every x in A.
3. Each element of the group has a (two-sided) **inverse**: for every x in A, there is a y in A such that $x * y = y * x = e$, where e is some identity element.

The words "together with" in this definition are a typical bit of jargon. Technically, a group is an ordered pair $(A, *)$ such that $*$ is a binary operation on A and the three listed conditions hold. So a group is a type of first-order structure in the sense of Chapter 5. More precisely, a group is a model of the theory consisting of conditions (1), (2), and (3), in the first-order language with a single binary function symbol.

Note that the set A is not a group by itself. In practice, mathematicians are often sloppy about this usage. For example, a reference to "the group of integers" would be understood to be about the group $(\mathbb{Z}, +)$, since $(\mathbb{Z}, \cdot)$ is not a group. Also, when we refer to an element of a group G, where $G = (A, *)$, we really mean an element of A.

Proposition D.1. *In any group, the identity element is unique, and the inverse of each element is unique.*

In most abstract algebra texts, this proposition is the first thing proved about groups. Because we have uniqueness, we can refer to *the* identity, and *the* inverse of any element, in a given group. Uniqueness also justifies the use of special symbols for the identity and inverses. There are two common conventions for this. In **multiplicative notation** for a group, the binary operation is written as $x \cdot y$ or simply xy, the identity is denoted 1, and the inverse of an element x is denoted x^{-1}. In **additive notation** for a group, the binary operation is written as $x + y$, the identity is denoted 0, and the inverse of x is denoted $-x$.

Furthermore, when multiplicative notation (or even the symbol $*$) is used for a group, it is common to use exponents: x^2 for $x \cdot x$, x^3 for $x \cdot x \cdot x$, and also x^{-2} for $x^{-1} \cdot x^{-1}$, etc. On the other hand, when additive notation is used, one writes $2x$ for $x + x$, $3x$ for $x + x + x$, $-2x$ for $(-x) + (-x)$, etc. Note that these integer exponents and coefficients do not denote elements of the group!

Example 1. The sets $\mathbb{R}$, $\mathbb{Q}$, and $\mathbb{Z}$, with ordinary addition as the binary operation, are all groups. So are the sets $\mathbb{R} - \{0\}$, $\mathbb{Q} - \{0\}$, $\mathbb{R}^+$ (the positive reals) and $\mathbb{Q}^+$ under multiplication. We have to exclude 0 in these last four groups since 0 has no multiplicative inverse.

There is a great variety of groups and special types of groups. Here is the most important category of them:

Definition. A group is called **abelian** if it satisfies the **commutative law**: $x * y = y * x$, for all elements x and y.

It would seem logical to refer to such groups as commutative groups, and this usage, although uncommon, would generally be understood. The term "abelian" honors the Norwegian mathematician Niels Abel, one of the pioneers of group theory. By the way, it's customary to use additive notation for a group only when the group is known to be abelian.

Example 2. The five groups mentioned in the previous example are all abelian. To come up with non-abelian groups, it's necessary to get away from familiar number systems. For example, let S be any set. Then the set of all bijections on S (one to one functions from S *onto* S), also known as **permutations** on S, forms a group with composition as the group operation. (Associativity of composition is very easy to show, and the usual identity function and inverse functions serve as the identity and inverses in this group.) This group is nonabelian as long as S has at least three members. For instance, suppose $S = \mathbb{R}$, $f(x) = x + 1$ and $g(x) = 2x$. Then f and g are permutations on S, but $f \circ g \neq g \circ f$. Thus the group of permutations on $\mathbb{R}$ is not abelian.

Example 3. Along with permutation groups, the simplest nonabelian groups are groups of matrices. Let n be any positive integer. Then any two $n \times n$ matrices (with real coefficients, say) can be multiplied, and this operation is known to be associative. Furthermore, there is an $n \times n$ identity matrix, with 1's down the main diagonal (top left to bottom right) and 0's everywhere else. Not every $n \times n$ matrix has an inverse, but the set of *invertible* $n \times n$ matrices forms a group under multiplication. If $n > 1$, this group is not abelian. To test this, choose two 2×2 matrices A and B at random. In all likelihood, you will find that $AB \neq BA$.

Abelian groups are much more well-behaved and easy to understand than nonabelian groups. Another important distinction among groups is the distinction between finite and infinite groups. All of the groups mentioned so far in this appendix are infinite, except for the group of permutations on a finite set S. Finite groups are not necessarily easier to work with than infinite groups; in fact, the classification of finite nonabelian groups has been one of the thorniest problems of modern algebra. However, finite abelian groups are rather simple, as we will see shortly. The number of elements in a finite group is called its **order**.

Example 4. The most straightforward way to construct finite abelian groups is by using "clock arithmetic," or **modular arithmetic** as it is more precisely called. Imagine an ordinary dial clock, except that 0 rather than 12 appears at the top. (Logically, it makes at least as much sense to say the day begins at zero o'clock as twelve o'clock.) If the time now is nine, we know that the time five hours from now will be two o'clock, not fourteen o'clock. In other words, in clock addition we add numbers in the usual way, but then we subtract twelve if necessary to make sure the answer we get is a number that appears on the clock.

A standard clock has twelve numbers, but this idea can be generalized to clocks with any number of numbers. So, for each natural number n, let $A_n = \{0, 1, \ldots, n-1\}$. The group of **integers modulo** n, denoted $\mathbb{Z}_n$, is the set A_n together with the operation of "clock addition" described above. For instance, in $\mathbb{Z}_7$, $2 + 2 = 4$, $4 + 3 = 0$, and

$6 + 4 = 3$. It is not hard to show that $\mathbb{Z}_n$ is an abelian group of order n. The identity of $\mathbb{Z}_n$ is 0. The inverse of 0 is 0, while the inverse of any other element x is $n - x$.

To state the main classification theorem for finite abelian groups, we need to define several important notions that are also discussed in Chapter 5, in the context of general algebraic structures:

Definitions. Let $G_i = (A_i, *_i)$ be a group, for $i = 1, 2$. A **homomorphism** from G_1 to G_2 is a function ϕ from A_1 to A_2 such that $\phi(x *_1 y) = \phi(x) *_2 \phi(y)$, for all x and y in A_1. A one-to-one onto homomorphism is called an **isomorphism**. Finally, two groups are called **isomorphic** if there is an isomorphism from one of them to the other.

A homomorphism is a "structure-preserving" function between groups or other algebraic structures. So an isomorphism is a one-to-one correspondence between two groups (more precisely, between their universes) that is structure-preserving. If two groups are isomorphic, they may be viewed as being "the same group, except possibly for how their elements are named." Isomorphic groups have exactly the same *mathematical* properties.

Definition. Let $G = (A, *)$ be a group and $B \subseteq A$. We say that B defines a **subgroup** of G if B contains e and is closed under $*$ and the inverse operation. Again, the subgroup isn't technically B but rather B together with the restriction of $*$ to $B \times B$, but it's common to be imprecise about this.

Example 5. For each fixed integer n, let $n\mathbb{Z} = \{kn \mid k \in \mathbb{Z}\}$ (not to be confused with $\mathbb{Z}_n$). Then it is easy to show that $n\mathbb{Z}$ defines a subgroup of $(\mathbb{Z}, +)$. In fact, these are the only subgroups of $(\mathbb{Z}, +)$. In contrast, it is not easy to describe all the subgroups of $(\mathbb{R}, +)$.

Exercise 1. Prove that all groups of the form $n\mathbb{Z}$, with $n \neq 0$, are isomorphic to each other.

Definition. Let $G = (A, *)$ be a group and $S \subseteq A$. Then there are three ways of defining what is meant by the subgroup of G **generated**

by S. It's the intersection of all subgroups of G that contain S, and also the smallest subgroup of G that contains S. More concretely, it's also the set of all elements of A that can be the interpretation of some term of the language of group theory (with symbols for $*$, the identity, and the inverse operation), with free variables assigned to elements of S. The equivalence of these definitions is implied by Corollaries 5.16 and 5.17 in Section 5.4.

Definitions. A group is said to be **finitely generated** (respectively, **cyclic**) if it is generated by some finite (respectively, one-element) set of its elements. The same terminology is applied to subgroups.

Proposition D.2. *Every cyclic group is isomorphic to the group of integers or to one of the groups* $\mathbb{Z}_n$.

Thus, cyclic groups are rather simple. In particular, $\mathbb{Z}$ is the unique infinite cyclic group, "up to isomorphism." Furthermore, we will soon see that cyclic groups are the main "building blocks" for a rather large category of groups.

Definition. Let $G_i = (A_i, *_i)$ be a group, for $i = 1, 2$. Their **direct product** $G_1 \times G_2$ is the group $(A_1 \times A_2, *)$, where $(u, v) * (x, y)$ is defined to be $(u *_1 x, v *_2 y)$. Similarly, we can define the direct product of three or more groups, or even an infinite collection of groups.

For example, if $G_1 = (\mathbb{Z}, +)$ and $G_2 = (\mathbb{R} - \{0\}, \cdot)$, then in $G_1 \times G_2$ we would have $(-4, 0.3) * (7, -5) = (3, -1.5)$.

Theorem D.3 (Classification of finitely generated abelian groups). *Every finitely generated abelian group is isomorphic to a direct product of a finite number of cyclic groups, with n being a power of a prime in each finite factor* $\mathbb{Z}_n$. *This representation is unique, except for the order of the factor groups.*

As a special case, every finite abelian group is isomorphic to a direct product of finite cyclic groups, with the same restriction on n and the same uniqueness condition as in the theorem.

Example 6. Suppose G is an abelian group of order 6. Then, since 2×3 is the only factorization of 6 into *powers* of primes, G must be isomorphic to $\mathbb{Z}_2 \times \mathbb{Z}_3$. The groups $\mathbb{Z}_3 \times \mathbb{Z}_2$ and $\mathbb{Z}_6$ are also isomorphic to G. By the way, the group of permutations on a set of three elements is a nonabelian group of order 6, and it is structurally the only nonabelian group of order 6. So there are exactly two groups of order 6, up to isomorphism.

Now suppose G is an abelian group of order 4. Then, according to the classification theorem, G could be isomorphic to $\mathbb{Z}_2 \times \mathbb{Z}_2$ or $\mathbb{Z}_4$. These two groups are not isomorphic, since $u + u$ is the identity for every u in $\mathbb{Z}_2 \times \mathbb{Z}_2$, but in $\mathbb{Z}_4$ we have $1 + 1 = 2 \neq 0$. (This is the typical sort of reasoning that shows groups are not isomorphic.) Thus, structurally, there are exactly two abelian groups of order 4. It is also simple to show that there are no nonabelian groups of this order.

Similarly, an abelian group of order 60 must be isomorphic to exactly one of $\mathbb{Z}_3 \times \mathbb{Z}_5 \times \mathbb{Z}_4$ or $\mathbb{Z}_3 \times \mathbb{Z}_5 \times \mathbb{Z}_2 \times \mathbb{Z}_2$. For instance, $\mathbb{Z}_{60}$ looks like the former of these, while $\mathbb{Z}_{30} \times \mathbb{Z}_2$ is isomorphic to the latter.

The classification theorem for finitely generated abelian groups provides a very precise, clear way of describing these groups. When applied to finite abelian groups, it is very reminiscent of the fundamental theorem of arithmetic, except that here the "factoring" is based on powers of primes, not simply on primes.

Here are a few other concepts of group theory that are mentioned in the text: Let $G = (A, \cdot)$ be a group and $x \in A$. If there is a positive integer n such that $x^n = 1$, then the smallest such n is called the **order** of x. In this case, the cyclic subgroup of G generated by x consists of $\{1, x, x^2, \dots, x^{n-1}\}$ and is isomorphic to $\mathbb{Z}_n$, so the order of x equals the order of the subgroup it generates. If there is no such n, x is said to have infinite order (and x generates a subgroup isomorphic to $\mathbb{Z}$).

A group in which every element has finite order is called a **torsion** group. Every finite group is torsion, as is every finite direct product of torsion groups. A group in which every element except the identity has infinite order is said to be **torsion-free**. The groups $\mathbb{Z}$, $\mathbb{Q}$, and $\mathbb{R}$

under addition are torsion-free, as is every direct product of torsion-free groups.

Finally, a group G is called **divisible** if "every element has an nth root, for every positive integer n." Symbolically, this can be written

$$\forall n \in \mathbb{Z}^+ \forall x \in G \exists y \in G(y^n = x).$$

But it's important to realize that this symbolic statement is not within the first-order language of a group, because n is not a variable for a group element, and group theory does not have exponential terms y^n. So an infinite axiom schema is required to express divisibility. A similar situation holds for the obvious attempts to axiomatize torsion groups and torsion-free groups. These limitations are discussed further in Section 5.7.

The term "divisible group" is more apt when the group operation is written additively, for then y^n becomes ny, and "has an nth root" becomes "is divisible by n." The additive groups $\mathbb{Q}$ and $\mathbb{R}$ are divisible, as is the multiplicative group $\mathbb{R}^+$. The additive group $\mathbb{Z}$ and the multiplicative group $\mathbb{Q}^+$ are not.

Rings and fields

Groups are the most important type of algebraic structure with a single binary operation. For the remainder of this appendix, we consider algebraic structures with two binary operations:

Definition. A **ring** is a set together with two binary operations on it (more formally, an ordered triple $(A, +, \cdot)$) satisfying these conditions:

1. The structure $(A, +)$ is an abelian group.
2. The operation $\cdot$ is associative.
3. The **distributive laws** hold: $x \cdot (y + z) = (x \cdot y) + (x \cdot z)$ and $(y + z) \cdot x = (y \cdot x) + (z \cdot x)$, for all x, y, and z in A.

Since a group has only one operation, the question of whether to use multiplicative or additive notation for a group is often a matter of taste. But a ring has two operations, and so it is almost universal to call them addition and multiplication, as in this definition. Note that the distributive laws are the only properties that connect the two operations. Also, note that the definition of a ring requires much more of addition (all four conditions needed to be an abelian group) than it does of multiplication. For this reason, most of the particular types of rings that are considered are based on putting more conditions on multiplication. Here are the definitions of some of these types of rings:

Definitions. A **commutative ring** is a ring in which multiplication is commutative: $xy = yx$ for all x and y in A.

A **ring with unity** is a ring with a nonzero multiplicative identity: an element 1 such that $1 \neq 0$ and $x \cdot 1 = 1 \cdot x = x$, for all x in A. As with groups, this identity element is easily shown to be unique.

A **division ring** is a ring with unity in which every nonzero element has a multiplicative inverse or **reciprocal**: in symbols, $\forall x \neq 0 \, \exists y(xy = yx = 1)$. Whenever an element x in a ring with unity has a reciprocal, that reciprocal is unique and we denote it x^{-1}.

A **field** is a commutative division ring.

Definitions. If x and y are nonzero elements of a ring such that $xy = 0$, then x and y are (each) called **zero-divisors**. An **integral domain** is a commutative ring with unity, with no zero-divisors.

Example 7. Most familiar number systems which have addition and multiplication operations are rings. Examples include $\mathbb{R}$, $\mathbb{Q}$, $\mathbb{C}$, and $\mathbb{Z}$. The first three of these are fields, which also implies that they are integral domains. On the other hand, $\mathbb{Z}$ is an integral domain but not a field. In fact, 1 and -1 are the only elements of $\mathbb{Z}$ that have reciprocals.

The number system $\mathbb{N}$ has addition and multiplication, but it is not a ring because some (in fact, almost all) elements in it don't have additive inverses.

We have called addition and multiplication the two main operations in a ring. But in grade school, we learn that there are four basic operations of arithmetic: addition, subtraction, multiplication, and division. In higher mathematics, subtraction and division are viewed as offshoots of the two basic operations and the inverse properties that relate to them. In other words, in any ring, $x - y$ means $x + (-y)$, and in a commutative ring with unity, x/y means $x \cdot y^{-1}$, provided that y^{-1} exists.

Example 8. The simplest examples of noncommutative rings are probably rings of matrices. Let R be a ring with unity. For each natural number n, the set of $n \times n$ matrices whose coefficients are in R, with the usual operations of matrix addition and multiplication, is also a ring with unity. (See the related discussion in Example 3.) But this ring is not commutative if $n > 1$. These noncommutative rings are also not division rings, since there are many nonzero square matrices that are not invertible.

Example 9. The simplest example of a ring without unity is the ring of even integers, with the usual operations. More generally, for each integer $n > 1$ the ring $n\mathbb{Z}$, whose elements are all multiples of n, is a commutative ring without unity. These rings also have no zero-divisors, so they are "almost" integral domains.

Exercise 2. Prove that the rings $m\mathbb{Z}$ and $n\mathbb{Z}$ are isomorphic if and only if $m = \pm n$. (Compare this exercise to the similar one about the groups $n\mathbb{Z}$.)

Exercise 3. Find a noncommutative ring without unity.

Example 10. The groups $\mathbb{Z}_n$ were defined in Example 4. These number systems become rings if we include the operation of **multiplication modulo** n that is analogous to addition modulo n. For instance, if we start at midnight and go five ten-hour periods into the future, the time will not be fifty o'clock. It will be two o'clock. We can determine this by first computing that $5 \times 10 = 50$, and then computing the *remainder* when 50 is divided by 12. So, in $\mathbb{Z}_{12}$, $5 \cdot 10 = 2$ (while $5 + 10 = 3$). Similarly, in $\mathbb{Z}_8$, $3 \cdot 7 = 5$. Arithmetic modulo n may also be viewed as "units place" arithmetic in base n.

It is easy to show that $\mathbb{Z}_n$ is a commutative ring with unity for $n > 1$. Beyond that, two different cases arise. If n is composite, $\mathbb{Z}_n$

cannot be an integral domain. For example, $2 \cdot 2 = 0$ in $\mathbb{Z}_4$, and $2 \cdot 3 = 0$ in $\mathbb{Z}_6$.

On the other hand, $\mathbb{Z}_p$ must be an integral domain if p is prime. For instance, zero-divisors in $\mathbb{Z}_7$ would be a pair of positive integers less than 7 whose product is a multiple of 7. This would clearly contradict the primality of 7. Furthermore, one of the important basic theorems of number theory says, in essence, that every nonzero element of $\mathbb{Z}_p$ (when p is prime) has a reciprocal. For instance, in $\mathbb{Z}_7$ we have $1^{-1} = 1$, $2^{-1} = 4$, $3^{-1} = 5$, and $6^{-1} = 6$. In other words, these rings $\mathbb{Z}_p$ are actually fields, and they are the simplest examples of finite fields.

In particular, consider $\mathbb{Z}_2$. This structure has only two elements, 0 and 1, with completely standard addition and multiplication except that $1 + 1 = 0$. Yet, somehow, this number system satisfies all the standard properties of addition, subtraction, multiplication, and division.

The fields $\mathbb{Z}_p$ have **finite characteristic**, meaning that some integer multiple of 1 (formally obtained by adding 1 to itself repeatedly) equals 0. For instance, $\mathbb{Z}_2$ has characteristic 2, since $1+1 = 0$ in it. Similarly, $\mathbb{Z}_p$ has characteristic p. It is easy to show that the characteristic of such a field—the smallest integer multiple of 1 that equals 0—must be prime. Trivially, every finite field has finite characteristic. But there are also infinite fields of finite characteristic.

Fields that do not have finite characteristic are said, perhaps illogically, to have **characteristic zero**. The familiar fields $\mathbb{Q}$, $\mathbb{R}$, and $\mathbb{C}$ are of this type.

The notions of **subrings** and **subfields** are defined in the same way as subgroups. These terms can be used in "hybrid" ways: one can refer to a subring of a field, or a subfield of a ring, or even a subgroup of the additive group of a ring, etc.

Every field K has a smallest subfield, called its **prime field**. It is, as usual, the intersection of all subfields of K. If K has characteristic p (respectively, 0), then its prime field is the unique subfield of K that is isomorphic to $\mathbb{Z}_p$ (respectively, $\mathbb{Q}$).

Let F be a subfield of a field K. An element x of K is called **algebraic** over F if it is a root (or "zero") of some nonzero polynomial with coefficients in F. Otherwise, x is **transcendental** over F. If every x in K is algebraic over F, then K is an **algebraic extension** of F.

When these terms are applied to complex (including real) numbers, the words "over $\mathbb{Q}$" are generally understood. So all rational numbers are algebraic, but so are all numbers that can be written using rational numbers and radicals. For instance, $\sqrt{5}$ is algebraic because it is a root of $x^2 - 5$. It's less obvious that a number like $\sqrt{7} + \sqrt[3]{10}$ is algebraic, but it is. Abel ushered in the modern age of algebra by proving that the converse of this is false: not every algebraic number is expressible by radicals. As Chapter 8 mentions, it took well into the nineteenth century to show that transcendental numbers exist. The most famous ones are π and e.

A field K is said to be **algebraically closed** if every polynomial with coefficients in K has a root in K. Finally, if K is algebraically closed and it is an algebraic extension of some subfield F, then K is called an **algebraic closure** of F. Here is one of the most important results of field theory, whose proof requires the axiom of choice:

Theorem D.4. *Every field F has an algebraic closure, which is unique "up to isomorphism over F." In other words, if K_1 and K_2 are both algebraic closures of F, then there is an isomorphism between K_1 and K_2 that is the identity on F.*

Example 11. The fundamental theorem of algebra says precisely that the field $\mathbb{C}$ is algebraically closed. Furthermore, it is easy to show that $\mathbb{C}$ is an algebraic extension of $\mathbb{R}$; in fact, every complex number is a root of a polynomial, with real coefficients, of degree at most 2. Therefore, $\mathbb{C}$ is the algebraic closure of $\mathbb{R}$.

On the other hand, $\mathbb{C}$ is an not algebraic extension of $\mathbb{Q}$, since $\mathbb{C}$ includes transcendental numbers. The algebraic closure of $\mathbb{Q}$ is, almost by definition, the field of complex algebraic numbers.

Ordered algebraic structures

In the basic algebra of the real numbers and other familiar number systems, one considers inequalities as well as equations. In other words, these algebraic structures have orderings defined on them. It is fruitful to generalize this idea:

Definition. An **ordered group** is a triple $(A, +, <)$ satisfying these conditions:

1. $(A, +)$ is an abelian group.
2. $<$ is an (irreflexive) total ordering on A.
3. Whenever x, y, and z are in A and $x < y$, then $x + z < y + z$.

Note that condition (3) is familiar from high-school algebra: an inequality is preserved if you add the same number to both sides of it.

Definition. An **ordered ring** is a 4-tuple $(A, +, \cdot, <)$ satisfying these conditions:

1. $(A, +, \cdot)$ is a commutative ring.
2. $(A, +, <)$ is an ordered group.
3. Whenever x, y, and z are in A, $x < y$, and $z > 0$, then $xz < yz$.

Here we see another elementary property: an inequality is preserved if you multiply both sides by the same positive number.

Example 12.

(a) $\mathbb{Z}$, $\mathbb{Q}$, and $\mathbb{R}$ are ordered rings, with the usual operations and ordering. The last two are also fields, so they are called **ordered fields**.

(b) It is easy to show that there is no ordering on a finite group or ring that will turn it into an ordered group or ring. So, for example, the groups $\mathbb{Z}_n$ cannot be ordered.

(c) The field $\mathbb{C}$ also cannot be ordered: it is not hard to show that, in any ordered field, $x^2 \geq 0$ for every x. Therefore $1 = 1^2 > 0$, and so $-1 < 0$. But in $\mathbb{C}$, $i^2 = -1$. (See the discussion of formally real fields in Section 5.5.)

Bibliography

[Ark] L. Arkeryd, N. Cutland, and C. W. Henson, eds., *Nonstandard Analysis: Theory and Applications*, Kluwer Academic Publishers, Dordrecht and Boston, 1997.

[Ax] J. Ax, "The elementary theory of finite fields," *Annals of Math.* 88 (1968), 239–271.

[Bar73] J. Barwise, "Back and forth through infinitary logic," in: M. D. Morley, ed., *Studies in Model Theory*, MAA Studies in Mathematics, Vol. 8, 1973.

[Bar77] J. Barwise, ed., *Handbook of Mathematical Logic*, North-Holland, Amsterdam, 1977.

[Bee] M. Beeson, *Foundations of Constructive Mathematics*, Springer-Verlag, Berlin, 1985.

[Bell] E. T. Bell, *Men of Mathematics*, Simon and Schuster, New York, 1937.

[BP] P. Benacerraf and H. Putnam, *Philosophy of Mathematics*, Prentice-Hall, Englewood Cliffs, 1964.

[BJ] G. Berkeley and D. M. Jesseph, "The analyst, or a discourse addressed to an infidel mathematician," in: *De Motu and the Analyst*, Kluwer Academic Publishers, Dordrecht and Boston, 1992.

[BB] E. Bishop and D. Bridges, *Constructive Analysis*, Springer-Verlag, Berlin, 1985.

[BDK] G. Brewka, J. Dix, and K. Konolige, *Nonmonotonic Reasoning: An Overview*, CSLI Publications, Stanford, 1997.

[Bro] F. Browder, ed., *Mathematical Developments Arising from Hilbert Problems*, Proceedings of Symposia in Pure Mathematics No. 28, AMS Publications, Providence, 1974.

[Bur] D. Burton, *The History of Mathematics - An Introduction*, 5th ed., McGraw-Hill, New York, 2002.

[CK] C. C. Chang and H. J. Keisler, *Model Theory*, 3rd ed., North-Holland, Amsterdam, 1990.

[Coh] P. Cohen, *Set Theory and the Continuum Hypothesis*, W. A. Benjamin, New York, 1966.

[Cut] N. Cutland, ed., *Nonstandard Analysis and its Applications*, Cambridge Univ. Press, Cambridge, 1988.

[Dav] M. Davis, ed., *The Undecidable*, Dover, New York, 2004.

[DSW] M. Davis, R. Sigal, and E. Weyuker, *Computability, Complexity, and Languages*, 2nd ed., Academic Press, New York, 1994.

[Daw] J. Dawson, *Logical Dilemmas: The Life and Work of Kurt Gödel*, A. K. Peters Ltd., Wellesley, 1997.

[Dra] F. Drake, *Set Theory: An Introduction to Large Cardinals*, North-Holland, Amsterdam, 1974.

[Eas] W. Easton, "Powers of regular cardinals," *Annals Math. Logic* 1 (1970), 139–178.

[Edw] C. H. Edwards, Jr., *The Historical Development of the Calculus*, Springer-Verlag, Berlin, 1979.

[End] H. Enderton, *A Mathematical Introduction to Logic*, Academic Press, New York, 1972.

[EGH] P. Erdös, L. Gillman and M. Henriksen, "An isomorphism theorem for real closed fields," *Annals of Math. Series 2*, 61 (1955), 542–554.

[Eves] H. Eves, *An Introduction to the History of Mathematics*, 6th ed., Harcourt Brace, Orlando, 1990.

[Fef60] S. Feferman, "Arithmetization of metamathematics in a general setting," *Fundamenta Mathematicae* 49 (1960), 35–92.

[Fef99] S. Feferman, "Does mathematics need new axioms?," *American Mathematical Monthly* 106 (1999), 99–111.

[GG] D. Gabbay and F. Guenther, eds., *Handbook of Philosophical Logic*, three volumes, D. Reidel, Dordrecht, 1983–1986.

[Go31] K. Gödel, "Über formal unentscheidbare Sätze der *Principia Mathematica* und verwandter Systeme I," *Monatshefte Math. Physik* 38 (1931), 173–198. (English translations in [Go86] and [Dav].)

[Go38] K. Gödel, "The consistency of the axiom of choice and of the generalized continuum hypothesis," *Proc. Nat. Acad. Sci., U.S.A.* 24 (1938), 556–557. (Reprinted in [Go86].)

[Go86] K. Gödel, *Collected Works*, (S. Feferman et al., eds.), three volumes, Oxford Univ. Press, New York, 1986–1995.

[Gol] D. Goldrei, *Classic Set Theory - for Guided Independent Study*, Chapman & Hall/CRC Press LLC, Boca Raton, 1996.

[GS] D. Gale and F. M. Stewart, "Infinite games with perfect information," *Contributions to the Theory of Games*, Annals of Mathematics Studies No. 28, Princeton Univ. Press, Princeton, 1953.

[Har] L. A. Harrington et al, eds., *Harvey Friedman's Research on the Foundations of Mathematics*, North-Holland, Amsterdam, 1985.

[HPS] D. Haskell, A. Pillay, and C. Steinhorn, eds., *Model Theory, Algebra, and Geometry*, Cambridge Univ. Press, Cambridge, 2000.

[Hei] J. van Heijenoort, *From Frege to Gödel: A Source Book in Mathematical Logic, 1879–1931*, Harvard Univ. Press, Cambridge, 1967.

[Hen] J. Henle, *An outline of Set Theory*, Springer-Verlag, Berlin, 1986.

[HK] J. Henle and E. Kleinberg, *Infinitesimal Calculus*, MIT Press, Cambridge, Cambridge, 1979.

[Hey] A. Heyting, *Intuitionism: An Introduction*, 2nd ed., North-Holland, Amsterdam, 1966.

[Hod] W. Hodges, *A Shorter Model Theory*, Cambridge Univ. Press, Cambridge, 1997.

[Hof] D. Hofstadter, *Gödel, Escher, Bach: An Eternal Golden Braid*, Harper-Collins, New York, 1999.

[Jech73] T. Jech, *The Axiom of Choice*, North-Holland, Amsterdam, 1973.

[Jech78] T. Jech, *Set Theory*, Academic Press, New York, 1978.

[JW] W. Just and M. Weese, *Discovering Modern Set Theory: Set-Theoretic Tools for Every Mathematician*, I and II, AMS Publications, Providence, 1996.

[Kan] A. Kanamori, *The Higher Infinite*, Springer-Verlag, Berlin, 1995.

[Kec] A. Kechris, *Classical Descriptive Set Theory*, Springer-Verlag, Berlin, 1995.

[Kei76] H. J. Keisler, *Foundations of Infinitesimal Calculus*, Prindle, Weber, and Schmidt, Boston, 1976.

[Kei86] H. J. Keisler, *Elementary Calculus: An Infinitesimal Approach*, 2nd ed., Prindle, Weber, and Schmidt, Boston, 1986.

[KV] S. Kleene and R. Vesley, *The Foundations of Intuitionistic Mathematics, Especially in Relation to Recursive Functions*, North-Holland, Amsterdam, 1965.

[Kun] K. Kunen, *Set Theory, An Introduction to Independence Proofs*, North-Holland, Amsterdam, 1983.

[LS] A. Lévy and R. Solovay, "Measurable cardinals and the continuum hypothesis," *Israel J. Math.* 5 (1967), 234–248.

[LP] H. Lewis and C. Papadimitriou, *Elements of the Theory of Computation*, 2nd ed., Prentice-Hall, Englewood Cliffs, 1998.

[LW] P. Loeb and M. Wolff, eds., *Nonstandard Analysis for the Working Mathematician*, Kluwer Academic Publishers, Dordrecht and Boston, 2000.

[Mad] P. Maddy, "Believing the axioms," I, *J. Symbolic Logic* 53 (1988), 481–511, and II, *ibid.*, 736–764.

[Mar] D. Marker, *Model Theory: an Introduction*, Springer-Verlag, Berlin, 2002.

[Mart] D. A. Martin, "Borel Determinacy," *Annals of Mathematics* Series 2, vol. 102 (1975), 363-371.

[Men] E. Mendelson, *Introduction to Mathematical Logic*, 4th ed., Lewis, Boca Raton, 1997.

[Moo] G. H. Moore, *Zermelo's Axiom of Choice*, Springer-Verlag, Berlin, 1982.

[Odi] P. Odifreddi, *Classical Recursion Theory*, North-Holland, Amsterdam, 1989.

[PH] J. Paris and L. Harrington, "A mathematical incompleteness in Peano arithmetic," in [Bar77].

[Robe] A. Robert, *Nonstandard Analysis*, John Wiley & Sons, New York, 1988.

[Robi] A. Robinson, *Non-standard Analysis*, Princeton Univ. Press, Princeton, 1996.

[Rog] H. Rogers, *Recursion Theory*, McGraw-Hill, New York, 1967.

[Roi] J. Roitman, *Introduction to Modern Set Theory*, Wiley, New York, 1990.

[Ross] K. A. Ross and C. R. B. Wright, *Discrete Mathematics*, 3rd ed., Prentice-Hall, Englewood Cliffs, 1992.

[RW] B. Russell and A. N. Whitehead, *Principia Mathematica*, Cambridge Univ. Press, Cambridge, 1989.

[Sho] J. Shoenfield, *Mathematical Logic*, A. K. Peters Ltd., Wellesley, 2001.

[Smo] C. Smorynski, *Self-Reference and Modal Logic*, Springer-Verlag, Berlin, 1985.

[Smu] R. Smullyan, *Gödel's Incompleteness Theorems*, Oxford Univ. Press, New York, 1992.

[SEP] *Stanford Encyclopedia of Philosophy*, online at http://plato.stanford.edu

[Str] D. J. Struik, *A Source Book in Mathematics 1200-1800*, Princeton Univ. Press, Princeton, 1986.

[Sup] P. Suppes, *Axiomatic Set Theory*, Dover, New York, 1960.

[TZ] G. Takeuti and W. M. Zaring, *Introduction to Axiomatic Set Theory*, Springer-Verlag, Berlin, 1971.

[Tod] T. D. Todorov, "Back to classics: teaching limits through infinitesimals," *International Journal of Mathematical Education in Science and Technology* 32 (2001), 1–20. (Online at www.tandf.co.uk/journals)

[Tro] A. S. Troelstra, "Proof theory and constructive mathematics," in [Bar77].

[TvD] A. S. Troelstra and D. van Dalen, *Constructivism in Mathematics: An Introduction*, North-Holland, Amsterdam, 1988.

[Vau] R. Vaught, *Set Theory, An Introduction*, Birkhäuser, Boston, 1985.

[Wag] S. Wagon, *The Banach-Tarski Paradox*, Cambridge Univ. Press, Cambridge, 1985.

[Wolf] R. Wolf, *Proof, Logic, and Conjecture: The Mathematician's Toolbox*, W. H. Freeman, New York, 1998.

[Woo] W. H. Woodin, "The Continuum Hypothesis," I, *Notices of the A.M.S.* 48 (2001), 567–576, and II, *ibid.*, 681–690.

[YOTN] R.R. Yager, S. Ovchinnikov, R. M. Tong, and H. T. Nguyen, eds., *Fuzzy Sets and Applications: Selected Papers by L. A. Zadeh*, John Wiley and Sons, New York, 1987.

[Zad] L. A. Zadeh, "Fuzzy sets," *Information and Control* 8 (1965), 338–353.

Symbols and Notation

This list contains most of the specialized notation used in the book. The exceptions are some notations that are used in only one section and are not germane to the rest of the book. When a notation is introduced on one page and defined more rigorously later, but with the same meaning, both page numbers are given on the same line. When a notation is used with two or more quite different meanings, it appears in separate entries in this list, with cross-referenced pages in brackets.

S	successor (of a natural number or ordinal)	32, 78
$\in$	is an element of (set membership)	36
Σ_n, Π_n	Sigma-*n* and Pi-*n* formulas	41
$\Box$	necessarily (modal operator)	50
$\Diamond$	possibly (modal operator)	51
$\preceq, \sim, \prec$	relations comparing cardinality of sets	64, [182]
$\mathcal{P}$	power set	66
$\subseteq$	is a subset of	69, 183
$\{x, y\}$	unordered pair (set)	69
$\bigcup$	union of a collection of sets	69
$\emptyset$	empty set	70
$\cup, \cap$	union and intersection (of two sets)	70
$\{x \mid \mathrm{P}(x)\}$	set-builder notation	71
$x - y$	relative complement	75
$x \Delta y$	symmetric difference	75
(x, y)	ordered pair	75, 349
$A \times B$	*A* cross *B* (Cartesian product)	75, 349
B^A	*B* to the *A* (set of functions from *A* to *B*)	75, 349
Suc, Lim	successor and limit ordinals	78
$<$	is less than (for ordinals; same as $\in$)	79, [355]
Ord	class of all ordinals	81
ω	omega (first infinite ordinal)	81
$\alpha + \beta, \alpha \cdot \beta$	ordinal addition and multiplication	82, [356]
α^β	ordinal exponentiation	84, [356]
$\omega_\alpha, \aleph_\alpha$	omega-alpha, aleph-alpha	91, 358
V_α	V-alpha (cumulative hierarchy)	92
TC	transitive closure of a set	92
V	"universe" or proper class of all sets	93
$\vec{x}$	sequence of objects $(x_1, \ldots, x_k)$	98
$A^{<\omega}$	set of finite sequences of elements of *A*	101, 356

Ψ	universal function for PR functions	105
M_w	Turing machine encoded by the number w	114
$\varphi_w^{(k)}$	partial recursive function on $\mathbb{N}^k$ computed by M_w	114
$\Phi^{(k)}$	universal function for partial recursive functions	114
T	Kleene's T-predicate	115
U	upshot function	115
μ	the least (mu operator)	116, [290]
P	set of all problems decidable in polynomial time	126
NP	set of all problems decidable in non-deterministic polynomial time	129
#P	Gödel number of formula P	138
A_n	formula encoded by n	138
$\overline{n}$	numeral for n	141
Sub	substitution function	141
Tr	truth predicate	149
Prf_K, Prov_K	provability predicates	150
Con_K, $\mathrm{Con}(K)$	consistency of K	152, 252
$A^{(n)}$	set of all n-element subsets of A	160
$j \to (k)^n_m$	Ramsey's combinatorial property	160
$\mathfrak{A}$, $\mathfrak{B}$	first-order structure	167
$\lvert\mathfrak{A}\rvert$	universe of $\mathfrak{A}$	167
g, g^i_x	assignment	168
$\hat{g}$	interpretation of terms	168
$\models$	satisfies	169
$\mathfrak{N}$	standard model of arithmetic	171
$\equiv$	is elementarily equivalent to	174
$\cong$	is isomorphic to	182
$\subseteq$	is a submodel or substructure of	182, [69]
$\preceq$	is an elementary submodel of	183, [64]

Symbol	Meaning	Page
η_α-set	eta-alpha set	196
$\overline{z}$	complex conjugate of z	197
Σ_n, Π_n, Δ_n	Sigma-n, Pi-n, and Delta-n sets	205
Σ_n^0, Π_n^0, Δ_n^0	arithmetical hierarchy	207
Mod	class of models of a theory	214
Th	theory of a class of models	214
$\overline{Thm}$	closed theory	217
L_α, L	L-alpha and L (constructible hierarchy)	229
P^L	relativization of P to L	231
$\mathfrak{M}$	minimal model of set theory	236, [295]
$\mathfrak{M}[G]$	generic extension of $\mathfrak{M}$	236
$\Vdash$	forces	237
$\Vdash^{\mathrm{C}}$	strongly forces (Cohen's forcing)	242
cf	cofinality of a limit ordinal	243
Σ_n^1, Π_n^1, Δ_n^1	analytical hierarchy	255
$\boldsymbol{\Sigma}_n^1$, $\boldsymbol{\Pi}_n^1$, $\boldsymbol{\Delta}_n^1$	projective hierarchy	256
σ-algebra	sigma-algebra	257
$\boldsymbol{\Sigma}_\alpha^0$, $\boldsymbol{\Pi}_\alpha^0$, $\boldsymbol{\Delta}_\alpha^0$	Borel hierarchy	258
$0^\#$	zero-sharp	267
G_A	infinite game with winning set A	269
$H(\kappa)$	sets that are hereditarily smaller than κ	274
$\approx$	is close to	290, 309
μ	monad	290, 309, [116]
$\mathbb{R}[t]$	ring of polynomials over the reals	293
$\mathbb{R}(t)$	field of rational functions over the reals	293
$V_n(S)$, $V(S)$	superstructure	294
$\mathfrak{M}$	standard model of analysis	295, [236]
$^*\mathfrak{M}$	nonstandard extension of $\mathfrak{M}$	296
$*$	canonical embedding of $\mathfrak{M}$ into $^*\mathfrak{M}$	297
$^\sigma A$	image of A under $*$	299

St	standart part function	303
$\#$	is apart from	334
A^n	set of n-tuples of elements of A	349
xRy	abbreviation for $(x, y) \in R$	350
Dom	domain	350
Rng	range	350
R^{-1}	R inverse (inverse relation)	350
$[a, b], [a, b)$, etc.	interval notation	352
$f(x)$	f of x (function notation)	353
$f : A \to B$	f is a function from A to B	353
$[x]_R, [x]$	equivalence class of x (under R)	354
$Card$	cardinality	355
$<, \leq$	ordering on cardinals	355, [79]
$+, \cdot, \kappa_1^{\kappa_2}$	cardinal arithmetic	356, [82]
$\coprod$	formal disjoint union	356
Max	maximum	356
B_k, B	bijections from $\mathbb{N}^k$ to $\mathbb{N}$ and $\mathbb{N}^{<\omega}$ to $\mathbb{N}$	357
$\prod_{i\in I} A_i$	infinitary Cartesian product	358
$\sum_{i\in I} \kappa_i$, $\prod_{i\in I} \kappa_i$	infinitary cardinal arithmetic	358
κ^+	least cardinal that is greater than κ	358
$\beth_\alpha$	beth-alpha	358
$\mathbb{R}^+, \mathbb{Q}^+, \mathbb{Z}^+$	positive reals, rationals, integers	363
$\mathbb{Z}_n$	group/ring of integers modulo n	364, 370
$n\mathbb{Z}$	group/ring of integer multiples of n	365, 370

Index

A boldface page number after a person's name indicates a biography. A boldface page number after any other entry indicates a primary definition or explanation that is not the first occurrence of that entry.